Engineering Tables and Data

First published 1972
by Chapman and Hall Ltd
11 New Fetter Lane, London EC4P 4EE
Reprinted 1977, 1982, 1985

Published in the USA
by Chapman and Hall
29 West 35th Street, New York, NY 10001

© *1972 A. M. Howatson, P. G. Lund, J. D. Todd*

Printed in Great Britain by
J. W. Arrowsmith Ltd, Bristol
ISBN 0 412 11550 6

Engineering Tables and Data

A. M. HOWATSON
P. G. LUND
J. D. TODD

*Department of Engineering Science,
University of Oxford*

LONDON NEW YORK

CHAPMAN AND HALL

Preface

This book brings together information which is used by engineers, and needed especially by students of engineering, but difficult to find in a collected form. In this respect engineering, perhaps because it is more often divided into separate branches, has so far been less well served than the other physical sciences; we hope to have in part redressed the balance. The contents are designed chiefly for engineering students of all kinds in universities and colleges, but they should also prove useful to practising engineers as a general reference.

There was some difficulty in choosing numerical values for parts of the section *Properties of Matter.* Information was culled from a range of sources which sometimes show an alarming lack of consistency. Given a choice, we have used values which are either average or more likely to be reliable. The degree of tolerance required varies very widely between, for example, the precision to which thermodynamic properties of steam are known and the uncertainty in those mechanical properties of solids which depend strongly on quality and preparation.

The tables on pages 4-12 inclusive are reproduced from *S.M.P. Advanced Tables* by permission of Cambridge University Press.

The tables on pages 35 and 36 are reproduced from *Elementary Statistical Tables:* Lindley and Miller, by permission of Cambridge University Press.

The tables on pages 37 and 38 are reproduced by permission of the Biometrika Trustees.

The tables on pages 66 and 67, the upper table on page 68 and the thermochemical data on pages 69-71 inclusive are reproduced from *Thermodynamic Tables:* Haywood by permission of Cambridge University Press.

The chart on page 75 is reproduced from *Engineering Thermodynamics Work and Heat Transfer:* Rogers and Mayhew, by permission of Longman.

Tables 1-3 on pages 76-85 inclusive and the charts on pages 104 and 105 are reproduced from *Elements of Gasdynamics:* Liepmann and Roshko, by permission of John Wiley.

Tables 4 and 5 on pages 85 to 103 inclusive are reproduced from *Introduction to Gas Dynamics:* Rotty, by permission of John Wiley.

The chart on page 106 and the table on page 107 are based on a corresponding diagram and table in *Fluid Mechanics:* Pao, by agreement with John Wiley.

The table on page 112 is reproduced from *Linear Structural Analysis:* Morice, by permission of Thames and Hudson.

The charts on page 121 are reproduced from *Moment Distribution:* Lightfoot, by permission of E. & F. N. Spon.

The tables on pages 123-143 inclusive are reproduced from the *Handbook on Structural Steelwork,* by permission of the British Constructional Steelwork Association Ltd., and the Constructional Steelwork Research and Development Organization.

The graphs on pages 147-149 inclusive are reproduced from *Vibration Theory and Applications:* Thomson, by permission of George Allen & Unwin and Prentice-Hall Inc.

The chart on page 154 is reproduced from *Transmission and Propagation:* Glazier and Lamont, by permission of the Controller of copyright, H.M.S.O.

The properties of water and steam tabulated on pages 48-65 inclusive are based on *U.K. Steam Tables in S.I. Units* published by Edward Arnold.

We are grateful to these publishers and authors for their collaboration and to our colleagues at Oxford who helped with advice and information. Our thanks are also due to Mr Jerome Davidson, who carried out the computation needed for certain tables. Inevitably, some mistakes will have escaped notice during preparation and checking; we should be grateful to hear of any which may be found.

A.M.H.
P.G.L.
J.D.T.

Oxford 1972.

Contents

Thermodynamics and fluid mechanics

Elasticity and structures

Mechanics

Electricity

General

The Greek alphabet

α	A	alpha		ν	N	nu
β	B	beta		ξ	Ξ	xi
γ	Γ	gamma		o	O	omicron
δ	Δ	delta		π	Π	pi
ϵ	E	epsilon		ρ	P	rho
ζ	Z	zeta		σ,ς	Σ	sigma
η	H	eta		τ	T	tau
θ	Θ	theta		υ	Υ	upsilon
ι	I	iota		ϕ	Φ	phi
κ	K	kappa		χ	X	chi
λ	Λ	lambda		ψ	Ψ	psi
μ	M	mu		ω	Ω	omega

SI units

Quantity	Unit	Symbol	Equivalent
Base-units			
Length	metre	m	—
Mass	kilogramme	kg	—
Time	second	s	—
Electric current	ampere	A	—
Temperature	kelvin	K	—
Luminous intensity	candela	cd	—
Supplementary units			
Plane angle	radian	rad	—
Solid angle	steradian	sr	—
Derived units			
Frequency	hertz	Hz	s^{-1}
Force	newton	N	$kg\,m/s^2$
Work, energy, quantity of heat	joule	J	Nm
Power	watt	W	J/s
Electric charge	coulomb	C	As
Electric potential, electromotive force	volt	V	W/A
Electric capacitance	farad	F	C/V
Electric resistance	ohm	Ω	V/A
Magnetic flux	weber	Wb	Vs
Magnetic flux density	tesla	T	Wb/m^2
Inductance	henry	H	Wb/A
Luminous flux	lumen	lm	cd sr
Illumination	lux	lx	lm/m^2

Other metric units

Quantity	Unit	Symbol	Equivalent
Length	Angstrom	Å	10^{-10} m
	micron	μm	10^{-6} m
Area	are	a	10^2 m^2
Volume	litre	l	10^{-3} m^3
Mass	tonne	t	10^3 kg
Force	dyne	—	10^{-5} N
Pressure	pascal	Pa	1 N/m^2
	bar	bar	10^5 N/m^2
Energy	erg	—	10^{-7} J
Viscosity (dynamic)	poise	P	10^{-1} Ns/m^2
Viscosity (kinetic)	stokes	St	10^{-4} m^2/s
Electrical conductance	siemens	S	1 Ω^{-1}
Magnetic field strength	oersted	—	$10^3/4\pi$ A/m
Magnetomotive force	gilbert	—	$10/4\pi$ A
Magnetic flux density	gauss	—	10^{-4} T
Magnetic flux	maxwell (or line)	—	10^{-8} Wb
Luminance	stilb	—	10^4 cd/m^2

Multiples and sub-multiples

Factor	Prefix	Symbol
10^{12}	tera	T
10^9	giga	G
10^6	mega	M
10^3	kilo	k
10^2	hecto	h
10	deca	da
10^{-1}	deci	d
10^{-2}	centi	c
10^{-3}	milli	m
10^{-6}	micro	μ
10^{-9}	nano	n
10^{-12}	pico	p

Conversion factors

Length

1 mil = 25·4 μm (1 mil = 10^{-3} inch)
1 foot = 0·3048 m
1 yard = 0·9144 m
1 fathom = 1·829 m (1 fathom = 6 feet)
1 furlong = 0·2012 km (1 furlong = 220 yards)
1 mile = 1·609 km
1 nautical mile = 1·852 km (1 nautical mile = 1·15 miles)

Area

1 square inch = 645·2 mm^2
1 square foot = 929·0 cm^2
1 square yard = 0·8361 m^2
1 acre = 4047 m^2 (1 acre = 4840 sq. yd)
1 square mile = 2·590 km^2

Volume

1 cubic inch = 16·39 cm^3
1 fluid ounce = 28·41 cm^3
1 pint = 568·3 cm^3
1 Imperial gallon = 4·546 l
1 U.S. gallon = 3·785 l
1 cubic foot = 28·32 l (1 cu. ft = 6·24 gallons)
1 cubic yard = 0·7646 m^3

Speed

1 revolution/minute = 0·1047 rad/s
1 mile/hour = 0·4470 m/s
1 knot = 0·5148 m/s (1 knot = 1 nautical mile/h)

Mass

1 ounce = 28·35 g
1 pound (lb) = 0·4536 kg
1 slug = 14·59 kg (1 slug = 32·17 lb)
1 hundredweight (cwt) = 50·8 kg (1 cwt = 112 lb)
1 ton = 1·016 t (1 ton = 2240 lb)

Density

1 lb/in^3 = 27·68 g/cm^3
1 lb/ft^3 = 16·02 kg/m^3
1 lb/gallon = 99·78 kg/m^3
1 ton/yd^3 = 1·329 t/m^3

Moments of mass

1 lb ft = 0·1383 kg m
1 lb ft^2 = 421·4 kg cm^2

Force

1 lb force (lbf) = 4·448 N (standard gravity)
1 poundal (pdl) = 0·1383 N (1 pdl = 0·0311 lbf)
1 ton force = 9·964 kN

Pressure (or stress)

1 lb/ft^2 = 47·88 N/m^2
1 lb/in^2 (psi) = 6·895 kN/m^2
1 ton/ft^2 = 107·3 kN/m^2
1 ton/in^2 = 15·44 N/mm^2
1 in Hg (0°C) = 3·386 kN/m^2 (1 in Hg = 0·491 psi)
1 ft H$_2$O (4°C) = 2·989 kN/m^2 (1 ft H$_2$O = 0·434 psi)
1 atmosphere (atm) = 1·013 bar

Energy, etc.

1 ft lbf = 1·356 J
1 ft pdl = 42·14 mJ
1 horse-power = 745·7 W (1 hp = 550 ft lbf/s)
1 electron volt (eV) = 1·602 x 10^{-19} J
1 degree Fahrenheit = $\frac{5}{9}$ K
* 1 calorie = 4·187 J
* 1 Btu = 1·055 kJ (1 Btu = 252 cal)
1 Chu = 1·899 kJ (1 Chu = $\frac{9}{5}$ Btu)
1 therm = 105·5 MJ (1 therm = 10^5 Btu)
1 Btu/hour = 0·293 W
1 Btu/lb = 2·325 kJ/kg

* The calorie and Btu were defined in terms of heating water, and their precise values depended on the temperature limits specified. The figures given are based on agreed international equivalents.

Mathematics

Logarithms, base 10

	0	1	2	3	4	5	6	7	8	9	1	2	3	4	5	6	7	8	9
1·0	·0000	0043	0086	0128	0170	0212	0253	0294	0334	0374	4	8	12	17	21	25	29	33	37
1·1	·0414	0453	0492	0531	0569	0607	0645	0682	0719	0755	4	8	11	15	19	23	27	30	34
1·2	·0792	0828	0864	0899	0934	0969	1004	1038	1072	1106	3	7	10	14	17	21	24	28	31
1·3	·1139	1173	1206	1239	1271	1303	1335	1367	1399	1430	3	6	10	13	16	19	23	26	29
1·4	·1461	1492	1523	1553	1584	1614	1644	1673	1703	1732	3	6	9	12	15	18	21	24	27
1·5	·1761	1790	1818	1847	1875	1903	1931	1959	1987	2014	3	6	8	11	14	17	20	22	25
1·6	·2041	2068	2095	2122	2148	2175	2201	2227	2253	2279	3	5	8	11	13	16	18	21	24
1·7	·2304	2330	2355	2380	2405	2430	2455	2480	2504	2529	2	5	7	10	12	15	17	20	22
1·8	·2553	2577	2601	2625	2648	2672	2695	2718	2742	2765	2	5	7	9	12	14	16	19	21
1·9	·2788	2810	2833	2856	2878	2900	2923	2945	2967	2989	2	4	7	9	11	13	16	18	20
2·0	·3010	3032	3054	3075	3096	3118	3139	3160	3181	3201	2	4	6	8	11	13	15	17	19
2·1	·3222	3243	3263	3284	3304	3324	3345	3365	3385	3404	2	4	6	8	10	12	14	16	18
2·2	·3424	3444	3464	3483	3502	3522	3541	3560	3579	3598	2	4	6	8	10	12	14	15	17
2·3	·3617	3636	3655	3674	3692	3711	3729	3747	3766	3784	2	4	6	7	9	11	13	15	17
2·4	·3802	3820	3838	3856	3874	3892	3909	3927	3945	3962	2	4	5	7	9	11	12	14	16
2·5	·3979	3997	4014	4031	4048	4065	4082	4099	4116	4133	2	3	5	7	9	10	12	14	15
2·6	·4150	4166	4183	4200	4216	4232	4249	4265	4281	4298	2	3	5	7	8	10	11	13	15
2·7	·4314	4330	4346	4362	4378	4393	4409	4425	4440	4456	2	3	5	6	8	9	11	13	14
2·8	·4472	4487	4502	4518	4533	4548	4564	4579	4594	4609	2	3	5	6	8	9	11	12	14
2·9	·4624	4639	4654	4669	4683	4698	4713	4728	4742	4757	1	3	4	6	7	9	10	12	13
3·0	·4771	4786	4800	4814	4829	4843	4857	4871	4886	4900	1	3	4	6	7	9	10	11	13
3·1	·4914	4928	4942	4955	4969	4983	4997	5011	5024	5038	1	3	4	6	7	8	10	11	12
3·2	·5051	5065	5079	5092	5105	5119	5132	5145	5159	5172	1	3	4	5	7	8	9	11	12
3·3	·5185	5198	5211	5224	5237	5250	5263	5276	5289	5302	1	3	4	5	6	8	9	10	12
3·4	·5315	5328	5340	5353	5366	5378	5391	5403	5416	5428	1	3	4	5	6	8	9	10	11
3·5	·5441	5453	5465	5478	5490	5502	5514	5527	5539	5551	1	2	4	5	6	7	9	10	11
3·6	·5563	5575	5587	5599	5611	5623	5635	5647	5658	5670	1	2	4	5	6	7	8	10	11
3·7	·5682	5694	5705	5717	5729	5740	5752	5763	5775	5786	1	2	3	5	6	7	8	9	10
3·8	·5798	5809	5821	5832	5843	5855	5866	5877	5888	5899	1	2	3	5	6	7	8	9	10
3·9	·5911	5922	5933	5944	5955	5966	5977	5988	5999	6010	1	2	3	4	6	7	8	9	10
4·0	·6021	6031	6042	6053	6064	6075	6085	6096	6107	6117	1	2	3	4	5	6	8	9	10
4·1	·6128	6138	6149	6160	6170	6180	6191	6201	6212	6222	1	2	3	4	5	6	7	8	9
4·2	·6232	6243	6253	6263	6274	6284	6294	6304	6314	6325	1	2	3	4	5	6	7	8	9
4·3	·6335	6345	6355	6365	6375	6385	6395	6405	6415	6425	1	2	3	4	5	6	7	8	9
4·4	·6435	6444	6454	6464	6474	6484	6493	6503	6513	6522	1	2	3	4	5	6	7	8	9
4·5	·6532	6542	6551	6561	6571	6580	6590	6599	6609	6618	1	2	3	4	5	6	7	8	9
4·6	·6628	6637	6646	6656	6665	6675	6684	6693	6702	6712	1	2	3	4	5	6	7	7	8
4·7	·6721	6730	6739	6749	6758	6767	6776	6785	6794	6803	1	2	3	4	5	5	6	7	8
4·8	·6812	6821	6830	6839	6848	6857	6866	6875	6884	6893	1	2	3	4	4	5	6	7	8
4·9	·6902	6911	6920	6928	6937	6946	6955	6964	6972	6981	1	2	3	4	4	5	6	7	8
5·0	·6990	6998	7007	7016	7024	7033	7042	7050	7059	7067	1	2	3	3	4	5	6	7	8
5·1	·7076	7084	7093	7101	7110	7118	7126	7135	7143	7152	1	2	3	3	4	5	6	7	8
5·2	·7160	7168	7177	7185	7193	7202	7210	7218	7226	7235	1	2	2	3	4	5	6	7	7
5·3	·7243	7251	7259	7267	7275	7284	7292	7300	7308	7316	1	2	2	3	4	5	6	7	7
	0	1	2	3	4	5	6	7	8	9	1	2	3	4	5	6	7	8	9

	0	1	2	3	4	5	6	7	8	9	1	2	3	4	5	6	7	8	9
5·4	·7324	7332	7340	7348	7356	7364	7372	7380	7388	7396	1	2	2	3	4	5	6	6	7
5·5	·7404	7412	7419	7427	7435	7443	7451	7459	7466	7474	1	2	2	3	4	5	5	6	7
5·6	·7482	7490	7497	7505	7513	7520	7528	7536	7543	7551	1	2	2	3	4	5	5	6	7
5·7	·7559	7566	7574	7582	7589	7597	7604	7612	7619	7627	1	2	2	3	4	5	5	6	7
5·8	·7634	7642	7649	7657	7664	7672	7679	7686	7694	7701	1	1	2	3	4	4	5	6	7
5·9	·7709	7716	7723	7731	7738	7745	7752	7760	7767	7774	1	1	2	3	4	4	5	6	7
6·0	·7782	7789	7796	7803	7810	7818	7825	7832	7839	7846	1	1	2	3	4	4	5	6	6
6·1	·7853	7860	7868	7875	7882	7889	7896	7903	7910	7917	1	1	2	3	4	4	5	6	6
6·2	·7924	7931	7938	7945	7952	7959	7966	7973	7980	7987	1	1	2	3	3	4	5	6	6
6·3	·7993	8000	8007	8014	8021	8028	8035	8041	8048	8055	1	1	2	3	3	4	5	5	6
6·4	·8062	8069	8075	8082	8089	8096	8102	8109	8116	8122	1	1	2	3	3	4	5	5	6
6·5	·8129	8136	8142	8149	8156	8162	8169	8176	8182	8189	1	1	2	3	3	4	5	5	6
6·6	·8195	8202	8209	8215	8222	8228	8235	8241	8248	8254	1	1	2	3	3	4	5	5	6
6·7	·8261	8267	8274	8280	8287	8293	8299	8306	8312	8319	1	1	2	3	3	4	5	5	6
6·8	·8325	8331	8338	8344	8351	8357	8363	8370	8376	8382	1	1	2	3	3	4	4	5	6
6·9	·8388	8395	8401	8407	8414	8420	8426	8432	8439	8445	1	1	2	3	3	4	4	5	6
7·0	·8451	8457	8463	8470	8476	8482	8488	8494	8500	8506	1	1	2	2	3	4	4	5	6
7·1	·8513	8519	8525	8531	8537	8543	8549	8555	8561	8567	1	1	2	2	3	4	4	5	5
7·2	·8573	8579	8585	8591	8597	8603	8609	8615	8621	8627	1	1	2	2	3	4	4	5	5
7·3	·8633	8639	8645	8651	8657	8663	8669	8675	8681	8686	1	1	2	2	3	4	4	5	5
7·4	·8692	8698	8704	8710	8716	8722	8727	8733	8739	8745	1	1	2	2	3	4	4	5	5
7·5	·8751	8756	8762	8768	8774	8779	8785	8791	8797	8802	1	1	2	2	3	3	4	5	5
7·6	·8808	8814	8820	8825	8831	8837	8842	8848	8854	8859	1	1	2	2	3	3	4	5	5
7·7	·8865	8871	8876	8882	8887	8893	8899	8904	8910	8915	1	1	2	2	3	3	4	4	5
7·8	·8921	8927	8932	8938	8943	8949	8954	8960	8965	8971	1	1	2	2	3	3	4	4	5
7·9	·8976	8982	8987	8993	8998	9004	9009	9015	9020	9025	1	1	2	2	3	3	4	4	5
8·0	·9031	9036	9042	9047	9053	9058	9063	9069	9074	9079	1	1	2	2	3	3	4	4	5
8·1	·9085	9090	9096	9101	9106	9112	9117	9122	9128	9133	1	1	2	2	3	3	4	4	5
8·2	·9138	9143	9149	9154	9159	9165	9170	9175	9180	9186	1	1	2	2	3	3	4	4	5
8·3	·9191	9196	9201	9206	9212	9217	9222	9227	9232	9238	1	1	2	2	3	3	4	4	5
8·4	·9243	9248	9253	9258	9263	9269	9274	9279	9284	9289	1	1	2	2	3	3	4	4	5
8·5	·9294	9299	9304	9309	9315	9320	9325	9330	9335	9340	1	1	2	2	3	3	4	4	5
8·6	·9345	9350	9355	9360	9365	9370	9375	9380	9385	9390	1	1	2	2	3	3	4	4	5
8·7	·9395	9400	9405	9410	9415	9420	9425	9430	9435	9440	0	1	1	2	2	3	3	4	4
8·8	·9445	9450	9455	9460	9465	9469	9474	9479	9484	9489	0	1	1	2	2	3	3	4	4
8·9	·9494	9499	9504	9509	9513	9518	9523	9528	9533	9538	0	1	1	2	2	3	3	4	4
9·0	·9542	9547	9552	9557	9562	9566	9571	9576	9581	9586	0	1	1	2	2	3	3	4	4
9·1	·9590	9595	9600	9605	9609	9614	9619	9624	9628	9633	0	1	1	2	2	3	3	4	4
9·2	·9638	9643	9647	9652	9657	9661	9666	9671	9675	9680	0	1	1	2	2	3	3	4	4
9·3	·9685	9689	9694	9699	9703	9708	9713	9717	9722	9727	0	1	1	2	2	3	3	4	4
9·4	·9731	9736	9741	9745	9750	9754	9759	9763	9768	9773	0	1	1	2	2	3	3	4	4
9·5	·9777	9782	9786	9791	9795	9800	9805	9809	9814	9818	0	1	1	2	2	3	3	4	4
9·6	·9823	9827	9832	9836	9841	9845	9850	9854	9859	9863	0	1	1	2	2	3	3	4	4
9·7	·9868	9872	9877	9881	9886	9890	9894	9899	9903	9908	0	1	1	2	2	3	3	4	4
9·8	·9912	9917	9921	9926	9930	9934	9939	9943	9948	9952	0	1	1	2	2	3	3	4	4
9·9	·9956	9961	9965	9969	9974	9978	9983	9987	9991	9996	0	1	1	2	2	3	3	3	4
	0	1	2	3	4	5	6	7	8	9	1	2	3	4	5	6	7	8	9

Natural cosines

Deg	·0	·1	·2	·3	·4	·5	·6	·7	·8	·9	1·0	(89−deg)	1	2	3	4	5	6	7	8	9
44°	·6947	[illegible]	[illegible]	[illegible]	[illegible]	[illegible]	[illegible]	7034	7046	7059	7071	45	[illegible]	[illegible]	[illegible]	[illegible]	[illegible]	[illegible]	[illegible]	[illegible]	[illegible]
45	·7071	[illegible]	[illegible]	[illegible]	[illegible]	[illegible]	[illegible]	7157	7169	7181	7193	44	[illegible]	[illegible]	[illegible]	[illegible]	[illegible]	[illegible]	[illegible]	[illegible]	[illegible]
46	·7193	[illegible]	[illegible]	[illegible]	[illegible]	[illegible]	[illegible]	7278	7290	7302	7314	43	[illegible]	[illegible]	[illegible]	[illegible]	[illegible]	[illegible]	[illegible]	[illegible]	[illegible]
47	·7314	[illegible]	[illegible]	[illegible]	[illegible]	[illegible]	[illegible]	7396	7408	7420	7431	42	[illegible]	[illegible]	[illegible]	[illegible]	[illegible]	[illegible]	[illegible]	[illegible]	[illegible]
48	·7431	[illegible]	[illegible]	[illegible]	[illegible]	[illegible]	[illegible]	7513	7524	7536	7547	41	[illegible]	[illegible]	[illegible]	[illegible]	[illegible]	[illegible]	[illegible]	[illegible]	[illegible]
49	·7547	[illegible]	[illegible]	[illegible]	[illegible]	[illegible]	[illegible]	7627	7638	7649	7660	40	[illegible]	[illegible]	[illegible]	[illegible]	[illegible]	[illegible]	[illegible]	[illegible]	[illegible]
50	·7660	[illegible]	[illegible]	[illegible]	[illegible]	[illegible]	[illegible]	7738	7749	7760	7771	39	[illegible]	[illegible]	[illegible]	[illegible]	[illegible]	[illegible]	[illegible]	[illegible]	[illegible]
51	·7771	[illegible]	[illegible]	[illegible]	[illegible]	[illegible]	[illegible]	7848	7859	7869	7880	38	[illegible]	[illegible]	[illegible]	[illegible]	[illegible]	[illegible]	[illegible]	[illegible]	[illegible]
52	·7880	[illegible]	[illegible]	[illegible]	[illegible]	[illegible]	[illegible]	7955	7965	7976	7986	37	[illegible]	[illegible]	[illegible]	[illegible]	[illegible]	[illegible]	[illegible]	[illegible]	[illegible]
53	·7986	[illegible]	[illegible]	[illegible]	[illegible]	[illegible]	[illegible]	8059	8070	8080	8090	36	[illegible]	[illegible]	[illegible]	[illegible]	[illegible]	[illegible]	[illegible]	[illegible]	[illegible]
54	·8090	[illegible]	[illegible]	[illegible]	[illegible]	[illegible]	[illegible]	8161	8171	8181	8192	35	[illegible]	[illegible]	[illegible]	[illegible]	[illegible]	[illegible]	[illegible]	[illegible]	[illegible]
55	·8192	[illegible]	[illegible]	[illegible]	[illegible]	[illegible]	[illegible]	8261	8271	8281	8290	34	[illegible]	[illegible]	[illegible]	[illegible]	[illegible]	[illegible]	[illegible]	[illegible]	[illegible]
56	·8290	[illegible]	[illegible]	[illegible]	[illegible]	[illegible]	[illegible]	8358	8368	8377	8387	33	[illegible]	[illegible]	[illegible]	[illegible]	[illegible]	[illegible]	[illegible]	[illegible]	[illegible]
57	·8387	[illegible]	[illegible]	[illegible]	[illegible]	[illegible]	[illegible]	8453	8462	8471	8480	32	[illegible]	[illegible]	[illegible]	[illegible]	[illegible]	[illegible]	[illegible]	[illegible]	[illegible]
58	·8480	[illegible]	[illegible]	[illegible]	[illegible]	[illegible]	[illegible]	8545	8554	8563	8572	31	[illegible]	[illegible]	[illegible]	[illegible]	[illegible]	[illegible]	[illegible]	[illegible]	[illegible]
59	·8572	[illegible]	[illegible]	[illegible]	[illegible]	[illegible]	[illegible]	8634	8643	8652	8660	30	[illegible]	[illegible]	[illegible]	[illegible]	[illegible]	[illegible]	[illegible]	[illegible]	[illegible]
60	·8660	[illegible]	[illegible]	[illegible]	[illegible]	[illegible]	[illegible]	8721	8729	8738	8746	29	[illegible]	[illegible]	[illegible]	[illegible]	[illegible]	[illegible]	[illegible]	[illegible]	[illegible]
61	·8746	[illegible]	[illegible]	[illegible]	[illegible]	[illegible]	[illegible]	8805	8813	8821	8829	28	[illegible]	[illegible]	[illegible]	[illegible]	[illegible]	[illegible]	[illegible]	[illegible]	[illegible]
62	·8829	[illegible]	[illegible]	[illegible]	[illegible]	[illegible]	[illegible]	8886	8894	8902	8910	27	[illegible]	[illegible]	[illegible]	[illegible]	[illegible]	[illegible]	[illegible]	[illegible]	[illegible]
63	·8910	[illegible]	[illegible]	[illegible]	[illegible]	[illegible]	[illegible]	8965	8973	8980	8988	26	[illegible]	[illegible]	[illegible]	[illegible]	[illegible]	[illegible]	[illegible]	[illegible]	[illegible]
64	·8988	[illegible]	[illegible]	[illegible]	[illegible]	[illegible]	[illegible]	9041	9048	9056	9063	25	[illegible]	[illegible]	[illegible]	[illegible]	[illegible]	[illegible]	[illegible]	[illegible]	[illegible]
65	·9063	[illegible]	[illegible]	[illegible]	9092	9100	9107	9114	9121	9128	9135	24	[illegible]	[illegible]	[illegible]	[illegible]	[illegible]	[illegible]	[illegible]	[illegible]	[illegible]
66	·9135	9143	9150	9157	9164	9171	9178	9184	9191	9198	9205	23	[illegible]	[illegible]	[illegible]	[illegible]	[illegible]	[illegible]	[illegible]	[illegible]	[illegible]
67	·9205	[illegible]	[illegible]	[illegible]	[illegible]	[illegible]	[illegible]	9252	9259	9265	9272	22	[illegible]	[illegible]	[illegible]	[illegible]	[illegible]	[illegible]	[illegible]	[illegible]	[illegible]
68	·9272	[illegible]	[illegible]	[illegible]	[illegible]	[illegible]	[illegible]	[illegible]	[illegible]	[illegible]	9336	21	[illegible]	[illegible]	[illegible]	[illegible]	[illegible]	[illegible]	[illegible]	[illegible]	[illegible]
69	·9336	[illegible]	[illegible]	[illegible]	[illegible]	[illegible]	[illegible]	[illegible]	[illegible]	[illegible]	9397	20	[illegible]	[illegible]	[illegible]	[illegible]	[illegible]	[illegible]	[illegible]	[illegible]	[illegible]
70	·9397	[illegible]	[illegible]	[illegible]	[illegible]	[illegible]	[illegible]	[illegible]	[illegible]	[illegible]	9455	19	[illegible]	[illegible]	[illegible]	[illegible]	[illegible]	[illegible]	[illegible]	[illegible]	[illegible]
71	·9455	[illegible]	[illegible]	[illegible]	[illegible]	[illegible]	[illegible]	[illegible]	[illegible]	[illegible]	9511	18	[illegible]	[illegible]	[illegible]	[illegible]	[illegible]	[illegible]	[illegible]	[illegible]	[illegible]
72	·9511	[illegible]	[illegible]	[illegible]	[illegible]	[illegible]	[illegible]	[illegible]	[illegible]	[illegible]	9563	17	[illegible]	[illegible]	[illegible]	[illegible]	[illegible]	[illegible]	[illegible]	[illegible]	[illegible]
73	·9563	[illegible]	[illegible]	[illegible]	[illegible]	[illegible]	[illegible]	[illegible]	[illegible]	[illegible]	9613	16	[illegible]	[illegible]	[illegible]	[illegible]	[illegible]	[illegible]	[illegible]	[illegible]	[illegible]
74	·9613	[illegible]	[illegible]	[illegible]	[illegible]	[illegible]	[illegible]	[illegible]	[illegible]	[illegible]	9659	15	[illegible]	[illegible]	[illegible]	[illegible]	[illegible]	[illegible]	[illegible]	[illegible]	[illegible]
75	·9659	[illegible]	[illegible]	[illegible]	[illegible]	[illegible]	[illegible]	[illegible]	[illegible]	[illegible]	9703	14	[illegible]	[illegible]	[illegible]	[illegible]	[illegible]	[illegible]	[illegible]	[illegible]	[illegible]
76	·9703	[illegible]	[illegible]	[illegible]	[illegible]	[illegible]	[illegible]	[illegible]	[illegible]	[illegible]	9744	13	[illegible]	[illegible]	[illegible]	[illegible]	[illegible]	[illegible]	[illegible]	[illegible]	[illegible]
77	·9744	[illegible]	[illegible]	[illegible]	[illegible]	[illegible]	[illegible]	[illegible]	[illegible]	[illegible]	9781	12	[illegible]	[illegible]	[illegible]	[illegible]	[illegible]	[illegible]	[illegible]	[illegible]	[illegible]
78	·9781	[illegible]	[illegible]	[illegible]	[illegible]	[illegible]	[illegible]	[illegible]	[illegible]	[illegible]	9816	11	[illegible]	[illegible]	[illegible]	[illegible]	[illegible]	[illegible]	[illegible]	[illegible]	[illegible]
79	·9816	[illegible]	[illegible]	[illegible]	[illegible]	[illegible]	[illegible]	[illegible]	[illegible]	[illegible]	9848	10	[illegible]	[illegible]	[illegible]	[illegible]	[illegible]	[illegible]	[illegible]	[illegible]	[illegible]
80	·9848	[illegible]	[illegible]	[illegible]	[illegible]	[illegible]	[illegible]	[illegible]	[illegible]	[illegible]	9877	9	[illegible]	[illegible]	[illegible]	[illegible]	[illegible]	[illegible]	[illegible]	[illegible]	[illegible]
81	·9877	[illegible]	[illegible]	[illegible]	[illegible]	[illegible]	[illegible]	[illegible]	[illegible]	[illegible]	9903	8	[illegible]	[illegible]	[illegible]	[illegible]	[illegible]	[illegible]	[illegible]	[illegible]	[illegible]
82	·9903	[illegible]	[illegible]	[illegible]	[illegible]	[illegible]	[illegible]	[illegible]	[illegible]	[illegible]	9925	7	[illegible]	[illegible]	[illegible]	[illegible]	[illegible]	[illegible]	[illegible]	[illegible]	[illegible]
83	·9925	[illegible]	[illegible]	[illegible]	[illegible]	[illegible]	[illegible]	[illegible]	[illegible]	[illegible]	9945	6	[illegible]	[illegible]	[illegible]	[illegible]	[illegible]	[illegible]	[illegible]	[illegible]	[illegible]
84	·9945	[illegible]	[illegible]	[illegible]	[illegible]	[illegible]	[illegible]	[illegible]	[illegible]	[illegible]	9962	5	[illegible]	[illegible]	[illegible]	[illegible]	[illegible]	[illegible]	[illegible]	[illegible]	[illegible]
85	·9962	[illegible]	[illegible]	[illegible]	[illegible]	[illegible]	[illegible]	[illegible]	[illegible]	[illegible]	9976	4	[illegible]	[illegible]	[illegible]	[illegible]	[illegible]	[illegible]	[illegible]	[illegible]	[illegible]
86	·9976	[illegible]	[illegible]	[illegible]	[illegible]	[illegible]	[illegible]	[illegible]	[illegible]	[illegible]	9986	3	[illegible]	[illegible]	[illegible]	[illegible]	[illegible]	[illegible]	[illegible]	[illegible]	[illegible]
87	·9986	[illegible]	[illegible]	[illegible]	[illegible]	[illegible]	[illegible]	[illegible]	[illegible]	[illegible]	9994	2	[illegible]	[illegible]	[illegible]	[illegible]	[illegible]	[illegible]	[illegible]	[illegible]	[illegible]
88	·9994	[illegible]	[illegible]	[illegible]	[illegible]	[illegible]	[illegible]	[illegible]	[illegible]	[illegible]	9998	1	[illegible]	[illegible]	[illegible]	[illegible]	[illegible]	[illegible]	[illegible]	[illegible]	[illegible]
89	·9998	[illegible]	[illegible]	[illegible]	[illegible]	[illegible]	[illegible]	[illegible]	[illegible]	[illegible]	0000	0	[illegible]	[illegible]	[illegible]	[illegible]	[illegible]	[illegible]	[illegible]	[illegible]	[illegible]

Natural sines

Deg	·0	·1	·2	·3	·4	·5	·6	·7	·8	·9	1·0	(89−deg)	1	2	3	4	5	6	7	8	9
0°	·0000	0017	0035	0052	0070	0087	0105	0122	0140	0157	0175	89°	[illegible]	[illegible]	[illegible]	[illegible]	[illegible]	[illegible]	[illegible]	[illegible]	[illegible]
1	·0175	0192	0209	0227	0244	0262	0279	0297	0314	0332	0349	88	[illegible]	[illegible]	[illegible]	[illegible]	[illegible]	[illegible]	[illegible]	[illegible]	[illegible]
2	·0349	0366	0384	0401	0419	0436	0454	0471	0488	0506	0523	87	[illegible]	[illegible]	[illegible]	[illegible]	[illegible]	[illegible]	[illegible]	[illegible]	[illegible]
3	·0523	0541	0558	0576	0593	0610	0628	0645	0663	0680	0698	86	[illegible]	[illegible]	[illegible]	[illegible]	[illegible]	[illegible]	[illegible]	[illegible]	[illegible]
4	·0698	0715	0732	0750	0767	0785	0802	0819	0837	0854	0872	85	[illegible]	[illegible]	[illegible]	[illegible]	[illegible]	[illegible]	[illegible]	[illegible]	[illegible]
5	·0872	0889	0906	0924	0941	0958	0976	0993	1011	1028	1045	84	[illegible]	[illegible]	[illegible]	[illegible]	[illegible]	[illegible]	[illegible]	[illegible]	[illegible]
6	·1045	1063	1080	1097	1115	1132	1149	1167	1184	1201	1219	83	[illegible]	[illegible]	[illegible]	[illegible]	[illegible]	[illegible]	[illegible]	[illegible]	[illegible]
7	·1219	1236	1253	1271	1288	1305	1323	1340	1357	1374	1392	82	[illegible]	[illegible]	[illegible]	[illegible]	[illegible]	[illegible]	[illegible]	[illegible]	[illegible]
8	·1392	1409	1426	1444	1461	1478	1495	1513	1530	1547	1564	81	[illegible]	[illegible]	[illegible]	[illegible]	[illegible]	[illegible]	[illegible]	[illegible]	[illegible]
9	·1564	1582	1599	1616	1633	1650	1668	1685	1702	1719	1736	80	[illegible]	[illegible]	[illegible]	[illegible]	[illegible]	[illegible]	[illegible]	[illegible]	[illegible]
10	·1736	1754	1771	1788	1805	1822	1840	1857	1874	1891	1908	79	[illegible]	[illegible]	[illegible]	[illegible]	[illegible]	[illegible]	[illegible]	[illegible]	[illegible]
11	·1908	1925	1942	1959	1977	1994	2011	2028	2045	2062	2079	78	[illegible]	[illegible]	[illegible]	[illegible]	[illegible]	[illegible]	[illegible]	[illegible]	[illegible]
12	·2079	2096	2113	2130	2147	2164	2181	2198	2215	2233	2250	77	[illegible]	[illegible]	[illegible]	[illegible]	[illegible]	[illegible]	[illegible]	[illegible]	[illegible]
13	·2250	2267	2284	2300	2317	2334	2351	2368	2385	2402	2419	76	[illegible]	[illegible]	[illegible]	[illegible]	[illegible]	[illegible]	[illegible]	[illegible]	[illegible]
14	·2419	2436	2453	2470	2487	2504	2521	2538	2554	2571	2588	75	[illegible]	[illegible]	[illegible]	[illegible]	[illegible]	[illegible]	[illegible]	[illegible]	[illegible]
15	·2588	2605	2622	2639	2656	2672	2689	2706	2723	2740	2756	74	[illegible]	[illegible]	[illegible]	[illegible]	[illegible]	[illegible]	[illegible]	[illegible]	[illegible]
16	·2756	2773	2790	2807	2823	2840	2857	2874	2890	2907	2924	73	[illegible]	[illegible]	[illegible]	[illegible]	[illegible]	[illegible]	[illegible]	[illegible]	[illegible]
17	·2924	2940	2957	2974	2990	3007	3024	3040	3057	3074	3090	72	[illegible]	[illegible]	[illegible]	[illegible]	[illegible]	[illegible]	[illegible]	[illegible]	[illegible]
18	·3090	3107	3123	3140	3156	3173	3190	3206	3223	3239	3256	71	[illegible]	[illegible]	[illegible]	[illegible]	[illegible]	[illegible]	[illegible]	[illegible]	[illegible]
19	·3256	3272	3289	3305	3322	3338	3355	3371	3387	3404	3420	70	[illegible]	[illegible]	[illegible]	[illegible]	[illegible]	[illegible]	[illegible]	[illegible]	[illegible]
20	·3420	3437	3453	3469	3486	3502	3518	3535	3551	3567	3584	69	[illegible]	[illegible]	[illegible]	[illegible]	[illegible]	[illegible]	[illegible]	[illegible]	[illegible]
21	·3584	3600	3616	3633	3649	3665	3681	3697	3714	3730	3746	68	[illegible]	[illegible]	[illegible]	[illegible]	[illegible]	[illegible]	[illegible]	[illegible]	[illegible]
22	·3746	3762	3778	3795	3811	3827	3843	3859	3875	3891	3907	67	[illegible]	[illegible]	[illegible]	[illegible]	[illegible]	[illegible]	[illegible]	[illegible]	[illegible]
23	·3907	3923	3939	3955	3971	3987	4003	4019	4035	4051	4067	66	[illegible]	[illegible]	[illegible]	[illegible]	[illegible]	[illegible]	[illegible]	[illegible]	[illegible]
24	·4067	4083	4099	4115	4131	4147	4163	4179	4195	4210	4226	65	[illegible]	[illegible]	[illegible]	[illegible]	[illegible]	[illegible]	[illegible]	[illegible]	[illegible]
25	·4226	4242	4258	4274	4289	4305	4321	4337	4352	4368	4384	64	[illegible]	[illegible]	[illegible]	[illegible]	[illegible]	[illegible]	[illegible]	[illegible]	[illegible]
26	·4384	4399	4415	4431	4446	4462	4478	4493	4509	4524	4540	63	[illegible]	[illegible]	[illegible]	[illegible]	[illegible]	[illegible]	[illegible]	[illegible]	[illegible]
27	·4540	4555	4571	4586	4602	4617	4633	4648	4664	4679	4695	62	[illegible]	[illegible]	[illegible]	[illegible]	[illegible]	[illegible]	[illegible]	[illegible]	[illegible]
28	·4695	4710	4726	4741	4756	4772	4787	4802	4818	4833	4848	61	[illegible]	[illegible]	[illegible]	[illegible]	[illegible]	[illegible]	[illegible]	[illegible]	[illegible]
29	·4848	4863	4879	4894	4909	4924	4939	4955	4970	4985	5000	60	[illegible]	[illegible]	[illegible]	[illegible]	[illegible]	[illegible]	[illegible]	[illegible]	[illegible]
30	·5000	5015	5030	5045	5060	5075	5090	5105	5120	5135	5150	59	[illegible]	[illegible]	[illegible]	[illegible]	[illegible]	[illegible]	[illegible]	[illegible]	[illegible]
31	·5150	5165	5180	5195	5210	5225	5240	5255	5270	5284	5299	58	[illegible]	[illegible]	[illegible]	[illegible]	[illegible]	[illegible]	[illegible]	[illegible]	[illegible]
32	·5299	5314	5329	5344	5358	5373	5388	5402	5417	5432	5446	57	[illegible]	[illegible]	[illegible]	[illegible]	[illegible]	[illegible]	[illegible]	[illegible]	[illegible]
33	·5446	5461	5476	5490	5505	5519	5534	5548	5563	5577	5592	56	[illegible]	[illegible]	[illegible]	[illegible]	[illegible]	[illegible]	[illegible]	[illegible]	[illegible]
34	·5592	5606	5621	5635	5650	5664	5678	5693	5707	5721	5736	55	[illegible]	[illegible]	[illegible]	[illegible]	[illegible]	[illegible]	[illegible]	[illegible]	[illegible]
35	·5736	5750	5764	5779	5793	5807	5821	5835	5850	5864	5878	54	[illegible]	[illegible]	[illegible]	[illegible]	[illegible]	[illegible]	[illegible]	[illegible]	[illegible]
36	·5878	5892	5906	5920	5934	5948	5962	5976	5990	6004	6018	53	[illegible]	[illegible]	[illegible]	[illegible]	[illegible]	[illegible]	[illegible]	[illegible]	[illegible]
37	·6018	6032	6046	6060	6074	6088	6101	6115	6129	6143	6157	52	[illegible]	[illegible]	[illegible]	[illegible]	[illegible]	[illegible]	[illegible]	[illegible]	[illegible]
38	·6157	6170	6184	6198	6211	6225	6239	6252	6266	6280	6293	51	[illegible]	[illegible]	[illegible]	[illegible]	[illegible]	[illegible]	[illegible]	[illegible]	[illegible]
39	·6293	6307	6320	6334	6347	6361	6374	6388	6401	6414	6428	50	[illegible]	[illegible]	[illegible]	[illegible]	[illegible]	[illegible]	[illegible]	[illegible]	[illegible]
40	·6428	6441	6455	6468	6481	6494	6508	6521	6534	6547	6561	49	[illegible]	[illegible]	[illegible]	[illegible]	[illegible]	[illegible]	[illegible]	[illegible]	[illegible]
41	·6561	6574	6587	6600	6613	6626	6639	6652	6665	6678	6691	48	[illegible]	[illegible]	[illegible]	[illegible]	[illegible]	[illegible]	[illegible]	[illegible]	[illegible]
42	·6691	6704	6717	6730	6743	6756	6769	6782	6794	6807	6820	47	[illegible]	[illegible]	[illegible]	[illegible]	[illegible]	[illegible]	[illegible]	[illegible]	[illegible]
43	·6820	6833	6845	6858	6871	6884	6896	6909	6921	6934	6947	46	[illegible]	[illegible]	[illegible]	[illegible]	[illegible]	[illegible]	[illegible]	[illegible]	[illegible]

Natural cotangents

Column headers (degrees · tenths), read left → right: 1·0, ·9, ·8, ·7, ·6, ·5, ·4, ·3, ·2, ·1, ·0. The last nine columns are the mean differences (headed 1–9). For angles above 66° (left-angle below 23°) the table prints: **Use Linear Interpolation**.

°	1·0	·9	·8	·7	·6	·5	·4	·3	·2	·1	·0	°	1	2	3	4	5	6	7	8	9
45	0000	9965	9930	9896	9861	9827	9793	9759	9725	9691	0·9657	44	3	7	10	14	17	21	24	27	31
44	0355	0319	0283	0247	0212	0176	0141	0105	0070	0035	1·0000	45	4	7	11	14	18	21	25	28	32
43	0724	0686	0649	0612	0575	0538	0501	0464	0428	0392	1·0355	46	4	7	11	15	18	22	26	29	33
42	1106	1067	1028	0990	0951	0913	0875	0837	0799	0761	1·0724	47	4	8	11	15	19	23	27	31	34
41	1504	1463	1423	1383	1343	1303	1263	1224	1184	1145	1·1106	48	4	8	12	16	20	24	28	32	36
40	1918	1875	1833	1792	1750	1708	1667	1626	1585	1544	1·1504	49	4	8	12	17	21	25	29	33	37
39	2349	2305	2261	2218	2174	2131	2088	2045	2002	1960	1·1918	50	4	9	13	17	22	26	30	35	39
38	2799	2753	2708	2662	2617	2572	2527	2482	2437	2393	1·2349	51	5	9	14	18	23	27	32	36	41
37	3270	3222	3175	3127	3079	3032	2985	2938	2892	2846	1·2799	52	5	9	14	19	24	28	33	38	42
36	3764	3713	3663	3613	3564	3514	3465	3416	3367	3319	1·3270	53	5	10	15	20	24	29	34	39	44
35	4281	4229	4176	4124	4071	4019	3968	3916	3865	3814	1·3764	54	5	10	16	21	26	31	36	41	47
34	4826	4770	4715	4659	4605	4550	4496	4442	4388	4335	1·4281	55	5	11	16	22	27	33	38	44	49
33	5399	5340	5282	5224	5166	5108	5051	4994	4938	4882	1·4826	56	6	11	17	23	29	34	40	46	52
32	6003	5941	5880	5818	5757	5697	5637	5577	5517	5458	1·5399	57	6	12	18	24	30	36	42	48	54
31	6643	6577	6512	6447	6383	6319	6255	6191	6128	6066	1·6003	58	6	13	19	26	32	38	45	51	58
30	7321	7251	7182	7113	7045	6977	6909	6842	6775	6709	1·6643	59	7	14	20	27	34	41	47	54	61
29	8040	7966	7893	7820	7747	7675	7603	7532	7461	7391	1·7321	60	7	14	22	29	36	43	50	58	65
28	8807	8728	8650	8572	8495	8418	8341	8265	8190	8115	1·8040	61	8	15	23	31	38	46	54	61	69
27	9626	9542	9458	9375	9292	9210	9128	9047	8967	8887	1·8807	62	8	16	25	33	41	49	57	66	74
26	0503	0413	0323	0233	0145	0057	9970	9883	9797	9711	1·9626	63	9	18	26	35	44	53	61	70	79
25	1445	1348	1251	1155	1060	0965	0872	0778	0686	0594	2·0503	64	9	19	28	38	47	57	66	75	85
24	2460	2355	2251	2148	2045	1943	1842	1742	1642	1543	2·1445	65	10	20	30	41	51	61	71	81	91
23	3559	3445	3332	3220	3109	2998	2889	2781	2673	2566	2·2460	66	11	22	33	44	55	66	77	88	99
22	4751	4627	4504	4383	4262	4142	4023	3906	3789	3673	2·3559	67									
21	6051	5916	5782	5649	5517	5386	5257	5129	5002	4876	2·4751	68									
20	7475	7326	7179	7034	6889	6746	6605	6464	6325	6187	2·6051	69									
19	9042	8878	8716	8556	8397	8239	8083	7929	7776	7625	2·7475	70									
18	0777	0595	0415	0237	0061	9887	9714	9544	9375	9208	2·9042	71									
17	2709	2506	2305	2106	1910	1716	1524	1334	1146	0961	3·0777	72									
16	4874	4646	4420	4197	3977	3759	3544	3332	3122	2914	3·2709	73									
15	7321	7062	6806	6554	6305	6059	5816	5576	5339	5105	3·4874	74									
14	0108	9812	9520	9232	8947	8667	8391	8118	7848	7583	3·7321	75									
13	3315	2972	2635	2303	1976	1653	1335	1022	0713	0408	4·0108	76									
12	7046	6646	6252	5864	5483	5107	4737	4373	4015	3662	4·3315	77									
11	1446	0970	0504	0045	9594	9152	8716	8288	7867	7453	4·7046	78									
10	6713	6140	5578	5026	4486	3955	3435	2924	2422	1929	5·1446	79									
9	6·314	6·243	6·174	6·107	6·041	5·976	5·912	5·850	5·789	5·730	5·671	80									
8	7·115	7·026	6·940	6·855	6·772	6·691	6·612	6·535	6·460	6·386	6·314	81									
7	8·144	8·028	7·916	7·806	7·700	7·596	7·495	7·396	7·300	7·207	7·115	82									
6	9·514	9·357	9·205	9·058	8·915	8·777	8·643	8·513	8·386	8·264	8·144	83									
5	11·430	11·205	10·988	10·780	10·579	10·385	10·199	10·019	9·845	9·677	9·514	84									
4	14·301	13·951	13·617	13·300	12·996	12·706	12·429	12·163	11·909	11·664	11·430	85									
3	19·081	18·464	17·886	17·343	16·832	16·350	15·895	15·464	15·056	14·669	14·301	86									
2	28·636	27·271	26·031	24·898	23·859	22·904	22·022	21·205	20·446	19·740	19·081	87									
1	57·290	52·081	47·740	44·066	40·917	38·188	35·801	33·694	31·821	30·145	28·636	88									
0		572·96	286·48	190·98	143·24	114·59	95·49	81·85	71·62	63·66	57·290	89									

Use Linear Interpolation

Natural tangents

Column headers (degrees · tenths), read left → right: 1·0, ·9, ·8, ·7, ·6, ·5, ·4, ·3, ·2, ·1, ·0. The last nine columns are the mean differences (headed 1–9).

°	1·0	·9	·8	·7	·6	·5	·4	·3	·2	·1	·0	°	1	2	3	4	5	6	7	8	9
0	0175	0157	0140	0122	0105	0087	0070	0052	0035	0017	0·0000	89	2	3	5	7	9	10	12	14	16
1	0349	0332	0314	0297	0279	0262	0244	0227	0209	0192	0·0175	88	2	3	5	7	9	11	12	14	16
2	0524	0507	0489	0472	0454	0437	0419	0402	0384	0367	0·0349	87	2	4	5	7	9	11	12	14	16
3	0699	0682	0664	0647	0629	0612	0594	0577	0559	0542	0·0524	86	2	4	5	7	9	11	13	14	16
4	0875	0857	0840	0822	0805	0787	0769	0752	0734	0717	0·0699	85	2	4	5	7	9	11	13	15	16
5	1051	1033	1016	0998	0981	0963	0945	0928	0910	0892	0·0875	84	2	4	5	7	9	11	13	15	16
6	1228	1210	1192	1175	1157	1139	1122	1104	1086	1069	0·1051	83	2	4	5	7	9	11	13	15	17
7	1405	1388	1370	1352	1334	1317	1299	1281	1263	1246	0·1228	82	2	4	5	7	9	11	13	15	17
8	1584	1566	1548	1530	1512	1495	1477	1459	1441	1423	0·1405	81	2	4	5	7	9	11	13	15	17
9	1763	1745	1727	1709	1691	1673	1655	1638	1620	1602	0·1584	80	2	4	5	7	9	11	13	15	17
10	1944	1926	1908	1890	1871	1853	1835	1817	1799	1781	0·1763	79	2	4	5	7	9	11	13	15	17
11	2126	2107	2089	2071	2053	2035	2016	1998	1980	1962	0·1944	78	2	4	5	7	9	11	13	15	17
12	2309	2290	2272	2254	2235	2217	2199	2180	2162	2144	0·2126	77	2	4	6	7	9	11	13	15	17
13	2493	2475	2456	2438	2419	2401	2382	2364	2345	2327	0·2309	76	2	4	6	8	9	11	13	15	17
14	2679	2661	2642	2623	2605	2586	2568	2549	2530	2512	0·2493	75	2	4	6	8	10	11	13	15	17
15	2867	2849	2830	2811	2792	2773	2754	2736	2717	2698	0·2679	74	2	4	6	8	10	12	13	15	17
16	3057	3038	3019	3000	2981	2962	2943	2924	2905	2886	0·2867	73	2	4	6	8	10	12	14	15	17
17	3249	3230	3211	3191	3172	3153	3134	3115	3096	3076	0·3057	72	2	4	6	8	10	12	14	16	18
18	3443	3424	3404	3385	3365	3346	3327	3307	3288	3269	0·3249	71	2	4	6	8	10	12	14	16	18
19	3640	3620	3600	3581	3561	3541	3522	3502	3482	3463	0·3443	70	2	4	6	8	10	12	14	16	18
20	3839	3819	3799	3779	3759	3739	3719	3699	3679	3659	0·3640	69	2	4	6	8	10	12	14	16	18
21	4040	4020	4000	3979	3959	3939	3919	3899	3879	3859	0·3839	68	2	4	6	8	10	12	15	17	19
22	4245	4224	4204	4183	4163	4142	4122	4101	4081	4061	0·4040	67	2	4	6	8	11	13	15	17	19
23	4452	4431	4411	4390	4369	4348	4327	4307	4286	4265	0·4245	66	2	4	6	9	11	13	15	17	19
24	4663	4642	4621	4599	4578	4557	4536	4515	4494	4473	0·4452	65	2	4	6	9	11	13	15	18	20
25	4877	4856	4834	4813	4791	4770	4748	4727	4706	4684	0·4663	64	2	4	6	9	11	13	16	18	20
26	5095	5073	5051	5029	5008	4986	4964	4942	4921	4899	0·4877	63	2	4	7	9	11	14	16	18	20
27	5317	5295	5272	5250	5228	5206	5184	5161	5139	5117	0·5095	62	2	5	7	9	11	14	16	18	21
28	5543	5520	5498	5475	5452	5430	5407	5384	5362	5340	0·5317	61	2	5	7	9	12	14	16	19	21
29	5774	5750	5727	5704	5681	5658	5635	5612	5589	5566	0·5543	60	2	5	7	9	12	14	17	19	21
30	6009	5985	5961	5938	5914	5890	5867	5844	5820	5797	0·5774	59	2	5	7	10	12	15	17	19	22
31	6249	6224	6200	6176	6152	6128	6104	6080	6056	6032	0·6009	58	2	5	7	10	12	15	17	20	22
32	6494	6469	6445	6420	6395	6371	6346	6322	6297	6273	0·6249	57	3	5	8	10	13	15	18	20	23
33	6745	6720	6694	6669	6644	6619	6594	6569	6544	6519	0·6494	56	3	5	8	10	13	15	18	21	23
34	7002	6976	6950	6924	6899	6873	6847	6822	6796	6771	0·6745	55	3	5	8	11	13	16	19	21	24
35	7265	7239	7212	7186	7159	7133	7107	7080	7054	7028	0·7002	54	3	5	8	11	14	16	19	22	24
36	7536	7508	7481	7454	7427	7400	7373	7346	7319	7292	0·7265	53	3	5	8	11	14	17	19	22	25
37	7813	7785	7757	7729	7701	7673	7646	7618	7590	7563	0·7536	52	3	6	8	11	14	17	20	23	25
38	8098	8069	8040	8012	7983	7954	7926	7898	7869	7841	0·7813	51	3	6	9	11	15	17	20	23	26
39	8391	8361	8332	8302	8273	8243	8214	8185	8156	8127	0·8098	50	3	6	9	12	15	18	21	23	26
40	8693	8662	8632	8601	8571	8541	8511	8481	8451	8421	0·8391	49	3	6	9	12	15	18	21	24	27
41	9004	8972	8941	8910	8878	8847	8816	8785	8754	8724	0·8693	48	3	6	9	12	16	19	22	25	28
42	9325	9293	9260	9228	9195	9163	9131	9099	9067	9036	0·9004	47	3	6	10	13	16	19	22	26	29
43	9657	9623	9590	9556	9523	9490	9457	9424	9391	9358	0·9325	46	3	7	10	13	17	20	23	27	30

Degrees to radians

Degrees	Radians	Degrees	Radians	Degrees	Radians	Degrees	Radians
0	0·0000	23	0·4014	46	0·8029	69	1·2043
1	0·0175	24	0·4189	47	0·8203	70	1·2217
2	0·0349	25	0·4363	48	0·8378	71	1·2392
3	0·0524	26	0·4538	49	0·8552	72	1·2566
4	0·0698	27	0·4712	50	0·8727	73	1·2741
5	0·0873	28	0·4887	51	0·8901	74	1·2915
6	0·1047	29	0·5061	52	0·9076	75	1·3090
7	0·1222	30	0·5236	53	0·9250	76	1·3265
8	0·1396	31	0·5411	54	0·9425	77	1·3439
9	0·1571	32	0·5585	55	0·9599	78	1·3614
10	0·1745	33	0·5760	56	0·9774	79	1·3788
11	0·1920	34	0·5934	57	0·9948	80	1·3963
12	0·2094	35	0·6109	58	1·0123	81	1·4137
13	0·2269	36	0·6283	59	1·0297	82	1·4312
14	0·2443	37	0·6458	60	1·0472	83	1·4486
15	0·2618	38	0·6632	61	1·0647	84	1·4661
16	0·2793	39	0·6807	62	1·0821	85	1·4835
17	0·2967	40	0·6981	63	1·0996	86	1·5010
18	0·3142	41	0·7156	64	1·1170	87	1·5184
19	0·3316	42	0·7330	65	1·1345	88	1·5359
20	0·3491	43	0·7505	66	1·1519	89	1·5533
21	0·3665	44	0·7679	67	1·1694	90	1·5708
22	0·3840	45	0·7854	68	1·1868		

Minutes to radians

Minutes	Radians	Minutes	Radians	Minutes	Radians
1	0·0003	6	0·0017	20	0·0058
2	0·0006	7	0·0020	30	0·0087
3	0·0009	8	0·0023	40	0·0116
4	0·0012	9	0·0026	50	0·0145
5	0·0015	10	0·0029	60	0·0175

To convert degrees and minutes to radians for angles other than those given in these tables, use linear interpolation.

Radians to degrees

Radians	Degrees	Radians	Degrees	Radians	Degrees	Radians	Degrees
0·0	0·0000	0·8	45·8366	1·6	91·6732	2·4	137·5099
0·1	5·7296	0·9	51·5662	1·7	97·4028	2·5	143·2394
0·2	11·4592	1·0	57·2958	1·8	103·1324	2·6	148·9690
0·3	17·1887	1·1	63·0254	1·9	108·8620	2·7	154·6986
0·4	22·9183	1·2	68·7549	2·0	114·5916	2·8	160·4282
0·5	28·6479	1·3	74·4845	2·1	120·3211	2·9	166·1578
0·6	34·3775	1·4	80·2141	2·2	126·0507	3·0	171·8873
0·7	40·1070	1·5	85·9437	2·3	131·7803	3·1	177·6169
						3·2	183·3465

To convert radians to degrees for angles other than those given in this table, use linear interpolation.

Logarithms of factorials

x	$\log_{10}x!$	x	$\log_{10}x!$	x	$\log_{10}x!$	x	$\log_{10}x!$
1	0·0000	26	26·6056	51	66·1906	76	111·2754
2	0·3010	27	28·0370	52	67·9066	77	113·1619
3	0·7782	28	29·4841	53	69·6309	78	115·0540
4	1·3802	29	30·9465	54	71·3633	79	116·9516
5	2·0792	30	32·4237	55	73·1037	80	118·8547
6	2·8573	31	33·9150	56	74·8519	81	120·7632
7	3·7024	32	35·4202	57	76·6077	82	122·6770
8	4·6055	33	36·9387	58	78·3712	83	124·5961
9	5·5598	34	38·4702	59	80·1420	84	126·5204
10	6·5598	35	40·0142	60	81·9202	85	128·4498
11	7·6012	36	41·5705	61	83·7055	86	130·3843
12	8·6803	37	43·1387	62	85·4979	87	132·3238
13	9·7943	38	44·7185	63	87·2972	88	134·2683
14	10·9404	39	46·3096	64	89·1034	89	136·2177
15	12·1165	40	47·9116	65	90·9163	90	138·1719
16	13·3206	41	49·5244	66	92·7359	91	140·1310
17	14·5511	42	51·1477	67	94·5619	92	142·0948
18	15·8063	43	52·7811	68	96·3945	93	144·0632
19	17·0851	44	54·4246	69	98·2333	94	146·0364
20	18·3861	45	56·0778	70	100·0784	95	148·0141
21	19·7083	46	57·7406	71	101·9297	96	149·9964
22	21·0508	47	59·4127	72	103·7870	97	151·9831
23	22·4125	48	61·0939	73	105·6503	98	153·9744
24	23·7927	49	62·7841	74	107·5196	99	155·9700
25	25·1906	50	64·4831	75	109·3946	100	157·9700

If x is greater than 100, factorials may be obtained from the series:

$$\log_{10}x! = 0·61624 + (x+\tfrac{1}{2})\,(\log_{10}x - 0·4342945) + 0·0362/x - \ldots .$$

Circular functions

x	sin x	cos x	tan x	x	sin x	cos x	tan x
0.00	0.0000	1.0000	0.0000	0.55	0.5227	0.8525	0.6131
0.01	0.0100	1.0000	0.0100	0.56	0.5312	0.8473	0.6269
0.02	0.0200	0.9998	0.0200	0.57	0.5396	0.8419	0.6410
0.03	0.0300	0.9996	0.0300	0.58	0.5480	0.8365	0.6552
0.04	0.0400	0.9992	0.0400	0.59	0.5564	0.8309	0.6696
0.05	0.0500	0.9988	0.0500	0.60	0.5646	0.8253	0.6841
0.06	0.0600	0.9982	0.0601	0.61	0.5729	0.8196	0.6989
0.07	0.0699	0.9976	0.0701	0.62	0.5810	0.8139	0.7139
0.08	0.0799	0.9968	0.0802	0.63	0.5891	0.8080	0.7291
0.09	0.0899	0.9960	0.0902	0.64	0.5972	0.8021	0.7445
0.10	0.0998	0.9950	0.1003	0.65	0.6052	0.7961	0.7602
0.11	0.1098	0.9940	0.1104	0.66	0.6131	0.7900	0.7761
0.12	0.1197	0.9928	0.1206	0.67	0.6210	0.7838	0.7923
0.13	0.1296	0.9916	0.1307	0.68	0.6288	0.7776	0.8087
0.14	0.1395	0.9902	0.1409	0.69	0.6365	0.7712	0.8253
0.15	0.1494	0.9888	0.1511	0.70	0.6442	0.7648	0.8423
0.16	0.1593	0.9872	0.1614	0.71	0.6518	0.7584	0.8595
0.17	0.1692	0.9856	0.1717	0.72	0.6594	0.7518	0.8771
0.18	0.1790	0.9838	0.1820	0.73	0.6669	0.7452	0.8949
0.19	0.1889	0.9820	0.1923	0.74	0.6743	0.7385	0.9131
0.20	0.1987	0.9801	0.2027	0.75	0.6816	0.7317	0.9316
0.21	0.2085	0.9780	0.2131	0.76	0.6889	0.7248	0.9505
0.22	0.2182	0.9759	0.2236	0.77	0.6961	0.7179	0.9697
0.23	0.2280	0.9737	0.2341	0.78	0.7033	0.7109	0.9893
0.24	0.2377	0.9713	0.2447	0.79	0.7104	0.7038	1.0092
0.25	0.2474	0.9689	0.2553	0.80	0.7174	0.6967	1.0296
0.26	0.2571	0.9664	0.2660	0.81	0.7243	0.6895	1.0505
0.27	0.2667	0.9638	0.2768	0.82	0.7311	0.6822	1.0717
0.28	0.2764	0.9611	0.2876	0.83	0.7379	0.6749	1.0934
0.29	0.2860	0.9582	0.2984	0.84	0.7446	0.6675	1.1156
0.30	0.2955	0.9553	0.3093	0.85	0.7513	0.6600	1.1383
0.31	0.3051	0.9523	0.3203	0.86	0.7578	0.6524	1.1616
0.32	0.3146	0.9492	0.3314	0.87	0.7643	0.6448	1.1853
0.33	0.3240	0.9460	0.3425	0.88	0.7707	0.6372	1.2097
0.34	0.3335	0.9428	0.3537	0.89	0.7771	0.6294	1.2346
0.35	0.3429	0.9394	0.3650	0.90	0.7833	0.6216	1.2602
0.36	0.3523	0.9359	0.3764	0.91	0.7895	0.6137	1.2864
0.37	0.3616	0.9323	0.3879	0.92	0.7956	0.6058	1.3133
0.38	0.3709	0.9287	0.3994	0.93	0.8016	0.5978	1.3409
0.39	0.3802	0.9249	0.4111	0.94	0.8076	0.5898	1.3692
0.40	0.3894	0.9211	0.4228	0.95	0.8134	0.5817	1.3984
0.41	0.3986	0.9171	0.4346	0.96	0.8192	0.5735	1.4284
0.42	0.4078	0.9131	0.4466	0.97	0.8249	0.5653	1.4592
0.43	0.4169	0.9090	0.4586	0.98	0.8305	0.5570	1.4910
0.44	0.4259	0.9048	0.4708	0.99	0.8360	0.5487	1.5237
0.45	0.4350	0.9004	0.4831	1.00	0.8415	0.5403	1.5574
0.46	0.4439	0.8961	0.4954	1.01	0.8468	0.5319	1.5922
0.47	0.4529	0.8916	0.5080	1.02	0.8521	0.5234	1.6281
0.48	0.4618	0.8870	0.5206	1.03	0.8573	0.5148	1.6652
0.49	0.4706	0.8823	0.5334	1.04	0.8624	0.5062	1.7036
0.50	0.4794	0.8776	0.5463	1.05	0.8674	0.4976	1.7433
0.51	0.4882	0.8727	0.5594	1.06	0.8724	0.4889	1.7844
0.52	0.4969	0.8678	0.5726	1.07	0.8772	0.4801	1.8270
0.53	0.5055	0.8628	0.5859	1.08	0.8820	0.4713	1.8712
0.54	0.5141	0.8577	0.5994	1.09	0.8866	0.4625	1.9171

x	sin x	cos x	tan x	x	sin x	cos x	tan x
1.10	0.8912	0.4536	1.9648	1.35	0.9757	0.2190	4.4552
1.11	0.8957	0.4447	2.0143	1.36	0.9779	0.2092	4.6734
1.12	0.9001	0.4357	2.0660	1.37	0.9799	0.1994	4.9131
1.13	0.9044	0.4267	2.1198	1.38	0.9819	0.1896	5.1774
1.14	0.9086	0.4176	2.1759	1.39	0.9837	0.1798	5.4707
1.15	0.9128	0.4085	2.2345	1.40	0.9854	0.1700	5.798
1.16	0.9168	0.3993	2.2958	1.41	0.9871	0.1601	6.165
1.17	0.9208	0.3902	2.3600	1.42	0.9887	0.1502	6.581
1.18	0.9246	0.3809	2.4273	1.43	0.9901	0.1403	7.056
1.19	0.9284	0.3717	2.4979	1.44	0.9915	0.1304	7.602
1.20	0.9320	0.3624	2.5722	1.45	0.9927	0.1205	8.238
1.21	0.9356	0.3530	2.6503	1.46	0.9939	0.1106	8.989
1.22	0.9391	0.3436	2.7328	1.47	0.9949	0.1006	9.887
1.23	0.9425	0.3342	2.8198	1.48	0.9959	0.0907	10.983
1.24	0.9458	0.3248	2.9119	1.49	0.9967	0.0807	12.35
1.25	0.9490	0.3153	3.0096	1.50	0.9975	0.0707	14.101
1.26	0.9521	0.3058	3.1133	1.51	0.9982	0.0608	16.428
1.27	0.9551	0.2963	3.2236	1.52	0.9987	0.0508	19.669
1.28	0.9580	0.2867	3.3413	1.53	0.9992	0.0408	24.498
1.29	0.9608	0.2771	3.4672	1.54	0.9995	0.0308	32.461
1.30	0.9636	0.2675	3.6021	1.55	0.9998	0.0208	48.079
1.31	0.9662	0.2579	3.7471	1.56	0.9999	0.0108	92.621
1.32	0.9687	0.2482	3.9033	1.57	1.0000	0.0008	1256
1.33	0.9711	0.2385	4.0723				
1.34	0.9735	0.2288	4.2556				

x	sin x	cos x	tan x
$\pi/2$	1	0	
π	0	−1	0
$3\pi/2$	−1	0	
2π	0	1	0

Exponential functions

x	e^x	e^{-x}	$\sinh x$	$\cosh x$
0·00	1·000	1·0000	0·00000	1·000
0·01	1·010	0·9900	0·01000	1·000
0·02	1·020	0·9802	0·02000	1·000
0·03	1·030	0·9704	0·03000	1·000
0·04	1·041	0·9608	0·04001	1·001
0·05	1·051	0·9512	0·05002	1·001
0·06	1·062	0·9418	0·06004	1·002
0·07	1·073	0·9324	0·07006	1·002
0·08	1·083	0·9231	0·08009	1·003
0·09	1·094	0·9139	0·09012	1·004
0·10	1·105	0·9048	0·1002	1·005
0·11	1·116	0·8958	0·1102	1·006
0·12	1·127	0·8869	0·1203	1·007
0·13	1·139	0·8781	0·1304	1·008
0·14	1·150	0·8694	0·1405	1·010
0·15	1·162	0·8607	0·1506	1·011
0·16	1·174	0·8521	0·1607	1·013
0·17	1·185	0·8437	0·1708	1·014
0·18	1·197	0·8353	0·1810	1·016
0·19	1·209	0·8270	0·1911	1·018
0·20	1·221	0·8187	0·2013	1·020
0·21	1·234	0·8106	0·2115	1·022
0·22	1·246	0·8025	0·2218	1·024
0·23	1·259	0·7945	0·2320	1·027
0·24	1·271	0·7866	0·2423	1·029
0·25	1·284	0·7788	0·2526	1·031
0·26	1·297	0·7711	0·2629	1·034
0·27	1·310	0·7634	0·2733	1·037
0·28	1·323	0·7558	0·2837	1·039
0·29	1·336	0·7483	0·2941	1·042
0·30	1·350	0·7408	0·3045	1·045
0·31	1·363	0·7334	0·3150	1·048
0·32	1·377	0·7261	0·3255	1·052
0·33	1·391	0·7189	0·3360	1·055
0·34	1·405	0·7118	0·3466	1·058
0·35	1·419	0·7047	0·3572	1·062
0·36	1·433	0·6977	0·3678	1·066
0·37	1·448	0·6907	0·3785	1·069
0·38	1·462	0·6839	0·3892	1·073
0·39	1·477	0·6771	0·4000	1·077
0·40	1·492	0·6703	0·4108	1·081
0·41	1·507	0·6637	0·4216	1·085
0·42	1·522	0·6570	0·4325	1·090
0·43	1·537	0·6505	0·4434	1·094
0·44	1·553	0·6440	0·4543	1·098
0·45	1·568	0·6376	0·4653	1·103
0·46	1·584	0·6313	0·4764	1·108
0·47	1·600	0·6250	0·4875	1·112
0·48	1·616	0·6188	0·4986	1·117
0·49	1·632	0·6126	0·5098	1·122

x	e^x	e^{-x}	$\sinh x$	$\cosh x$
0·50	1·649	0·6065	0·5211	1·128
0·51	1·665	0·6005	0·5324	1·133
0·52	1·682	0·5945	0·5438	1·138
0·53	1·699	0·5886	0·5552	1·144
0·54	1·716	0·5827	0·5666	1·149
0·55	1·733	0·5769	0·5782	1·155
0·56	1·751	0·5712	0·5897	1·161
0·57	1·768	0·5655	0·6014	1·167
0·58	1·786	0·5599	0·6131	1·173
0·59	1·804	0·5543	0·6248	1·179
0·60	1·822	0·5488	0·6367	1·185
0·61	1·840	0·5434	0·6485	1·192
0·62	1·859	0·5379	0·6605	1·198
0·63	1·878	0·5326	0·6725	1·205
0·64	1·896	0·5273	0·6846	1·212
0·65	1·916	0·5220	0·6967	1·219
0·66	1·935	0·5169	0·7090	1·226
0·67	1·954	0·5117	0·7213	1·233
0·68	1·974	0·5066	0·7336	1·240
0·69	1·994	0·5016	0·7461	1·248
0·70	2·014	0·4966	0·7586	1·255
0·71	2·034	0·4916	0·7712	1·263
0·72	2·054	0·4868	0·7838	1·271
0·73	2·075	0·4819	0·7966	1·278
0·74	2·096	0·4771	0·8094	1·287
0·75	2·117	0·4724	0·8223	1·295
0·76	2·138	0·4677	0·8353	1·303
0·77	2·160	0·4630	0·8484	1·311
0·78	2·181	0·4584	0·8615	1·320
0·79	2·203	0·4538	0·8748	1·329
0·80	2·226	0·4493	0·8881	1·337
0·81	2·248	0·4449	0·9015	1·346
0·82	2·270	0·4404	0·9150	1·355
0·83	2·293	0·4360	0·9286	1·365
0·84	2·316	0·4317	0·9423	1·374
0·85	2·340	0·4274	0·9561	1·384
0·86	2·363	0·4232	0·9700	1·393
0·87	2·387	0·4190	0·9840	1·403
0·88	2·411	0·4148	0·9981	1·413
0·89	2·435	0·4107	1·012	1·423
0·90	2·460	0·4066	1·027	1·433
0·91	2·484	0·4025	1·041	1·443
0·92	2·509	0·3985	1·055	1·454
0·93	2·535	0·3946	1·070	1·465
0·94	2·560	0·3906	1·085	1·475
0·95	2·586	0·3867	1·099	1·486
0·96	2·612	0·3829	1·114	1·497
0·97	2·638	0·3791	1·129	1·509
0·98	2·664	0·3753	1·145	1·520
0·99	2·691	0·3716	1·160	1·531

Exponential functions

x	e^x	e^{-x}	$\sinh x$	$\cosh x$
1·00	2·718	0·3679	1·175	1·543
1·01	2·746	0·3642	1·191	1·555
1·02	2·773	0·3606	1·206	1·567
1·03	2·801	0·3570	1·222	1·579
1·04	2·829	0·3535	1·238	1·591
1·05	2·858	0·3499	1·254	1·604
1·06	2·886	0·3465	1·270	1·616
1·07	2·915	0·3430	1·286	1·629
1·08	2·945	0·3396	1·303	1·642
1·09	2·974	0·3362	1·319	1·655
1·10	3·004	0·3329	1·336	1·669
1·11	3·034	0·3296	1·352	1·682
1·12	3·065	0·3263	1·369	1·696
1·13	3·096	0·3230	1·386	1·709
1·14	3·127	0·3198	1·403	1·723
1·15	3·158	0·3166	1·421	1·737
1·16	3·190	0·3135	1·438	1·752
1·17	3·222	0·3104	1·456	1·766
1·18	3·254	0·3073	1·474	1·781
1·19	3·287	0·3042	1·491	1·796
1·20	3·320	0·3012	1·509	1·811
1·21	3·353	0·2982	1·528	1·826
1·22	3·387	0·2952	1·546	1·841
1·23	3·421	0·2923	1·564	1·857
1·24	3·456	0·2894	1·583	1·872
1·25	3·490	0·2865	1·602	1·888
1·26	3·525	0·2837	1·621	1·905
1·27	3·561	0·2808	1·640	1·921
1·28	3·597	0·2780	1·659	1·937
1·29	3·633	0·2753	1·679	1·954
1·30	3·669	0·2725	1·698	1·971
1·31	3·706	0·2698	1·718	1·988
1·32	3·743	0·2671	1·738	2·005
1·33	3·781	0·2645	1·758	2·023
1·34	3·819	0·2618	1·779	2·040
1·35	3·857	0·2592	1·799	2·058
1·36	3·896	0·2567	1·820	2·076
1·37	3·935	0·2541	1·841	2·095
1·38	3·975	0·2516	1·862	2·113
1·39	4·015	0·2491	1·883	2·132
1·40	4·055	0·2466	1·904	2·151
1·41	4·096	0·2441	1·926	2·170
1·42	4·137	0·2417	1·948	2·189
1·43	4·179	0·2393	1·970	2·209
1·44	4·221	0·2369	1·992	2·229
1·45	4·263	0·2346	2·014	2·249
1·46	4·306	0·2322	2·037	2·269
1·47	4·349	0·2299	2·060	2·290
1·48	4·393	0·2276	2·083	2·310
1·49	4·437	0·2254	2·106	2·331

x	e^x	e^{-x}	$\sinh x$	$\cosh x$
1·50	4·482	0·2231	2·129	2·352
1·51	4·527	0·2209	2·153	2·374
1·52	4·572	0·2187	2·177	2·395
1·53	4·618	0·2165	2·201	2·417
1·54	4·665	0·2144	2·225	2·439
1·55	4·711	0·2122	2·250	2·462
1·56	4·759	0·2101	2·274	2·484
1·57	4·807	0·2080	2·299	2·507
1·58	4·855	0·2060	2·324	2·530
1·59	4·904	0·2039	2·350	2·554
1·60	4·953	0·2019	2·376	2·577
1·61	5·003	0·1999	2·401	2·601
1·62	5·053	0·1979	2·428	2·625
1·63	5·104	0·1959	2·454	2·650
1·64	5·155	0·1940	2·481	2·675
1·65	5·207	0·1920	2·507	2·700
1·66	5·259	0·1901	2·535	2·725
1·67	5·312	0·1882	2·562	2·750
1·68	5·366	0·1864	2·590	2·776
1·69	5·419	0·1845	2·617	2·802
1·70	5·474	0·1827	2·646	2·828
1·71	5·529	0·1809	2·674	2·855
1·72	5·585	0·1791	2·703	2·882
1·73	5·641	0·1773	2·732	2·909
1·74	5·697	0·1755	2·761	2·936
1·75	5·755	0·1738	2·790	2·964
1·76	5·812	0·1720	2·820	2·992
1·77	5·871	0·1703	2·850	3·021
1·78	5·930	0·1686	2·881	3·049
1·79	5·989	0·1670	2·911	3·078
1·80	6·050	0·1653	2·942	3·107
1·81	6·110	0·1637	2·973	3·137
1·82	6·172	0·1620	3·005	3·167
1·83	6·234	0·1604	3·037	3·197
1·84	6·297	0·1588	3·069	3·228
1·85	6·360	0·1572	3·101	3·259
1·86	6·424	0·1557	3·134	3·290
1·87	6·488	0·1541	3·167	3·321
1·88	6·554	0·1526	3·200	3·353
1·89	6·619	0·1511	3·234	3·385
1·90	6·686	0·1496	3·268	3·418
1·91	6·753	0·1481	3·303	3·451
1·92	6·821	0·1466	3·337	3·484
1·93	6·890	0·1451	3·372	3·517
1·94	6·959	0·1437	3·408	3·551
1·95	7·029	0·1423	3·443	3·585
1·96	7·099	0·1409	3·479	3·620
1·97	7·171	0·1395	3·516	3·655
1·98	7·243	0·1381	3·552	3·690
1·99	7·316	0·1367	3·589	3·726

Exponential functions

x	e^x	e^{-x}	sinh x	cosh x
2·00	7·389	0·1353	3·627	3·762
2·01	7·463	0·1340	3·665	3·799
2·02	7·538	0·1327	3·703	3·835
2·03	7·614	0·1313	3·741	3·873
2·04	7·691	0·1300	3·780	3·910
2·05	7·768	0·1287	3·820	3·948
2·06	7·846	0·1275	3·859	3·987
2·07	7·925	0·1262	3·899	4·026
2·08	8·004	0·1249	3·940	4·065
2·09	8·085	0·1237	3·981	4·104
2·10	8·166	0·1225	4·022	4·144
2·11	8·248	0·1212	4·064	4·185
2·12	8·331	0·1200	4·106	4·226
2·13	8·415	0·1188	4·148	4·267
2·14	8·499	0·1177	4·191	4·309
2·15	8·585	0·1165	4·234	4·351
2·16	8·671	0·1153	4·278	4·393
2·17	8·758	0·1142	4·322	4·436
2·18	8·846	0·1130	4·367	4·480
2·19	8·935	0·1119	4·412	4·524
2·20	9·025	0·1108	4·457	4·568
2·21	9·116	0·1097	4·503	4·613
2·22	9·207	0·1086	4·549	4·658
2·23	9·300	0·1075	4·596	4·704
2·24	9·393	0·1065	4·643	4·750
2·25	9·488	0·1054	4·691	4·797
2·26	9·583	0·1044	4·739	4·844
2·27	9·679	0·1033	4·788	4·891
2·28	9·777	0·1023	4·837	4·939
2·29	9·875	0·1013	4·887	4·988
2·30	9·974	0·1003	4·937	5·037
2·31	10·07	0·09926	4·988	5·087
2·32	10·18	0·09827	5·039	5·137
2·33	10·28	0·09730	5·090	5·188
2·34	10·38	0·09633	5·142	5·239
2·35	10·49	0·09537	5·195	5·290
2·36	10·59	0·09442	5·248	5·343
2·37	10·70	0·09348	5·302	5·395
2·38	10·80	0·09255	5·356	5·449
2·39	10·91	0·09163	5·411	5·503
2·40	11·02	0·09072	5·466	5·557
2·41	11·13	0·08982	5·522	5·612
2·42	11·25	0·08892	5·578	5·667
2·43	11·36	0·08804	5·635	5·723
2·44	11·47	0·08716	5·693	5·780
2·45	11·59	0·08629	5·751	5·837
2·46	11·70	0·08543	5·810	5·895
2·47	11·82	0·08458	5·869	5·954
2·48	11·94	0·08374	5·929	6·013
2·49	12·06	0·08291	5·989	6·072

x	e^x	e^{-x}	sinh x	cosh x
2·50	12·18	0·08208	6·050	6·132
2·51	12·30	0·08127	6·112	6·193
2·52	12·43	0·08046	6·174	6·255
2·53	12·55	0·07966	6·237	6·317
2·54	12·68	0·07887	6·300	6·379
2·55	12·81	0·07808	6·365	6·443
2·56	12·94	0·07730	6·429	6·507
2·57	13·07	0·07654	6·495	6·571
2·58	13·20	0·07577	6·561	6·636
2·59	13·33	0·07502	6·627	6·702
2·60	13·46	0·07427	6·695	6·769
2·61	13·60	0·07353	6·763	6·836
2·62	13·74	0·07280	6·831	6·904
2·63	13·87	0·07208	6·901	6·973
2·64	14·01	0·07136	6·971	7·042
2·65	14·15	0·07065	7·042	7·112
2·66	14·30	0·06995	7·113	7·183
2·67	14·44	0·06925	7·185	7·255
2·68	14·59	0·06856	7·258	7·327
2·69	14·73	0·06788	7·332	7·400
2·70	14·88	0·06721	7·406	7·473
2·71	15·03	0·06654	7·481	7·548
2·72	15·18	0·06587	7·557	7·623
2·73	15·33	0·06522	7·634	7·699
2·74	15·49	0·06457	7·711	7·776
2·75	15·64	0·06393	7·789	7·853
2·76	15·80	0·06329	7·868	7·932
2·77	15·96	0·06266	7·948	8·011
2·78	16·12	0·06204	8·028	8·091
2·79	16·28	0·06142	8·110	8·171
2·80	16·44	0·06081	8·192	8·253
2·81	16·61	0·06020	8·275	8·335
2·82	16·78	0·05961	8·359	8·418
2·83	16·95	0·05901	8·443	8·502
2·84	17·12	0·05843	8·529	8·587
2·85	17·29	0·05784	8·615	8·673
2·86	17·46	0·05727	8·702	8·759
2·87	17·64	0·05670	8·790	8·847
2·88	17·81	0·05613	8·879	8·935
2·89	17·99	0·05558	8·969	9·024
2·90	18·17	0·05502	9·060	9·115
2·91	18·36	0·05448	9·151	9·206
2·92	18·54	0·05393	9·244	9·298
2·93	18·73	0·05340	9·337	9·391
2·94	18·92	0·05287	9·431	9·484
2·95	19·11	0·05234	9·572	9·579
2·96	19·30	0·05182	9·623	9·675
2·97	19·49	0·05130	9·720	9·772
2·98	19·69	0·05079	9·819	9·869
2·99	19·89	0·05029	9·918	9·968

Exponential functions

x	e^x	e^{-x}	$\sinh x$	$\cosh x$
3·00	20·09	0·04979	10·02	10·07
3·01	20·29	0·04929	10·12	10·17
3·02	20·49	0·04880	10·22	10·27
3·03	20·70	0·04832	10·32	10·37
3·04	20·91	0·04783	10·43	10·48
3·05	21·12	0·04736	10·53	10·58
3·06	21·33	0·04689	10·64	10·69
3·07	21·54	0·04642	10·75	10·79
3·08	21·76	0·04596	10·86	10·90
3·09	21·98	0·04550	10·97	11·01
3·10	22·20	0·04505	11·08	11·12
3·11	22·42	0·04460	11·19	11·23
3·12	22·65	0·04416	11·30	11·35
3·13	22·87	0·04372	11·42	11·46
3·14	23·10	0·04328	11·53	11·57
3·15	23·34	0·04285	11·65	11·69
3·16	23·57	0·04243	11·76	11·81
3·17	23·81	0·04200	11·88	11·92
3·18	24·05	0·04159	12·00	12·04
3·19	24·29	0·04117	12·12	12·16
3·20	24·53	0·04076	12·25	12·29
3·21	24·78	0·04036	12·37	12·41
3·22	25·03	0·03996	12·49	12·53
3·23	25·28	0·03956	12·62	12·66
3·24	25·53	0·03916	12·75	12·79
3·25	25·79	0·03877	12·88	12·91
3·26	26·05	0·03839	13·01	13·04
3·27	26·31	0·03801	13·14	13·17
3·28	26·58	0·03763	13·27	13·31
3·29	26·84	0·03725	13·40	13·44
3·30	27·11	0·03688	13·54	13·57
3·31	27·39	0·03652	13·67	13·71
3·32	27·66	0·03615	13·81	13·85
3·33	27·94	0·03579	13·95	13·99
3·34	28·22	0·03544	14·09	14·13
3·35	28·50	0·03508	14·23	14·27
3·36	28·79	0·03474	14·38	14·41
3·37	29·08	0·03439	14·52	14·56
3·38	29·37	0·03405	14·67	14·70
3·39	29·67	0·03371	14·82	14·85
3·40	29·96	0·03337	14·97	15·00
3·41	30·27	0·03304	15·12	15·15
3·42	30·57	0·03271	15·27	15·30
3·43	30·88	0·03239	15·42	15·45
3·44	31·19	0·03206	15·58	15·61
3·45	31·50	0·03175	15·73	15·77
3·46	31·82	0·03143	15·89	15·92
3·47	32·14	0·03112	16·05	16·08
3·48	32·46	0·03081	16·21	16·25
3·49	32·79	0·03050	16·38	16·41

x	e^x	e^{-x}	$\sinh x$	$\cosh x$
3·50	33·12	0·03020	16·54	16·57
3·51	33·45	0·02990	16·71	16·74
3·52	33·78	0·02960	16·88	16·91
3·53	34·12	0·02930	17·05	17·08
3·54	34·47	0·02901	17·22	17·25
3·55	34·81	0·02872	17·39	17·42
3·56	35·16	0·02844	17·57	17·60
3·57	35·52	0·02816	17·74	17·77
3·58	35·87	0·02788	17·92	17·95
3·59	36·23	0·02760	18·10	18·13
3·60	36·60	0·02732	18·29	18·31
3·61	36·97	0·02705	18·47	18·50
3·62	37·34	0·02678	18·66	18·68
3·63	37·71	0·02652	18·84	18·87
3·64	38·09	0·02625	19·03	19·06
3·65	38·47	0·02599	19·22	19·25
3·66	38·86	0·02573	19·42	19·44
3·67	39·25	0·02548	19·61	19·64
3·68	39·65	0·02522	19·81	19·84
3·69	40·04	0·02497	20·01	20·03
3·70	40·45	0·02472	20·21	20·24
3·71	40·85	0·02448	20·41	20·44
3·72	41·26	0·02423	20·62	20·64
3·73	41·68	0·02399	20·83	20·85
3·74	42·10	0·02375	21·04	21·06
3·75	42·52	0·02352	21·25	21·27
3·76	42·95	0·02328	21·46	21·49
3·77	43·38	0·02305	21·68	21·70
3·78	43·82	0·02282	21·90	21·92
3·79	44·26	0·02260	22·12	22·14
3·80	44·70	0·02237	22·34	22·36
3·81	45·15	0·02215	22·56	22·59
3·82	45·60	0·02193	22·79	22·81
3·83	46·06	0·02171	23·02	23·04
3·84	46·53	0·02149	23·25	23·27
3·85	46·99	0·02128	23·49	23·51
3·86	47·47	0·02107	23·72	23·74
3·87	47·94	0·02086	23·96	23·98
3·88	48·42	0·02065	24·20	24·22
3·89	48·91	0·02045	24·45	24·47
3·90	49·40	0·02024	24·69	24·71
3·91	49·90	0·02004	24·94	24·96
3·92	50·40	0·01984	25·19	25·21
3·93	50·91	0·01964	25·44	25·46
3·94	51·42	0·01945	25·70	25·72
3·95	51·94	0·01925	25·96	25·98
3·96	52·46	0·01906	26·22	26·24
3·97	52·98	0·01887	26·48	26·50
3·98	53·52	0·01869	26·75	26·77
3·99	54·05	0·01850	27·02	27·04
4·00	54·60	0·01832	27·29	27·31

Constants

Constant	Value	Log_{10}	Log_e
π	3·141 59	0·497 15	1·144 73
π^2	9·869 60	0·994 30	2·289 46
$1/\pi$	0·318 31	$\bar{1}$·502 85	−1·144 73
$1/\pi^2$	0·101 32	$\bar{1}$·005 70	−2·289 46
$\sqrt{\pi}$	1·772 45	0·248 57	0·572 36
e	2·718 28	0·434 29	1·000 00
γ (Euler's constant)	0·577 22	$\bar{1}$·761 34	−0·549 54
	2	0·301 03	0·693 15
	3	0·477 12	1·098 61
	10	1·000 00	2·302 59
$\sqrt{2}$	1·414 21	0·150 51	0·346 57
$\sqrt{3}$	1·732 05	0·238 56	0·549 31
$\sqrt{10}$	3·162 28	0·500 00	1·151 29
$180/\pi$	57·295 78	1·758 12	4·048 23

Binomial coefficients

$$\binom{n}{m} = \frac{n!}{(n-m)!\,m!} = \binom{n}{n-m}$$

	m						
n	0	1	2	3	4	5	6
0	1						
1	1	1					
2	1	2	1				
3	1	3	3	1			
4	1	4	6	4	1		
5	1	5	10	10	5	1	
6	1	6	15	20	15	6	1
7	1	7	21	35	35	21	7
8	1	8	28	56	70	56	28
9	1	9	36	84	126	126	84
10	1	10	45	120	210	252	210

$\left.\right\}$ etc.

Series

$$1 - \frac{1}{3} + \frac{1}{5} - \frac{1}{7} + \ldots = \frac{\pi}{4}$$

$$\frac{1}{1^2} + \frac{1}{2^2} + \frac{1}{3^2} + \ldots = \frac{\pi^2}{6}$$

$$\frac{1}{1^2} + \frac{1}{3^2} + \frac{1}{5^2} + \ldots = \frac{\pi^2}{8}$$

$$\frac{1}{1^3} - \frac{1}{3^3} + \frac{1}{5^3} - \ldots = \frac{\pi^3}{32}$$

$$1 + 2 + 3 + \ldots + n = \sum_1^n r = \frac{n(n+1)}{2}$$

$$1^2 + 2^2 + 3^2 + \ldots + n^2 = \sum_1^n r^2 = \frac{n(n+1)(2n+1)}{6}$$

$$1^3 + 2^3 + 3^3 + \ldots + n^3 = \sum_1^n r^3 = \frac{n^2(n+1)^2}{4}$$

Arithmetic

$$a + (a+d) + (a+2d) + \ldots + \{a + (n-1)d\}$$
$$= \frac{n}{2}\{2a + (n-1)d\}$$

Geometric

$$1 + x + x^2 + x^3 + \ldots + x^{n-1} = \frac{1-x^n}{1-x} \quad (x \neq 1)$$
$$= \frac{1}{1-x} \quad (|x| < 1, n \to \infty)$$

Binomial

$$(1+x)^n = 1 + nx + \frac{n(n-1)}{2!}x^2 + \frac{n(n-1)(n-2)}{3!}x^3$$
$$+ \ldots + \binom{n}{r}x^r + \ldots.$$
$$(|x| < 1, \text{ all real } n; \text{ all } x, n \text{ a positive integer})$$

Exponential and logarithmic

$$e^x = 1 + x + \frac{x^2}{2!} + \frac{x^3}{3!} + \ldots.$$

$$a^x = 1 + x \ln a + \frac{(x \ln a)^2}{2!} + \frac{(x \ln a)^3}{3!} + \ldots.$$

$$\ln(1+x) = x - \frac{x^2}{2} + \frac{x^3}{3} - \frac{x^4}{4} + \ldots. \quad (|x| < 1)$$

Trigonometric

$$\sin x = x - \frac{x^3}{3!} + \frac{x^5}{5!} - \ldots$$

$$\cos x = 1 - \frac{x^2}{2!} + \frac{x^4}{4!} - \ldots$$

$$\tan x = x + \frac{x^3}{3} + \frac{2x^5}{15} + \frac{17x^7}{315} + \ldots \quad \left(|x| < \frac{\pi}{2}\right)$$

$$\sin^{-1}x = x + \frac{x^3}{6} + \frac{1.3}{2.4}\frac{x^5}{5} + \frac{1.3.5}{2.4.6}\frac{x^7}{7} + \ldots \quad (|x| < 1)$$

$$\tan^{-1}x = x - \frac{x^3}{3} + \frac{x^5}{5} - \ldots \quad (|x| < 1)$$

$$\sinh x = x + \frac{x^3}{3!} + \frac{x^5}{5!} + \ldots$$

$$\cosh x = 1 + \frac{x^2}{2!} + \frac{x^4}{4!} + \ldots$$

$$\tanh x = x - \frac{x^3}{3} + \frac{2x^5}{15} - \frac{17x^7}{315} + \ldots \quad \left(|x| < \frac{\pi}{2}\right)$$

Maclaurin

$$f(x) = f(0) + xf'(0) + \frac{x^2}{2!}f''(0) + \frac{x^3}{3!}f'''(0) + \ldots$$

Taylor

$$f(a+h) = f(a) + hf'(a) + \frac{h^2}{2!}f''(a) + \frac{h^3}{3!}f'''(a) + \ldots$$

Fourier

$$f(x) = \frac{a_0}{2} + a_1 \cos\frac{2\pi x}{T} + a_2 \cos\frac{4\pi x}{T} + \ldots$$
$$+ b_1 \sin\frac{2\pi x}{T} + b_2 \sin\frac{4\pi x}{T} + \ldots \quad (0 < x < T)$$

where

$$a_n = \frac{2}{T}\int_0^T f(x) \cos\frac{2\pi nx}{T} \, dx$$

$$b_n = \frac{2}{T}\int_0^T f(x) \sin\frac{2\pi nx}{T} \, dx$$

If $f(x) = f(-x)$, f is *even* and $b_n = 0$.

If $f(x) = -f(-x)$, f is *odd* and $a_n = 0$.

If $f(x) = -f\left(x + \frac{T}{2}\right)$, f has half-wave symmetry and

$a_n = b_n = 0$ for all even n.

Fourier series for certain waveforms

The series below are expressed in terms of θ according to the substitution $\theta = 2\pi x/T$ so that the period T of $f(x)$ becomes simply 2π. The origin of θ is so chosen as to make the waveforms even functions ($b_n = 0$) wherever possible.

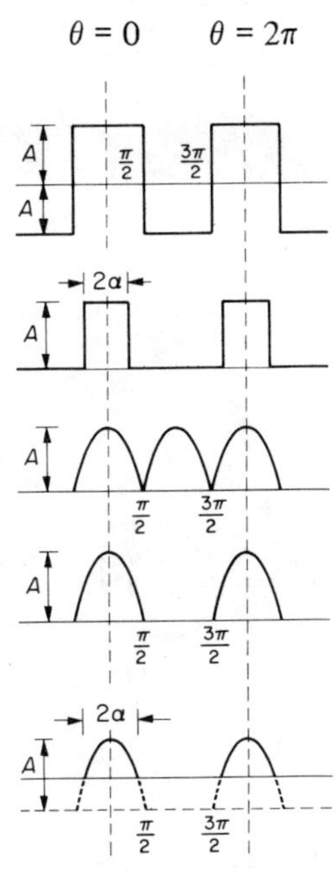

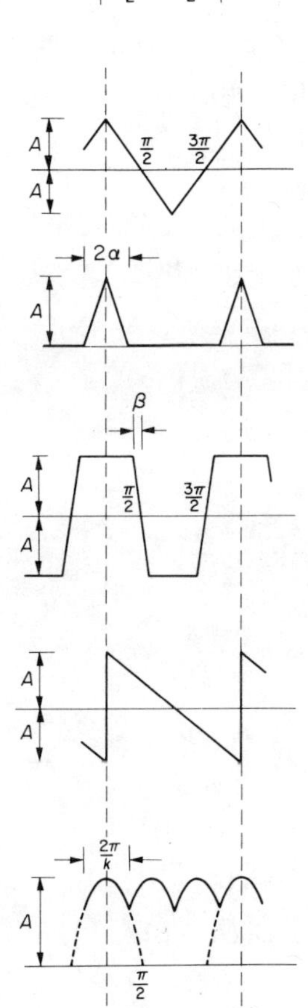

Series	Mean square value
$\dfrac{4A}{\pi}\{\cos\theta - \tfrac{1}{3}\cos 3\theta + \tfrac{1}{5}\cos 5\theta - \ldots\}$	A^2
$A\left\{\dfrac{\alpha}{\pi} + \dfrac{2}{\pi}(\sin\alpha\cos\theta + \tfrac{1}{2}\sin 2\alpha\cos 2\theta + \tfrac{1}{3}\sin 3\alpha\cos 3\theta + \ldots)\right\}$	$\dfrac{A^2\alpha}{\pi}$
$\dfrac{2A}{\pi}\{1 + \tfrac{2}{3}\cos 2\theta - \tfrac{2}{15}\cos 4\theta + \tfrac{2}{35}\cos 6\theta - \ldots\}$	$\dfrac{A^2}{2}$
$\dfrac{A}{\pi}\left\{1 + \dfrac{\pi}{2}\cos\theta + \tfrac{2}{3}\cos 2\theta - \tfrac{2}{15}\cos 4\theta + \ldots\right\}$	$\dfrac{A^2}{4}$
$\dfrac{A}{\pi}\{(\sin\alpha - \alpha\cos\alpha) + (\alpha - \tfrac{1}{2}\sin 2\alpha)\cos\theta$ $\quad + (\sin\alpha + \tfrac{1}{3}\sin 3\alpha - \cos\alpha\sin 2\alpha)\cos 2\theta$ $\quad + (\tfrac{1}{2}\sin 2\alpha + \tfrac{1}{4}\sin 4\alpha - \tfrac{2}{3}\cos\alpha\sin 3\alpha)\cos 3\theta + \ldots\}$	$\dfrac{A^2}{2\pi}\{\alpha - \tfrac{3}{2}\sin 2\alpha + 2\alpha\cos^2\alpha\}$
$\dfrac{8A}{\pi^2}\{\cos\theta + \tfrac{1}{9}\cos 3\theta + \tfrac{1}{25}\cos 5\theta + \ldots\}$	$\dfrac{A^2}{3}$
$\dfrac{A}{\pi\alpha}\left\{\dfrac{\alpha^2}{2} + 4\sin^2\dfrac{\alpha}{2}\cos\theta + \sin^2\alpha\cos 2\theta + \tfrac{4}{9}\sin^2\dfrac{3\alpha}{2}\cos 3\theta + \ldots\right\}$	$\dfrac{A^2\alpha}{3\pi}$
$\dfrac{4A}{\pi\beta}\{\sin\beta\cos\theta - \tfrac{1}{9}\sin 3\beta\cos 3\theta + \tfrac{1}{25}\sin 5\beta\cos 5\theta - \ldots\}$	$A^2\left(1 - \tfrac{4}{3}\dfrac{\beta}{\pi}\right)$
$\dfrac{2A}{\pi}\{\sin\theta + \tfrac{1}{2}\sin 2\theta + \tfrac{1}{3}\sin 3\theta + \ldots\}$	$\dfrac{A^2}{3}$
$\dfrac{Ak}{\pi}\sin\dfrac{\pi}{k}\left\{1 + \dfrac{2}{k^2 - 1}\cos k\theta - \dfrac{2}{4k^2 - 1}\cos 2k\theta + \dfrac{2}{9k^2 - 1}\cos 3k\theta - \ldots\right\}$	

Trigonometric, hyperbolic and exponential functions

$$e^{jx} = \cos x + j \sin x$$

$$e^x = \cosh x + \sinh x$$

$$\sin x = \frac{e^{jx} - e^{-jx}}{2j}; \quad \cos x = \frac{e^{jx} + e^{-jx}}{2}$$

$$\sinh x = \frac{e^x - e^{-x}}{2}; \quad \cosh x = \frac{e^x + e^{-x}}{2}$$

$$\sin jx = j \sinh x; \quad \cos jx = \cosh x$$

$$\sinh jx = j \sin x; \quad \cosh jx = \cos x$$

$$\sin^{-1} jx = j \sinh^{-1} x; \quad \cos^{-1} x = -j \cosh^{-1} x$$

$$\sinh^{-1} jx = j \sin^{-1} x; \quad \cosh^{-1} x = j \cos^{-1} x$$

$$\sinh^{-1} x = \ln\{x + \sqrt{(x^2 + 1)}\};$$

$$\cosh^{-1} x = \ln\{x + \sqrt{(x^2 - 1)}\}$$

$$\tanh^{-1} x = \tfrac{1}{2} \ln \frac{1 + x}{1 - x}$$

Trigonometric relations

$$\sin(-A) = -\sin A; \quad \cos(-A) = \cos A$$

$$\sin^2 A + \cos^2 A = 1$$

$$\tan^2 A + 1 = \sec^2 A$$

$$1 + \cot^2 A = \mathrm{cosec}^2 A$$

$$\sin(A \pm B) = \sin A \cos B \pm \cos A \sin B$$

$$\cos(A \pm B) = \cos A \cos B \mp \sin A \sin B$$

$$\tan(A \pm B) = \frac{\tan A \pm \tan B}{1 \mp \tan A \tan B}$$

$$\sin^2 A = \tfrac{1}{2}(1 - \cos 2A)$$

$$\cos^2 A = \tfrac{1}{2}(1 + \cos 2A)$$

$$\tan A = \frac{\sin 2A}{1 + \cos 2A}$$

$$a \sin A + b \cos A = (a^2 + b^2)^{1/2} \sin\left(A + \tan^{-1}\frac{b}{a}\right)$$

$$\sin A \pm \sin B = 2 \sin \frac{A \pm B}{2} \cos \frac{A \mp B}{2}$$

$$\cos A + \cos B = 2 \cos \frac{A + B}{2} \cos \frac{A - B}{2}$$

$$\cos A - \cos B = -2 \sin \frac{A + B}{2} \sin \frac{A - B}{2}$$

$$2 \sin A \cos B = \sin(A + B) + \sin(A - B)$$

$$2 \cos A \cos B = \cos(A + B) + \cos(A - B)$$

$$2 \sin A \sin B = \cos(A - B) - \cos(A + B)$$

$$\left. \begin{aligned} \frac{a}{\sin A} &= \frac{b}{\sin B} = \frac{c}{\sin C} \\ a^2 &= b^2 + c^2 - 2bc \cos A \end{aligned} \right\} \text{ in a triangle ABC}$$

Hyperbolic relations

$$\sinh(-A) = -\sinh A; \quad \cosh(-A) = \cosh A$$

$$\cosh^2 A - \sinh^2 A = 1$$

$$1 - \tanh^2 A = \mathrm{sech}^2 A$$

$$\coth^2 A - 1 = \mathrm{cosech}^2 A$$

$$\sinh(A \pm B) = \sinh A \cosh B \pm \cosh A \sinh B$$

$$\cosh(A \pm B) = \cosh A \cosh B \pm \sinh A \sinh B$$

$$\tanh(A \pm B) = \frac{\tanh A \pm \tanh B}{1 \pm \tanh A \tanh B}$$

$$\sinh^2 A = \tfrac{1}{2}(\cosh 2A - 1)$$

$$\cosh^2 A = \tfrac{1}{2}(\cosh 2A + 1)$$

$$\tanh A = \frac{\sinh 2A}{\cosh 2A + 1}$$

$$\sinh A \pm \sinh B = 2 \sinh \frac{A \pm B}{2} \cosh \frac{A \mp B}{2}$$

$$\cosh A + \cosh B = 2 \cosh \frac{A + B}{2} \cosh \frac{A - B}{2}$$

$$\cosh A - \cosh B = 2 \sinh \frac{A + B}{2} \sinh \frac{A - B}{2}$$

$$2 \sinh A \cosh B = \sinh(A + B) + \sinh(A - B)$$

$$2 \cosh A \cosh B = \cosh(A + B) + \cosh(A - B)$$

$$2 \sinh A \sinh B = \cosh(A + B) - \cosh(A - B)$$

Differentials

In the following, u, v and w are functions of x and a is a constant.

$f(x)$	$f'(x)$		$f(x)$	$f'(x)$
uv	$u\dfrac{\mathrm{d}v}{\mathrm{d}x} + v\dfrac{\mathrm{d}u}{\mathrm{d}x}$		$\sin x$	$\cos x$
			$\cos x$	$-\sin x$
			$\tan x$	$\sec^2 x$
uvw	$uv\dfrac{\mathrm{d}w}{\mathrm{d}x} + vw\dfrac{\mathrm{d}u}{\mathrm{d}x} + wu\dfrac{\mathrm{d}v}{\mathrm{d}x}$		$\operatorname{cosec} x$	$-\cot x \operatorname{cosec} x$
			$\sec x$	$\tan x \sec x$
			$\cot x$	$-\operatorname{cosec}^2 x$
$\dfrac{u}{v}$	$\dfrac{1}{v^2}\left(v\dfrac{\mathrm{d}u}{\mathrm{d}x} - u\dfrac{\mathrm{d}v}{\mathrm{d}x}\right)$		$\sinh x$	$\cosh x$
			$\cosh x$	$\sinh x$
			$\tanh x$	$\operatorname{sech}^2 x$
$f(u, v)$	$\dfrac{\partial f}{\partial u}\dfrac{\mathrm{d}u}{\mathrm{d}x} + \dfrac{\partial f}{\partial v}\dfrac{\mathrm{d}v}{\mathrm{d}x}$		$\operatorname{cosech} x$	$-\coth x \operatorname{cosech} x$
			$\operatorname{sech} x$	$-\tanh x \operatorname{sech} x$
			$\coth x$	$-\operatorname{cosech}^2 x$
ax^n	anx^{n-1}		$\sin^{-1}(x/a)$	$(a^2 - x^2)^{-1/2}$
			$\cos^{-1}(x/a)$	$-(a^2 - x^2)^{-1/2}$
e^{ax}	$a\,\mathrm{e}^{ax}$		$\tan^{-1}(x/a)$	$a/(a^2 + x^2)$
			$\operatorname{cosec}^{-1}(x/a)$	$-a/x(x^2 - a^2)^{1/2}$
a^x	$a^x \ln a$		$\sec^{-1}(x/a)$	$a/x(x^2 - a^2)^{1/2}$
			$\cot^{-1}(x/a)$	$-a/(a^2 + x^2)$
x^x	$x^x(1 + \ln x)$		$\sinh^{-1}(x/a)$	$(a^2 + x^2)^{-1/2}$
			$\cosh^{-1}(x/a)$	$(x^2 - a^2)^{-1/2}$
$\ln x$	$1/x$		$\tanh^{-1}(x/a)$	$a/(a^2 - x^2)$
			$\operatorname{cosech}^{-1}(x/a)$	$-a/x(x^2 + a^2)^{1/2}$
$\log_a x$	$\dfrac{1}{x}\log_a \mathrm{e}$		$\operatorname{sech}^{-1}(x/a)$	$-a/x(a^2 - x^2)^{1/2}$
			$\coth^{-1}(x/a)$	$-a/(x^2 - a^2)$

Indefinite integrals

The constant of integration is omitted in each case.

$f(x)$	$\int f(x)\,\mathrm{d}x$		$f(x)$	$\int f(x)\,\mathrm{d}x$
$(a + bx)^n \;(n \neq -1)$	$\dfrac{(a + bx)^{n+1}}{b(n + 1)}$		$\ln x$	$x(\ln x - 1)$
$(a + bx)^{-1}$	$\dfrac{1}{b}\ln(a + bx)$		$(\ln x)^2$	$x\{(\ln x)^2 - 2\ln x + 2\}$
$\dfrac{x}{ax + b}$	$\dfrac{ax + b - b\ln(ax + b)}{a^2}$		$x^n \ln ax \;(n \neq -1)$	$\dfrac{x^{n+1}}{n + 1}\left(\ln ax - \dfrac{1}{n + 1}\right)$
a^x	$a^x/\ln a$		$(\ln ax)^n/x \;(n \neq -1)$	$(\ln ax)^{n+1}/(n + 1)$
xa^x	$\dfrac{a^x x}{\ln a} - \dfrac{a^x}{(\ln a)^2}$		$1/x \ln x$	$\ln(\ln x)$
$x\,\mathrm{e}^{ax}$	$\mathrm{e}^{ax}(ax - 1)/a^2$			
$(a + b\,\mathrm{e}^{cx})^{-1}$	$\dfrac{x}{a} - \dfrac{1}{ac}\ln(a + b\,\mathrm{e}^{cx})$			

$f(x)$	$\int f(x)\,\mathrm{d}x$	$f(x)$	$\int f(x)\,\mathrm{d}x$
$(x^2 + a^2)^{-1}$	$\dfrac{1}{a}\,\tan^{-1}(x/a)$		
$(x^2 - a^2)^{-1}$	$\dfrac{1}{2a}\ln\left\|\dfrac{x-a}{x+a}\right\|$	$(x^2 + a^2)^{-1/2}$	$\sinh^{-1}(x/a)$
		$(x^2 - a^2)^{-1/2}$	$\cosh^{-1}(x/a)$
$(x^2 + a^2)^{1/2}$	$\frac{1}{2}\{x(x^2+a^2)^{1/2} + a^2\sinh^{-1}(x/a)\}$	$(a^2 - x^2)^{-1/2}$	$\sin^{-1}(x/a)$
$(x^2 - a^2)^{1/2}$	$\frac{1}{2}\{x(x^2-a^2)^{1/2} - a^2\cosh^{-1}(x/a)\}$	$a/x(x^2-a^2)^{1/2}$	$\sec^{-1}(x/a)$
$(a^2 - x^2)^{1/2}$	$\frac{1}{2}\{x(a^2-x^2)^{1/2} + a^2\sin^{-1}(x/a)\}$		

$f(x)$	$\int f(x)\,\mathrm{d}x$	$f(x)$	$\int f(x)\,\mathrm{d}x$
$\sin x$	$-\cos x$	$\sinh x$	$\cosh x$
$\cos x$	$\sin x$	$\cosh x$	$\sinh x$
$\tan x$	$-\ln\cos x$	$\tanh x$	$\ln\cosh x$
$\operatorname{cosec} x$	$\ln\tan(x/2)$	$\operatorname{cosech} x$	$\ln\tanh(x/2)$
$\sec x$	$\ln(\sec x + \tan x)$	$\operatorname{sech} x$	$\tan^{-1}(\sinh x)$
$\cot x$	$\ln\sin x$	$\coth x$	$\ln\sinh x$
$\sin^2 x$	$\frac{1}{2}(x - \frac{1}{2}\sin 2x)$	$\sinh^2 x$	$\frac{1}{2}(-x + \frac{1}{2}\sinh 2x)$
$\cos^2 x$	$\frac{1}{2}(x + \frac{1}{2}\sin 2x)$	$\cosh^2 x$	$\frac{1}{2}(x + \frac{1}{2}\sinh 2x)$
$\tan^2 x$	$\tan x - x$	$\tanh^2 x$	$x - \tanh x$
$\operatorname{cosec}^2 x$	$-\cot x$	$\operatorname{cosech}^2 x$	$-\coth x$
$\sec^2 x$	$\tan x$	$\operatorname{sech}^2 x$	$\tanh x$
$\cot^2 x$	$-x - \cot x$	$\coth^2 x$	$x - \coth x$
$x\sin x$	$\sin x - x\cos x$	$x\sinh x$	$x\cosh x - \sinh x$
$x\cos x$	$\cos x + x\sin x$	$x\cosh x$	$x\sinh x - \cosh x$
$\sin^{-1} x$	$x\sin^{-1} x + (1-x^2)^{1/2}$	$\sinh^{-1} x$	$x\sinh^{-1} x - (x^2+1)^{1/2}$
$\cos^{-1} x$	$x\cos^{-1} x - (1-x^2)^{1/2}$	$\cosh^{-1} x$	$x\cosh^{-1} x - (x^2-1)^{1/2}$
$\tan^{-1} x$	$x\tan^{-1} x - \frac{1}{2}\ln(1+x^2)$	$\tanh^{-1} x$	$x\tanh^{-1} x + \frac{1}{2}\ln(1-x^2)$

Definite integrals

Legendre's normal *elliptic integrals* include:

$$F(\theta, k) = \int_0^\theta \frac{\mathrm{d}\theta}{(1 - k^2\sin^2\theta)^{1/2}} \quad \text{(first kind)}$$

$$E(\theta, k) = \int_0^\theta (1 - k^2\sin^2\theta)^{1/2}\,\mathrm{d}\theta \quad \text{(second kind)}$$

The 'complete' form of these is:

$$F(\pi/2, k) = K(k) = \int_0^{\pi/2} \frac{\mathrm{d}\theta}{(1 - k^2\sin^2\theta)^{1/2}}$$

$$= \frac{\pi}{2}\left\{1 + (\tfrac{1}{2})^2 k^2 + \left(\frac{1\cdot 3}{2\cdot 4}\right)^2 k^4 + \ldots\right\}(k^2 < 1)$$

$$E(\pi/2, k) = E(k) = \int_0^{\pi/2} (1 - k^2\sin^2\theta)^{1/2}\,\mathrm{d}\theta$$

$$= \frac{\pi}{2}\left\{1 - (\tfrac{1}{2})^2 k^2 - \left(\frac{1\cdot 3}{2\cdot 4}\right)^2 \frac{k^4}{3} - \ldots\right\}(k^2 < 1)$$

The *error function* is:

$$\operatorname{erf} x = \frac{2}{\sqrt{\pi}}\int_0^x e^{-u^2}\,\mathrm{d}u = \frac{2}{\sqrt{\pi}}\left(x - \frac{x^3}{3} + \frac{1}{2!}\frac{x^5}{5} - \frac{1}{3!}\frac{x^7}{7} + \ldots\right)$$

The *gamma function* is:

$$\Gamma(n) = \int_0^\infty x^{n-1} e^{-x}\,\mathrm{d}x = \int_0^1 \left(\ln\frac{1}{x}\right)^{n-1}\,\mathrm{d}x \quad (\operatorname{Re} n > 0)$$

If n is a positive integer,

$$\Gamma(n) = (n - 1)!$$

Also

$$\Gamma(\tfrac{1}{2}) = \sqrt{\pi}$$

The *beta function* is:

$$B(p,q) = \int_0^1 x^{p-1}(1 - x)^{q-1}\,dx = \frac{\Gamma(p)\Gamma(q)}{\Gamma(p+q)} \quad (\mathrm{Re}\,p, q > 0)$$

Stirling's formula can be expressed as:

$$\ln\Gamma(n) \sim (n - \tfrac{1}{2})\ln n - n + \ln(\sqrt{2\pi})$$

Euler's constant is:

$$\gamma = -\int_0^\infty e^{-x}\ln x\,dx = 0{\cdot}577\,21$$

Other definite integrals

$$\int_0^\infty \frac{x^{m-1}}{1 + x^n}\,dx = (\pi/n)\operatorname{cosec}(m\pi/n) \quad (0 < m < n)$$

$$\int_0^\infty e^{-ax}\,dx = \frac{1}{a} \quad (\mathrm{Re}\,a > 0)$$

$$\int_0^\infty x^n e^{-ax}\,dx = \frac{\Gamma(n+1)}{a^{n+1}} \quad (n > -1, \mathrm{Re}\,a > 0)$$

$$\int_0^\infty x\, e^{-x^2}\,dx = \tfrac{1}{2}$$

$$\int_0^\infty x^2 e^{-x^2}\,dx = \frac{\sqrt{\pi}}{4}$$

$$\int_0^\infty \frac{\sin ax}{x}\,dx = \begin{cases} \pi/2 & (a > 0) \\ 0 & (a = 0) \\ -\pi/2 & (a < 0) \end{cases} \quad (a\ \text{real})$$

$$\int_0^\infty \frac{\tan x}{x}\,dx = \frac{\pi}{2}$$

$$\int_0^\infty e^{-ax}\sin bx\,dx = \frac{b}{a^2 + b^2} \quad (a > 0)$$

$$\int_0^\infty e^{-ax}\cos bx\,dx = \frac{a}{a^2 + b^2} \quad (a > 0)$$

$$\int_0^\pi \sin ax \cos ax\,dx = 0$$

$$\int_0^\pi \sin ax \sin bx\,dx = \int_0^\pi \cos ax \cos bx\,dx = 0$$

$$(a, b\ \text{integers};\ a \neq b)$$

$$\int_0^{\pi/2} \sin^m x \cos^n x\,dx =$$

$$\frac{(m-1)(m-3)\ldots(2\ \text{or}\ 1)\,(n-1)(n-3)\ldots(2\ \text{or}\ 1)}{(m+n)(m+n-2)\ldots(2\ \text{or}\ 1)} \times C$$

$$(m, n\ \text{integers};\ C = \frac{\pi}{2}\ \text{for}\ m\ \text{and}\ n\ \text{even,}$$

$$C = 1\ \text{for}\ m\ \text{or}\ n\ \text{odd})$$

$$\int_0^\pi \frac{\cos nu\,du}{\cos u - \cos x} = \frac{\pi \sin nx}{\sin x} \quad (n = 1, 2, 3\ldots)$$
$$(\text{principal value})$$

Fourier transform

The Fourier transform of a function $f(t)$ may be written

$$F(\omega) = \int_{-\infty}^\infty f(t)\, e^{-j\omega t}\,dt,$$

and the inverse transform of $F(\omega)$ is then

$$f(t) = \frac{1}{2\pi}\int_{-\infty}^\infty F(\omega)e^{j\omega t}\,d\omega.$$

(The constants unity and $1/2\pi$ preceding the two integrals may be rewritten in any other way which retains their product as $1/2\pi$.)

If $f(t)$ is even, $F(\omega)$ becomes the *Fourier cosine transform*,

$$F_c(\omega) = 2\int_0^\infty f(t)\cos\omega t\,dt$$

and

$$f(t) = \frac{1}{\pi}\int_0^\infty F_c(\omega)\cos\omega t\,d\omega,$$

and if $f(t)$ is odd, $F(\omega)$ becomes the *Fourier sine transform*,

$$F_s(\omega) = 2\int_0^\infty f(t)\sin\omega t\,dt$$

and

$$f(t) = \frac{1}{\pi} \int_0^\infty F_s(\omega) \sin \omega t \, d\omega.$$

Convolution theorem for Fourier transforms

If $F(\omega)$, $G(\omega)$ are the Fourier transforms of $f(t)$, $g(t)$ then $F(\omega)G(\omega)$ is the transform of the convolution of f and g, i.e. of

$$f * g = \int_{-\infty}^\infty f(\tau)g(t - \tau) \, d\tau.$$

Fourier transforms of various functions

Function	Transform
$f(t - \tau)$	$e^{-j\omega\tau} F(\omega)$
Unit impulse at $t = 0$, $\delta(t)$	1
Unit step function at $t = 0$,	
$\quad u(t)$	$\pi\delta(\omega) + 1/j\omega$
1	$2\pi\delta(\omega)$
$e^{j\omega_0 t}$	$2\pi\delta(\omega - \omega_0)$
$\cos \omega_0 t$	$\pi\{\delta(\omega - \omega_0) + \delta(\omega + \omega_0)\}$
$\sin \omega_0 t$	$j\pi\{-\delta(\omega - \omega_0) + \delta(\omega + \omega_0)\}$

Function	Transform
Sign function $\epsilon(t)$:	
$\quad 1 \ (t > 0)$	$\dfrac{2}{j\omega}$
$\quad -1 \ (t < 0)$	
Pulses	
Rectangular:	
$\quad 1 \ (\lvert t \rvert < \tau)$	$\dfrac{2 \sin \omega\tau}{\omega}$
$\quad 0 \ (\lvert t \rvert > \tau)$	
Triangular:	
$\quad 1 - t\epsilon(t)/\tau \ \ (\lvert t \rvert < \tau)$	$\dfrac{4 \sin^2 (\omega\tau/2)}{\omega^2 \tau}$
$\quad 0 \qquad\qquad (\lvert t \rvert > \tau)$	
Carrier wave:	
$\quad \cos \omega_0 t \ \ (\lvert t \rvert < \tau)$	$\dfrac{\sin (\omega_0 + \omega)\tau}{\omega_0 + \omega} + \dfrac{\sin (\omega_0 - \omega)\tau}{\omega_0 - \omega}$
$\quad 0 \qquad (\lvert t \rvert > \tau)$	
Gaussian:	
$\quad e^{-t^2/\sigma^2} \ (\sigma + \text{ve real})$	$\sqrt{\pi}\sigma \, e^{-\sigma^2\omega^2/4}$
Cosine-squared:	
$\quad \cos^2 \pi t/2\tau \ \ (\lvert t \rvert < \tau)$	$\dfrac{\pi^2 \sin \omega\tau}{\omega(\pi^2 - \omega^2\tau^2)}$
$\quad 0 \qquad\qquad (\lvert t \rvert > \tau)$	

Laplace transform

The Laplace transform of a function $f(t)$ is

$$F(s) = \int_0^\infty f(t) \, e^{-st} \, dt$$

and the inverse transform of $F(s)$ for $t > 0$ is then

$$f(t) = \frac{1}{2\pi j} \int_{\sigma - j\infty}^{\sigma + j\infty} F(s) \, e^{st} \, ds$$

where σ is a real constant greater than the real part of each singularity of $F(s)$.

Convolution theorem for Laplace transforms

If $F(s)$, $G(s)$ are the Laplace transforms of $f(t)$, $g(t)$ then $F(s)G(s)$ is the transform of the convolution of f and g from 0 to t, i.e. of

$$f * g = \int_0^t f(\tau)g(t - \tau) \, d\tau.$$

Laplace transforms of various functions

Function	Transform
$f(t - \tau)$	$e^{-s\tau} F(s)$
$f'(t)$	$sF(s) - f(0+)$
$f''(t)$	$s^2 F(s) - sf(0+) - f'(0+)$
$\quad f^{(n)}(t)$	$s^n F(s) - s^{n-1}f(0+)$
	$\qquad - s^{n-2}f'(0+) - \ldots$
	$\qquad - f^{(n-1)}(0+)$
$\displaystyle\int_0^t f(t) \, dt$	$F(s)/s$
Unit impulse at $t = 0$, $\delta(t)$	1
Unit step function at $t = 0$, $u(t)$	$1/s$
$t^{n-1}/(n - 1)!$	$1/s^n$
e^{-at}	$1/(s + a)$
$(1/a)(1 - e^{-at})$	$1/\{s(s + a)\}$
$\cos at$	$s/(s^2 + a^2)$

Function	Transform
$\cosh at$	$s/(s^2 - a^2)$
$\sin at$	$a/(s^2 + a^2)$
$\sinh at$	$a/(s^2 - a^2)$
$(1/a^2)(1 - \cos at)$	$1/\{s(s^2 + a^2)\}$
$(1/a^3)(at - \sin at)$	$1/\{s^2(s^2 + a^2)\}$
te^{-at}	$1/(s + a)^2$

Function	Transform
$e^{-at}(1 - at)$	$s/(s + a)^2$
$(1/2a^3)(\sin at - at\cos at)$	$1/(s^2 + a^2)^2$
$(t/2a)\sin at$	$s/(s^2 + a^2)^2$
$t\cos at$	$(s^2 - a^2)/(s^2 + a^2)^2$
$e^{-at}\cos bt$	$(s + a)/\{(s + a)^2 + b^2\}$
$e^{-at}\sin bt$	$b/\{(s + a)^2 + b^2\}$

Complex variable

Cauchy-Riemann relations

If $z = x + \mathrm{j}y$ and the function $f(z) = u + \mathrm{j}v$, then for $f(z)$ to be analytic it is necessary that

$$\frac{\partial u}{\partial x} = \frac{\partial v}{\partial y}; \quad \frac{\partial u}{\partial y} = -\frac{\partial v}{\partial x}.$$

Cauchy's theorem

If $f(z)$ is analytic in a closed region bounded by a contour C,

$$\oint_C f(z)\,\mathrm{d}z = 0.$$

Cauchy's integral

If a is a point inside C, $f(z)$ being analytic within and on C,

$$f(a) = \frac{1}{2\pi\mathrm{j}} \oint_C \frac{f(z)}{z - a}\,\mathrm{d}z.$$

Also,

$$f^{(n)}(a) = \frac{n!}{2\pi\mathrm{j}} \oint_C \frac{f(z)}{(z - a)^{n+1}}\,\mathrm{d}z.$$

Residue theorem

If $f(z)$ is analytic within and on C except at poles a, b, c . . . enclosed by C,

$$\oint_C f(z)\,\mathrm{d}z = 2\pi\mathrm{j}(A + B + C \ldots)$$

where A, B, C . . . are the residues of the poles.

The Nyquist criterion

A consequence of the residue theorem is the following: If $f(z)$ is analytic within and on C except for P poles and Z zeros (a pole or zero of order n being counted n times) within C, then

$$\theta_c = 2\pi(Z - P),$$

where θ_c is the change in the argument of $f(z)$ for one circuit of C; or

$$N = Z - P,$$

where N is the number of times $f(z)$ encircles its origin counter-clockwise for one counter-clockwise circuit of C. This is the basis of the Nyquist stability criterion.

Algebraic equations

The *quadratic* equation $ax^2 + bx + c = 0$ has roots

$$x = \frac{-b \pm \sqrt{(b^2 - 4ac)}}{2a}$$

which are:

real and unequal if $b^2 > 4ac$,

real, equal and given by $-b/2a$ if $b^2 = 4ac$,

complex and conjugate if $b^2 < 4ac$.

The *cubic* equation $x^3 + ax^2 + bx + c = 0$ is reduced to the form $y^3 + py + q = 0$, in which

$$p = -\frac{a^2}{3} + b; \quad q = 2\left(\frac{a}{3}\right)^3 - \frac{ab}{3} + c$$

by the substitution $x = y - a/3$. The roots are obtained from

$$y = A + B, \quad -\tfrac{1}{2}(A + B) \pm \mathrm{j}\frac{\sqrt{3}}{2}(A - B)$$

where

$$A = \sqrt[3]{\left\{-\frac{q}{2} + \sqrt{\left(\frac{q^2}{4} + \frac{p^3}{27}\right)}\right\}};$$

$$B = \sqrt[3]{\left\{-\frac{q}{2} - \sqrt{\left(\frac{q^2}{4} + \frac{p^3}{27}\right)}\right\}}.$$

If

$$\frac{q^2}{4} + \frac{p^3}{27} > 0,\text{ one root is real and two are complex and}$$
$$\text{conjugate,}$$
$$= 0,\text{ all roots are real and two are equal,}$$
$$< 0,\text{ all roots are real and different.}$$

The general equation

$$a_0 x^n + a_1 x^{n-1} + \ldots + a_{n-1} x + a_n = 0$$

of degree n has n roots (real, or complex and conjugate in pairs) of which at least one is real for odd n. Their sum is $-a_1/a_0$ and the sum of their products taken r at a time is $(-1)^r a_r/a_0$.

The Routh-Hurwitz criterion

The number of roots of the general equation which are positive or have positive real parts is the number of sign changes in the sequence

$$D_0, D_1, D_1 D_2, D_2 D_3, \ldots ,\text{ where}$$

Differential equations

Bessel's equation of order v is

$$z^2 \frac{d^2 w}{dz^2} + z \frac{dw}{dz} + (z^2 - v^2) w = 0$$

and its solutions include the following Bessel functions.

First kind, order v:

$$J_v(z) = \sum_{r=0}^{\infty} (-1)^r \left(\frac{z}{2}\right)^{v+2r} / r! \, \Gamma(v + r + 1).$$

First kind, order n (an integer):

$$J_n(z) = \sum_{r=0}^{\infty} (-1)^r \left(\frac{z}{2}\right)^{n+2r} / r!(n + r)!$$

Second kind, order v (non-integral):

$$Y_v(z) = \{\cos v\pi \, J_v(z) - J_{-v}(z)\} / \sin v\pi.$$

Second kind, order n (an integer):

$$Y_n(z) = \left[\frac{\partial}{\partial v} \{\cos v\pi \, J_v(z) - J_{-v}(z)\} \middle/ \frac{\partial}{\partial v} \sin v\pi\right]_{v=n}.$$

Third kind (Hankel functions), order v:

$$H_v^{(1)}(z) = J_v(z) + jY_v(z)$$
$$H_v^{(2)}(z) = J_v(z) - jY_v(z)$$

Complete solutions may take the form, for any v,

$$w = AJ_v(z) + BY_v(z)$$

$$D_0, D_1, D_2, D_3, \ldots$$
$$= a_0, a_1, \begin{vmatrix} a_1 & a_0 \\ a_3 & a_2 \end{vmatrix}, \begin{vmatrix} a_1 & a_0 & 0 \\ a_3 & a_2 & a_1 \\ a_5 & a_4 & a_3 \end{vmatrix}, \ldots$$

All roots have negative real parts if there is no sign change and no coefficient is zero.

Simultaneous linear equations

The set of n equations in n unknowns

$$\sum_{k=1}^{n} a_{ik} x_k = b_i \quad (i = 1, 2, 3, \ldots n)$$

has a unique solution if the determinant of the coefficients $\det[a_{ik}]$ or Δ is non-zero; the solution is given by *Cramer's rule*:

$$x_k = \Delta_k/\Delta \quad (k = 1, 2, 3, \ldots n)$$

where Δ_k is obtained by replacing the kth column of Δ by the column of b_i.

or

$$w = AH_v^{(1)}z + BH_v^{(2)}(z)$$

where A and B are constants.

Legendre's equation of degree n has the form

$$(1 - z^2)\frac{d^2 w}{dz^2} - 2z \frac{dw}{dz} + n(n + 1)w = 0$$

and its solutions are Legendre functions of the first and second kinds; for positive integral n the first are Legendre polynomials. Associated Legendre functions of degree n and order m are solutions to equations of the form

$$(1 - z^2)\frac{d^2 w}{dz^2} - 2z \frac{dw}{dz} + \left\{n(n + 1) - \frac{m^2}{1 - z^2}\right\} w = 0$$

Laguerre's equation of degree n has the form

$$z \frac{d^2 w}{dz^2} + (1 - z)\frac{dw}{dz} + nw = 0.$$

Its solutions for positive integral n are the Laguerre polynomials

$$L_n(z) = \frac{e^z}{n!} \frac{d^n(z^n e^{-z})}{dz^n}.$$

Chebyshev polynomials are solutions to equations of the form

$$(1 - z^2)\frac{d^2 w}{dz^2} - z\frac{dw}{dz} + n^2 w = 0$$

for positive integral n, including:

First kind $T_n(z) = \cos(n\cos^{-1} z)$
Second kind $U_n(z) = \sin(n\cos^{-1} z)$.

Mathieu's equation takes the form

$$\frac{d^2 w}{dz^2} + (a - 16q\cos 2z)w = 0$$

in which a, q are real numbers.

Riccati's equation has the general form

$$\frac{dy}{dx} = ay^2 + by + c$$

in which a, b, c may be functions of x.

Cauchy's equation has the form

$$x^2\frac{d^2 y}{dx^2} + ax\frac{dy}{dx} + by = 0$$

where a, b are constants; its solution is

$$y = Ax^{m_1} + Bx^{m_2}$$

where m_1, m_2 are the roots of

$$m^2 + (a - 1)m + b = 0.$$

The general form of the *wave equation* is

$$\nabla^2 y - a\frac{\partial^2 y}{\partial t^2} - b\frac{\partial y}{\partial t} - cy = 0$$

and has solutions in the form of attenuated travelling waves. If $b = c = 0$, the solutions are undamped travelling waves of phase velocity $1/\sqrt{a}$. If $a = c = 0$, there results the equation of *diffusion* or *heat conduction* which has solutions of exponential form. If $a = b = c = 0$, the equation becomes

$$\nabla^2 y = 0$$

which is *Laplace's equation*; its solutions give the spatial variation of a potential y whose gradient is a vector field of zero divergence.

Poisson's equation is

$$\nabla^2 y = \rho$$

in which ρ may be a function of position; its solutions give the potential y of a field whose divergence at any point is ρ.

Vector analysis

For two vectors $\mathbf{A}, \mathbf{B}$ with angle θ between them:

Scalar product $= \mathbf{A}.\mathbf{B} = \mathbf{B}.\mathbf{A} = AB\cos\theta$
Vector product $= \mathbf{A} \times \mathbf{B} = -\mathbf{B} \times \mathbf{A}$

The vector product has magnitude $AB\sin\theta$ and is normal to the plane containing $\mathbf{A}$ and $\mathbf{B}$.

For unit vectors $\mathbf{i}, \mathbf{j}, \mathbf{k}$ on right-handed orthogonal axes:

$\mathbf{i}.\mathbf{i} = \mathbf{j}.\mathbf{j} = \mathbf{k}.\mathbf{k} = 1$
$\mathbf{i}.\mathbf{j} = \mathbf{j}.\mathbf{k} = \mathbf{k}.\mathbf{i} = 0$
$\mathbf{i} \times \mathbf{i} = \mathbf{j} \times \mathbf{j} = \mathbf{k} \times \mathbf{k} = 0$
$\mathbf{i} \times \mathbf{j} = \mathbf{k} = -(\mathbf{j} \times \mathbf{i})$ etc.

In *Cartesian coordinates:*

$\mathbf{A}.\mathbf{B} = A_x B_x + A_y B_y + A_z B_z$
$\mathbf{A} \times \mathbf{B} = \mathbf{i}(A_y B_z - B_y A_z) + \mathbf{j}(A_z B_x - B_z A_x)$
$\qquad\qquad + \mathbf{k}(A_x B_y - B_x A_y)$

$$= \begin{vmatrix} \mathbf{i} & \mathbf{j} & \mathbf{k} \\ A_x & A_y & A_z \\ B_x & B_y & B_z \end{vmatrix}$$

$$\nabla \equiv \mathbf{i}\frac{\partial}{\partial x} + \mathbf{j}\frac{\partial}{\partial y} + \mathbf{k}\frac{\partial}{\partial z}$$

(In the following, V is a scalar field, $\mathbf{F}$ a vector field.)

$$\text{grad } V = \nabla V = \mathbf{i}\frac{\partial V}{\partial x} + \mathbf{j}\frac{\partial V}{\partial y} + \mathbf{k}\frac{\partial V}{\partial z}$$

$$\text{div } \mathbf{F} = \nabla.\mathbf{F} = \frac{\partial F_x}{\partial x} + \frac{\partial F_y}{\partial y} + \frac{\partial F_z}{\partial z}$$

$$\text{curl } \mathbf{F} = \nabla \times \mathbf{F} = \mathbf{i}\left(\frac{\partial F_z}{\partial y} - \frac{\partial F_y}{\partial z}\right) + \mathbf{j}\left(\frac{\partial F_x}{\partial z} - \frac{\partial F_z}{\partial x}\right)$$
$$+ \mathbf{k}\left(\frac{\partial F_y}{\partial x} - \frac{\partial F_x}{\partial y}\right)$$

$$\nabla^2 \equiv \frac{\partial^2}{\partial x^2} + \frac{\partial^2}{\partial y^2} + \frac{\partial^2}{\partial z^2}$$

$$\text{div grad } V = \nabla^2 V = \frac{\partial^2 V}{\partial x^2} + \frac{\partial^2 V}{\partial y^2} + \frac{\partial^2 V}{\partial z^2}.$$

In *spherical coordinates* (unit vectors $\mathbf{u}_r, \mathbf{u}_\theta, \mathbf{u}_\phi$):

$$\text{grad } V = \mathbf{u}_r\frac{\partial V}{\partial r} + \mathbf{u}_\theta\frac{1}{r}\frac{\partial V}{\partial \theta} + \mathbf{u}_\phi\frac{1}{r\sin\theta}\frac{\partial V}{\partial \phi}$$

$$\text{div } \mathbf{F} = \frac{1}{r^2}\frac{\partial}{\partial r}(r^2 F_r) + \frac{1}{r\sin\theta}\frac{\partial}{\partial \theta}(F_\theta\sin\theta) + \frac{1}{r\sin\theta}\frac{\partial F_\phi}{\partial \phi}$$

$$\text{curl } \mathbf{F} = \mathbf{u}_r \, \frac{1}{r \sin \theta} \left[\frac{\partial}{\partial \theta} (F_\phi \sin \theta) - \frac{\partial F_\theta}{\partial \phi} \right]$$

$$+ \mathbf{u}_\theta \, \frac{1}{r} \left[\frac{1}{\sin \theta} \frac{\partial F_r}{\partial \phi} - \frac{\partial}{\partial r} (rF_\phi) \right] + \mathbf{u}_\phi \, \frac{1}{r} \left[\frac{\partial}{\partial r} (rF_\theta) - \frac{\partial F_r}{\partial \theta} \right]$$

$$\nabla^2 V = \frac{1}{r^2} \frac{\partial}{\partial r} \left(r^2 \frac{\partial V}{\partial r} \right) + \frac{1}{r^2 \sin \theta} \frac{\partial}{\partial \theta} \left(\sin \theta \frac{\partial V}{\partial \theta} \right)$$

$$+ \frac{1}{r^2 \sin^2 \theta} \frac{\partial^2 V}{\partial \phi^2}.$$

In *cylindrical coordinates* (unit vectors $\mathbf{u}_r$, $\mathbf{u}_\phi$, $\mathbf{u}_z$):

$$\text{grad } V = \mathbf{u}_r \, \frac{\partial V}{\partial r} + \mathbf{u}_\phi \, \frac{1}{r} \frac{\partial V}{\partial \phi} + \mathbf{u}_z \, \frac{\partial V}{\partial z}$$

$$\text{div } \mathbf{F} = \frac{1}{r} \frac{\partial}{\partial r} (rF_r) + \frac{1}{r} \frac{\partial F_\phi}{\partial \phi} + \frac{\partial F_z}{\partial z}$$

$$\text{curl } \mathbf{F} = \mathbf{u}_r \left(\frac{1}{r} \frac{\partial F_z}{\partial \phi} - \frac{\partial F_\phi}{\partial z} \right) + \mathbf{u}_\phi \left(\frac{\partial F_r}{\partial z} - \frac{\partial F_z}{\partial r} \right)$$

$$+ \mathbf{u}_z \, \frac{1}{r} \left[\frac{\partial}{\partial r} (rF_\phi) - \frac{\partial F_r}{\partial \phi} \right]$$

$$\nabla^2 V = \frac{1}{r} \frac{\partial}{\partial r} \left(r \frac{\partial V}{\partial r} \right) + \frac{1}{r^2} \frac{\partial^2 V}{\partial \phi^2} + \frac{\partial^2 V}{\partial z^2}.$$

General vector identities

$$\text{curl grad } V = \nabla \times \nabla V = 0$$

$$\text{div curl } \mathbf{F} = \nabla . \nabla \times \mathbf{F} = 0$$

$$\text{curl curl } \mathbf{F} = \nabla \times \nabla \times \mathbf{F} = \text{grad div } \mathbf{F} - \nabla^2 \mathbf{F}$$

$$\text{grad } (V_1 V_2) = V_1 \text{ grad } V_2 + V_2 \text{ grad } V_1$$

$$\text{div } (V\mathbf{F}) = V \text{ div } \mathbf{F} + \mathbf{F} . \text{ grad } V$$

$$\text{div } (\mathbf{F}_1 \times \mathbf{F}_2) = \mathbf{F}_2 . \text{ curl } \mathbf{F}_1 - \mathbf{F}_1 . \text{ curl } \mathbf{F}_2$$

$$\text{curl } (V\mathbf{F}) = V \text{ curl } \mathbf{F} - \mathbf{F} \times \text{ grad } V$$

$$\text{curl } (\mathbf{F}_1 \times \mathbf{F}_2) = \mathbf{F}_1 \text{ div } \mathbf{F}_2 - \mathbf{F}_2 \text{ div } \mathbf{F}_1 + (\mathbf{F}_2 . \nabla) \mathbf{F}_1 - (\mathbf{F}_1 . \nabla)\mathbf{F}_2.$$

A vector field of zero divergence is said to be *solenoidal.* If the line integral of $\mathbf{F}$ around any closed path is zero, then $\mathbf{F}$ has zero curl, can always be expressed as grad V and is said to be *lamellar, conservative* or *irrotational.*

Gauss's divergence theorem

If $\mathrm{d}\tau$ is an element of a volume T bounded by a surface S of which $\mathrm{d}\mathbf{S}$ is an element, then

$$\iiint_T \text{div } \mathbf{F} \, \mathrm{d}\tau = \iint_S \mathbf{F} . \, \mathrm{d}\mathbf{S}.$$

Stokes's theorem

If $\mathrm{d}\mathbf{S}$ is an element of a surface S bounded by a closed curve C of which $\mathrm{d}\mathbf{l}$ is an element, then

$$\iint_S \text{curl } \mathbf{F} . \, \mathrm{d}\mathbf{S} = \int_C \mathbf{F} . \, \mathrm{d}\mathbf{l}.$$

Matrices

If $A_{m,n}$ represents a matrix A of order $m \times n$ (i.e. having m rows and n columns) with the element a_{jk} in the jth row and kth column, then:

(i) $A_{m,n} + B_{m,n} = C_{m,n} = B_{m,n} + A_{m,n}$
and $c_{jk} = a_{jk} + b_{jk}$.
(Only matrices of the same order may be added or subtracted.)

(ii) $\lambda A_{m,n} = B_{m,n}$
and $b_{jk} = \lambda a_{jk}$, where λ is a scalar.

(iii) $A_{m,n} B_{n,p} = C_{m,p}$
and $c_{jk} = \sum_{l=1}^{n} a_{jl} b_{lk}$.
(Two matrices can be multiplied only if they are *conformable;* i.e. if the first has as many columns as the second has rows; in general $AB \neq BA$.)

(iv)   The *transpose* of $A_{m,n}$ is
$B_{n,m} = A^{\mathrm{T}}$
and $b_{jk} = a_{kj}$.
(It follows that $(A + B)^{\mathrm{T}} = A^{\mathrm{T}} + B^{\mathrm{T}}$
and $(AB)^{\mathrm{T}} = B^{\mathrm{T}} A^{\mathrm{T}}$.)

(v) A is a *square matrix* if $m = n$
A is a *row matrix* if $m = 1$
A is a *column matrix* if $n = 1$.

(vi)   A square matrix A is *symmetric* if $A^{\mathrm{T}} = A$
A square matrix A is *skew-symmetric* if $A^{\mathrm{T}} = -A$
A square matrix A is *diagonal* if $a_{jk} = 0 \; (j \neq k)$.
A unit matrix U is a diagonal matrix in which
$u_{jk} = 1 \; (j = k)$.

(vii)   If $AB = U$, then $B = A^{-1}$ is the *inverse* of A.
(The inverse of any product AB is $(AB)^{-1} = B^{-1} A^{-1}$.)

(viii)   If $A^{-1} = A^{\mathrm{T}}$, A is *orthogonal.*

Matrix representation of vectors

An n-dimensional vector may be represented as a row matrix of order $1 \times n$ or as a column matrix of order $n \times 1$. The column matrix for a vector $\mathbf{A}$ may be written $\{A\}$. The scalar product of $\mathbf{A}$ and $\mathbf{B}$ is then

$$\mathbf{A} . \mathbf{B} = \{A\}^{\mathrm{T}} \{B\} = \{B\}^{\mathrm{T}} \{A\}.$$

The vector product of two three-dimensional vectors $\mathbf{A}$ and $\mathbf{B}$ in Cartesian coordinates is

$$\mathbf{A} \times \mathbf{B} = [A]\{B\}$$

where $[A]$ is the skew-symmetric matrix

$$\begin{bmatrix} 0 & -A_z & A_y \\ A_z & 0 & -A_x \\ -A_y & A_x & 0 \end{bmatrix}.$$

Rotation of axes

If a vector is represented by $\{R\}$ in a system of Cartesian coordinates $OXYZ$, and by $\{r\}$ in a second system $Oxyz$ having the same origin O, then

$$\{r\} = [C]\{R\}$$

where $[C]$ is the *rotation matrix*

$$\begin{bmatrix} l_{xX} & l_{xY} & l_{xZ} \\ l_{yX} & l_{yY} & l_{yZ} \\ l_{zX} & l_{zY} & l_{zZ} \end{bmatrix}$$

in which $l_{xX} = \cos(xOX)$, xOX being the angle between Ox and OX, etc.

Properties of plane curves and figures

Pappus's theorems

(i) The surface area generated by a curve of length l revolving about an axis is

$$A = 2\pi l \bar{y}$$

where $\bar{y}$ is the perpendicular distance of the centroid of the curve from the axis.

(ii) The volume generated by a plane surface of area A rotating about an axis is

$$V = 2\pi A \bar{y}$$

where $\bar{y}$ is the perpendicular distance of the centroid of A from the axis.

Conic sections

	Circle	Ellipse	Hyperbola	Parabola
Cartesian equation	$x^2 + y^2 = a^2$	$\dfrac{x^2}{a^2} + \dfrac{y^2}{b^2} = 1$	$\dfrac{x^2}{a^2} - \dfrac{y^2}{b^2} = 1$	$y^2 = lx$
Eccentricity ϵ	0	$\left(1 - \dfrac{b^2}{a^2}\right)^{1/2} < 1$	$\left(1 + \dfrac{b^2}{a^2}\right)^{1/2} > 1$	1
Focal distance OF	0	$a\epsilon$	$a\epsilon$	$l/4$
Latus rectum l	$2a$	$2b^2/a$	$2b^2/a$	
Circumference	$2\pi a$	$4aE \approx 2\pi\sqrt{\left(\dfrac{a^2 + b^2}{2}\right)}$		
Enclosed area	πa^2	πab		$h^3/6l^*$
Polar equation, origin O	$r = a$	$r^2 = b^2/(1 - \epsilon^2 \cos^2\theta)$	$r^2 = -b^2/(1 - \epsilon^2 \cos^2\theta)$	$r = l\cos\theta/(1 - \cos^2\theta)$
Polar equation, origin F	$r = a$	$r = l/2(1 - \epsilon\cos\theta)$	$r = l/2(1 - \epsilon\cos\theta)$	$r = l/2(1 - \cos\theta)$

* Area enclosed by curve and vertical chord of length h.

Other curves

Catenary

$$y = a \cosh (x/a)$$

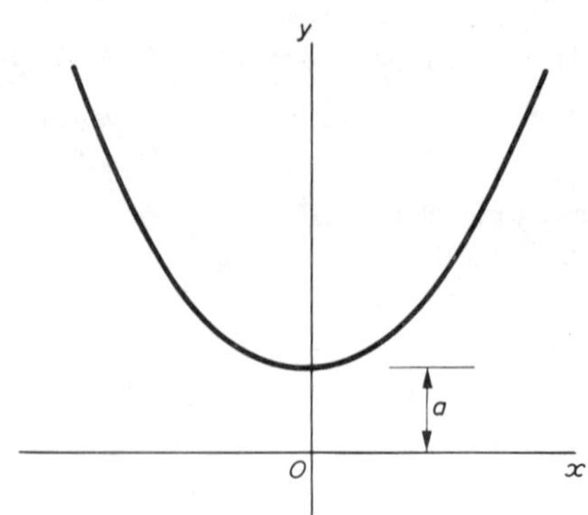

Cycloids

$$x = a\theta - b \sin \theta$$
$$y = a - b \cos \theta$$

(i) $a = b$ (arc $8a$, area $3\pi a^2$)

(ii) $a < b$ (prolate)

(iii) $a > b$ (curtate)

b is the generating radius on the circle of radius a.

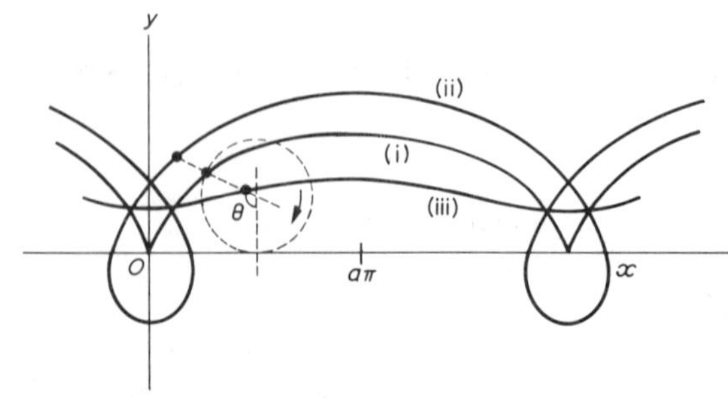

Epicycloids

$$x = (a + b) \cos \phi - b \cos \left(\frac{a + b}{b}\right) \phi$$

$$y = (a + b) \sin \phi - b \sin \left(\frac{a + b}{b}\right) \phi$$

(i) $0 < b < a$

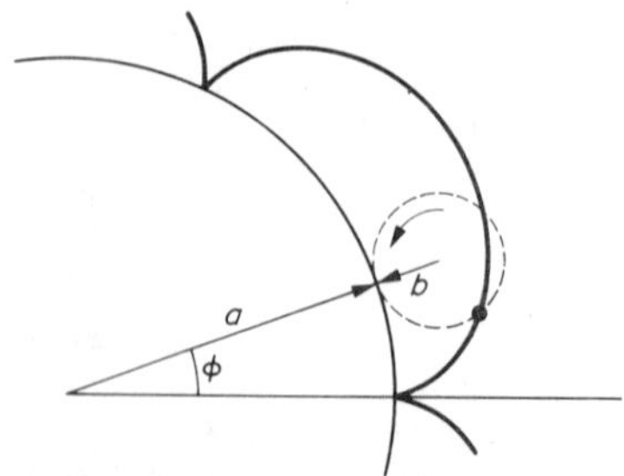

(ii) $a = b$: *cardioid*

polar equation: $r = 2a(1 - \cos \theta)$

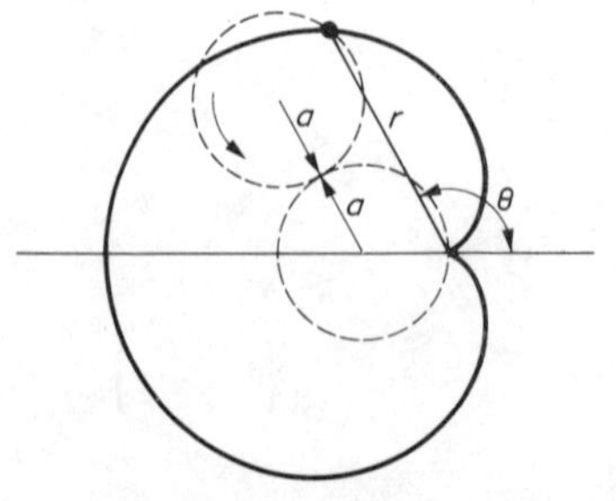

(iii) $b \rightarrow -b$: *hypocycloid*

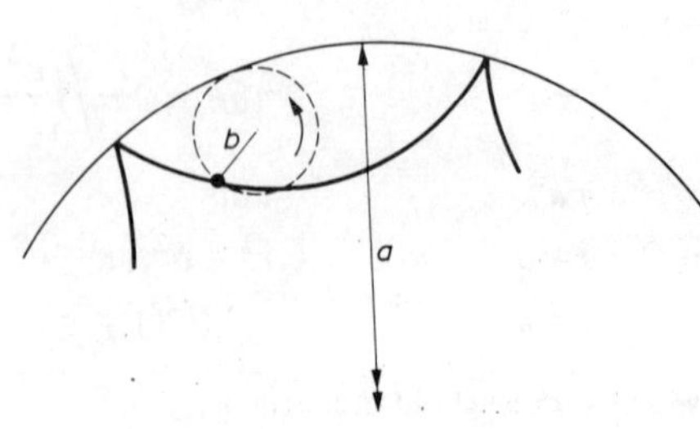

Logarithmic spiral

$$r = a\,e^{b\theta}$$

Area between radii $r_1, r_2 = \dfrac{r_2^2 - r_1^2}{4b}$

Length between radii $r_1, r_2 = \dfrac{(r_2 - r_1)\sqrt{(b^2 + 1)}}{b}$

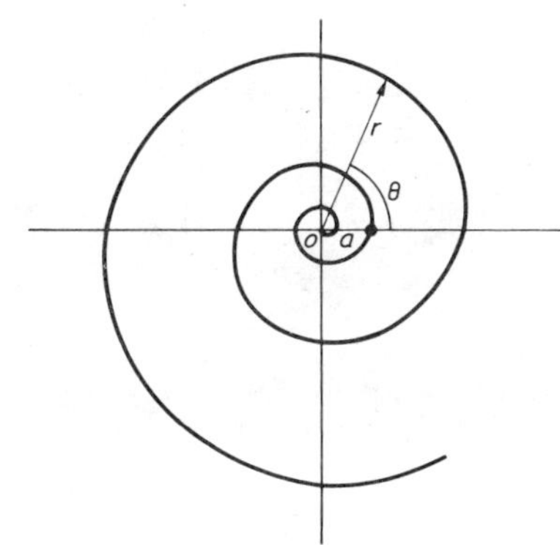

Archimedean spiral

$$r = a\theta$$
Area $= a^2\theta^3/6$
Length $\to a\theta^2/2$ for large θ

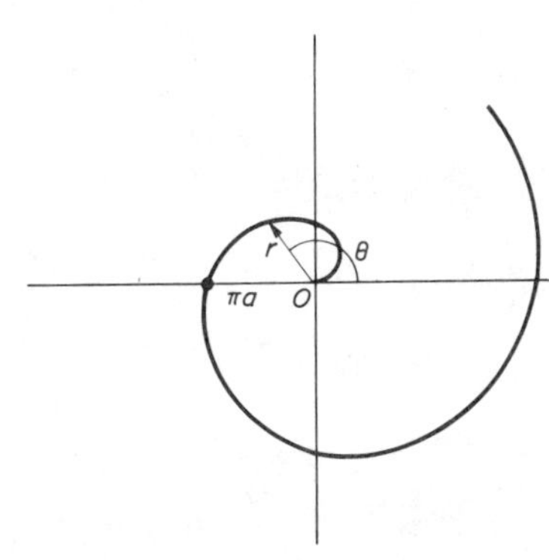

Areas, centroids and second moments of area

The second moment of a plane area A about an axis x is

$$I_{xx} = \int y^2\,\mathrm{d}A = Ak_x^2$$

where k_x is the radius of gyration about x. The product moment of area is

$$I_{xy} = \int xy\,\mathrm{d}A$$

and is zero if x and y are principal axes; an axis of symmetry is a principal axis for any origin lying on it. These moments are sometimes incorrectly called moment and product of inertia.

Parallel-axis theorem

If I is the second moment of area about any axis through the centroid of A, then that for a parallel axis at a perpendicular distance d is

$$I' = I + Ad^2$$

and if I is the product moment for any pair of Cartesian axes through the centroid, that for a parallel pair at distances a, b is

$$I' = I + Aab$$

Rotation of axes

If I_{xx}, I_{yy} and I_{xy} are values for axes x,y, then the corresponding values for axes x',y' at an angle α to x,y and having the same origin are

$$I_{x'x'} = \tfrac{1}{2}(I_{xx} + I_{yy}) + \tfrac{1}{2}(I_{xx} - I_{yy})\cos 2\alpha - I_{xy}\sin 2\alpha$$

$$I_{y'y'} = \tfrac{1}{2}(I_{xx} + I_{yy}) + \tfrac{1}{2}(I_{yy} - I_{xx})\cos 2\alpha + I_{xy}\sin 2\alpha$$

$$I_{x'y'} = \tfrac{1}{2}(I_{xx} - I_{yy})\sin 2\alpha + I_{xy}\cos 2\alpha.$$

Polar moment

The polar second moment for any point O in the plane of A is the second moment of A about an axis z through O normal to A. It is given by

$$I_{zz} = I_{xx} + I_{yy}$$

for any pair of orthogonal axes x, y in A with origin O.

In the following table, $\bar{x}, \bar{y}$ are the coordinates of the centroid C with respect to the origin O; I_{xx}, etc., are moments for axes *through C* in the directions x, y.

Figure and area, A	$\bar{x}, \bar{y}$	I_{xx}	I_{yy}	I_{xy}
bh rectangle	$\dfrac{b}{2}, \dfrac{h}{2}$	$\dfrac{Ah^2}{12}$	$\dfrac{Ab^2}{12}$	0
$ab \sin \theta$ parallelogram	$\dfrac{b + a \cos \theta}{2}, \dfrac{a \sin \theta}{2}$	$\dfrac{A}{12}(a \sin \theta)^2$	$\dfrac{A}{12}(b^2 + a^2 \cos^2 \theta)$	$\dfrac{A}{12}a^2 \sin \theta \cos \theta$
$\dfrac{bh}{2}$ triangle	$\dfrac{a + b}{3}, \dfrac{h}{3}$	$\dfrac{Ah^2}{18}$	$\dfrac{A}{18}(b^2 - ab + a^2)$	$\dfrac{Ah}{36}(2a - b)$
$\dfrac{h}{2}(a + b)$ trapezium	$\bar{y} = \dfrac{h(2a + b)}{3(a + b)}$	$\dfrac{Ah^2(a^2 + 4ab + b^2)}{18(a + b)^2}$		
πa^2 circle	a, a	$\dfrac{Aa^2}{4}$	$\dfrac{Aa^2}{4}$	0
$\dfrac{\pi a^2}{2}$ semicircle	$a, \dfrac{4a}{3\pi}$	$\dfrac{Aa^2(9\pi^2 - 64)}{36\pi^2}$	$\dfrac{Aa^2}{4}$	0
$a^2\theta$ sector of circle	$\dfrac{2a \sin \theta}{3\theta}, 0$	$\dfrac{Aa^2(\theta - \sin \theta \cos \theta)}{4\theta}$	$A\left\{\dfrac{a^2(\theta + \sin \theta \cos \theta)}{4\theta} - \bar{x}^2\right\}$	0

Shape	Area	Centroid		
segment of circle	$a^2(\theta - \frac{1}{2}\sin 2\theta)$	$\dfrac{2a\sin^3\theta}{3(\theta - \frac{1}{2}\sin 2\theta)}, 0$	$\dfrac{Aa^2}{4}\left\{1 - \dfrac{\sin^2\theta\,\sin 2\theta}{3(\theta - \frac{1}{2}\sin 2\theta)}\right\}$	$\dfrac{Aa^2}{4}\left\{1 + \dfrac{\sin^2\theta\,\sin 2\theta}{\theta - \frac{1}{2}\sin 2\theta}\right\} - A\bar{x}^2$	0
rectangle	bh	$\frac{1}{2}\sqrt{(b^2 + h^2)}, 0$	$\dfrac{Ab^2h^2}{6(h^2 + b^2)}$	$\dfrac{A(h^4 + b^4)}{12(h^2 + b^2)}$	$\dfrac{Abh(h^2 - b^2)}{12(h^2 + b^2)}$
ellipse	πab	a, b	$\dfrac{Ab^2}{4}$	$\dfrac{Aa^2}{4}$	0
semi-ellipse	$\dfrac{\pi ab}{2}$	$a, \dfrac{4b}{3\pi}$	$\dfrac{Ab^2(9\pi^2 - 64)}{36\pi^2}$	$\dfrac{Aa^2}{4}$	0
parabola	$\dfrac{4ab}{3}$	$\dfrac{3a}{5}, 0$	$\dfrac{Ab^2}{5}$	$\dfrac{12Aa^2}{175}$	0
$x = a\left(\frac{y}{b}\right)^n$	$\dfrac{nab}{n+1}$	$\dfrac{(n+1)a}{2n+1}, \dfrac{(n+1)b}{2(n+2)}$			
$y = b\left(\frac{x}{a}\right)^n$	$\dfrac{ab}{n+1}$	$\dfrac{(n+1)a}{n+2}, \dfrac{(n+1)b}{2(2n+1)}$			

Moments of inertia, etc., of rigid bodies

The three products of inertia for any set of principal axes are all zero.

If two Cartesian axes, say x and y, lie in a plane of mass symmetry, then only the product I_{xy} can be non-zero. If there are two orthogonal planes of symmetry, their intersection is a principal axis for any origin lying on it.

Parallel-axis theorem

If I is the moment of inertia about any axis through the centre of mass C of a body of mass m, then that for a parallel axis at a perpendicular distance d is

$$I' = I + md^2$$

and if I is the product of inertia for any Cartesian pair through C, that for a parallel pair at distances a, b is

$$I' = I + mab$$

Rotation of axes

If I is the inertia matrix for certain axes, the matrix for a new set having the same origin and a rotation matrix $[C]$ is

$$[I'] = [C]\,[I]\,[C]^T$$

For any origin, the sum of the moments $I_{xx} + I_{yy} + I_{zz}$ is invariant.

In the following table for homogeneous bodies $\bar{x}, \bar{y}, \bar{z}$ are the coordinates of the centre of mass C with respect to the origin O; I_{xx}, etc., are the (principal) moments for axes $\underline{through\ C}$ in the directions x, y, z; A is the area of external curved surfaces only and V is the volume.

Body	A	V	$\bar{x}, \bar{y}, \bar{z}$	I_{xx}	I_{yy}	I_{zz}
uniform rod	0	0	$\dfrac{l}{2}, 0, 0$	0	$\dfrac{ml^2}{12}$	$\dfrac{ml^2}{12}$
rectangular prism	—	abc	$\dfrac{a}{2}, \dfrac{b}{2}, \dfrac{c}{2}$	$\dfrac{m(b^2 + c^2)}{12}$	$\dfrac{m(c^2 + a^2)}{12}$	$\dfrac{m(a^2 + b^2)}{12}$
right rectangular pyramid	—	$\dfrac{abh}{3}$	$0, \dfrac{h}{4}, 0$	$\dfrac{m(4b^2 + 3h^2)}{80}$	$\dfrac{m(a^2 + b^2)}{20}$	$\dfrac{m(4a^2 + 3h^2)}{80}$
uniform hoop	0	0	$0, 0, 0$	$\dfrac{ma^2}{2}$	$\dfrac{ma^2}{2}$	ma^2
arc of hoop	0	0	$\dfrac{a \sin \theta}{\theta}, 0, 0$	$\dfrac{ma^2(\theta - \sin \theta \cos \theta)}{2\theta}$		$ma^2 \left(1 - \dfrac{\sin^2 \theta}{\theta^2} \right)$
spherical shell	$4\pi a^2$	0	$0, 0, 0$	$\dfrac{2ma^2}{3}$	$\dfrac{2ma^2}{3}$	$\dfrac{2ma^2}{3}$
hollow sphere	$4\pi a^2$	$\dfrac{4\pi}{3}(a^3 - b^3)$	$0, 0, 0$	$\dfrac{2m(a^5 - b^5)}{5(a^3 - b^3)}$	$\dfrac{2m(a^5 - b^5)}{5(a^3 - b^3)}$	$\dfrac{2m(a^5 - b^5)}{5(a^3 - b^3)}$
sphere	$4\pi a^2$	$\dfrac{4\pi a^3}{3}$	$0, 0, 0$	$\dfrac{2ma^2}{5}$	$\dfrac{2ma^2}{5}$	$\dfrac{2ma^2}{5}$

Body	A	V	$\bar{x}, \bar{y}, \bar{z}$	I_{xx}	I_{yy}	I_{zz}
hemisphere	$2\pi a^2$	$\dfrac{2\pi a^3}{3}$	$0, \dfrac{3a}{8}, 0$		$\dfrac{2ma^2}{5}$	
right circular cylinder	$2\pi ah$	$\pi a^2 h$	$0, \dfrac{h}{2}, 0$	$\dfrac{m(3a^2 + h^2)}{12}$	$\dfrac{ma^2}{2}$	$\dfrac{m(3a^2 + h^2)}{12}$
right circular cone	$\pi a\sqrt{a^2 + h^2}$	$\dfrac{\pi a^2 h}{3}$	$0, \dfrac{h}{4}, 0$	$\dfrac{3m(4a^2 + h^2)}{80}$	$\dfrac{3ma^2}{10}$	$\dfrac{3m(4a^2 + h^2)}{80}$
ellipsoid		$\dfrac{4\pi abc}{3}$	$0, 0, 0$	$\dfrac{m(b^2 + c^2)}{5}$	$\dfrac{m(c^2 + a^2)}{5}$	$\dfrac{m(a^2 + b^2)}{5}$
segment of spherical shell*	$2\pi rh$	0	$0, \dfrac{h}{2}, 0$			
segment of sphere*	$2\pi rh$	$\pi h^2\left(r - \dfrac{h}{3}\right)$	$0, \dfrac{h(4a - h)}{4(3a - h)}, 0$			

* radius r

Numerical analysis

Solution of algebraic equation $f(x) = 0$.

:(i) Newton's method:

$$x_{n+1} = x_n - f(x_n)/f'(x_n)$$

(ii) Secant method:

$$x_{n+1} = \frac{-x_n f(x_{n-1}) + x_{n-1} f(x_n)}{f(x_n) - f(x_{n-1})}$$

where x_n is the nth estimate.

Approximations to derivatives

$$f'(x) = \frac{f(x+h) - f(x-h)}{2h}$$

$$f''(x) = \frac{f(x+h) - 2f(x) + f(x-h)}{h^2}$$

$$f'''(x) = \frac{f(x+2h) - 3f(x+h) + 3f(x-h) - f(x-2h)}{2h^3}$$

$$f^{(4)}(x) = $$
$$\frac{f(x+2h) - 4f(x+h) + 6f(x) - 4f(x-h) + f(x-2h)}{h^4}$$

where h is an increment in x.

Numerical integration by equal intervals h

(i) Trapezoidal rule:

$$\int_{x_0}^{x_1} y(x)\,dx = \frac{h}{2}(y_0 + y_1) - 0(h^3 y_0''/12)$$

(ii) Simpson's rule:

$$\int_{x_0}^{x_2} y(x)\,dx = \frac{h}{3}(y_0 + 4y_1 + y_2) - 0(h^5 y_1^{(4)}/90)$$

where $x_n = x_0 + nh, y_n = y(x_n)$.

Everett's interpolation formula for a table of $y(x)$

If $x = x_0 + s(x_1 - x_0)$ and $p = 1 - s$, then

$$y(x) \approx \binom{p}{1} y_0 + \binom{p+1}{3} \delta^2 y_0 + \ldots$$
$$+ \binom{s}{1} y_1 + \binom{s+1}{3} \delta^2 y_1 + \ldots$$

where $\delta^2 y_0 = y_1 - 2y_0 + y_{-1}$, etc.

Smoothing

Third-order, five-point: a least-squares cubic for five successive points $y_{-2} \ldots y_2$ is fixed by the points

$$y_0^1 = y_0 - \tfrac{3}{35}\delta^4 y_0 \text{ (similarly } y_1^1, y_2^1, \ldots)$$
$$\left.\begin{aligned} y_{-1}^1 &= y_{-1} + \tfrac{2}{35}\delta^4 y_0 \\ y_{-2}^1 &= y_{-2} - \tfrac{1}{70}\delta^4 y_0 \end{aligned}\right\} \text{ end points}$$

Gaussian integration (second order)

$$\int_{-1}^{1} f(x)\,dx = f\left(-\frac{1}{\sqrt{3}}\right) + f\left(\frac{1}{\sqrt{3}}\right)$$

Integration of ordinary differential equations $\dfrac{dy}{dx} = f(x, y)$

(i) *Runge-Kutta*

2nd order: $y_{n+1} = y_n + \dfrac{h}{2}\,[f(x_n, y_n)$

$\qquad + f\{x_n + h, y_n + hf(x_n, y_n)\}]$

4th order: $y_{n+1} = y_n + \tfrac{1}{6}(k_1 + 2k_2 + 2k_3 + k_4)$

where

$$k_1 = hf(x_n, y_n)$$

$$k_2 = hf\left(x_n + \frac{h}{2}, y_n + \frac{k_1}{2}\right)$$

$$k_3 = hf\left(x_n + \frac{h}{2}, y_n + \frac{k_2}{2}\right)$$

$$k_4 = hf(x_n + h, y_n + k_3)$$

(ii) *Adams-Bashforth*

Predictor:

$$y_{n+1} = y_n + \frac{h}{24}\{55f(x_n, y_n) - 59f(x_{n-1}, y_{n-1})$$
$$+ 37f(x_{n-2}, y_{n-2}) - 9f(x_{n-3}, y_{n-3})\}$$

Corrector:

$$y_{n+1}^k = y_n + \frac{h}{24}\{9f(x_{n+1}, y_{n+1}^{k-1}) + 19f(x_n, y_n)$$
$$- 5f(x_{n-1}, y_{n-1}) + f(x_{n-2}, y_{n-2})\}$$

Here $x_n = x_0 + nh$, y_n^k is the kth estimate of $y(x_n)$.

Statistics

The variance of n values of a variable x is

$$s^2 = \frac{1}{n-1} \sum_{j=1}^{n} (x_j - \bar{x})^2$$

and the standard deviation is s.

If the value of a variable x has a probability density $f(x)$ then the mean of x is

$$\mu = \int_{-\infty}^{\infty} x f(x) \, \mathrm{d}x$$

the variance of the distribution of x is

$$\sigma^2 = \int_{-\infty}^{\infty} (x - \mu)^2 f(x) \, \mathrm{d}x$$

and σ is the standard deviation.

Distributions

If a certain event has a probability p of occurring in each of n independent trials, then the occurrence of x events has the probability

$$f(x) = \binom{n}{x} p^x (1 - p)^{n-x}$$

x has the mean $\mu = np$ and $f(x)$ has the variance $np(1 - p)$. This is the binomial distribution.

The limit of the binomial distribution for $p \to 0$, $n \to \infty$ is the *Poisson distribution* with probability function

$$f(x) = \frac{\mu^x}{x!} \, \mathrm{e}^{-\mu} \quad (x = 0, 1, 2 \ldots)$$

when np is defined as the mean, μ. The variance is then

$$\sigma^2 = \mu$$

The Gauss or *normal distribution* is defined by the probability function

$$f(x) = \frac{1}{\sigma\sqrt{(2\pi)}} \, \mathrm{e}^{-(x-\mu)^2/2\sigma^2}$$

in which μ is the mean of x and σ the standard deviation.

The probability that in any one trial the variable will assume a value $\leqslant x$ is the *distribution function*

$$F(x) = \int_{-\infty}^{x} f(t) \, \mathrm{d}t$$

For the normal distribution with $\mu = 0$, $\sigma^2 = 1$ this becomes

$$\Phi(x) = \frac{1}{\sqrt{(2\pi)}} \int_{-\infty}^{x} \mathrm{e}^{-t^2/2} \, \mathrm{d}t$$

corresponding to a probability function (or frequency curve)

$$\phi(x) = \frac{1}{\sqrt{(2\pi)}} \, \mathrm{e}^{-x^2/2}$$

The probability that $a < x < b$ for a variable x with mean μ and standard deviation σ is then

$$p(a < x < b) = \Phi\left(\frac{b - \mu}{\sigma}\right) - \Phi\left(\frac{a - \mu}{\sigma}\right)$$

the probability that $(\mu - n\sigma) < x < (\mu + n\sigma)$ is

$$p_{n\sigma} = \Phi(n) - \Phi(-n)$$

For $n = 2$ this gives $p_{2\sigma} = 0.955$, for $n = 3$ $p_{3\sigma} = 0.997$.

The normal distribution function

x	$\Phi(x)$	x	$\Phi(x)$	x	$\Phi(x)$	x	$\Phi(x)$	x	$\Phi(x)$
0·00	$0{\cdot}5000_{40}$	0·50	$0{\cdot}6915_{35}$	1·00	$0{\cdot}8413_{25}$	1·50	$0{\cdot}9332_{13}$	2·00	$0{\cdot}97725_{53}$
·01	$\cdot5040_{40}$	·51	$\cdot6950_{35}$	·01	$\cdot8438_{23}$	·51	$\cdot9345_{12}$	·01	$\cdot97778_{53}$
·02	$\cdot5080_{40}$	·52	$\cdot6985_{34}$	·02	$\cdot8461_{24}$	·52	$\cdot9357_{13}$	·02	$\cdot97831_{51}$
·03	$\cdot5120_{40}$	·53	$\cdot7019_{35}$	·03	$\cdot8485_{23}$	·53	$\cdot9370_{12}$	·03	$\cdot97882_{50}$
·04	$\cdot5160_{39}$	·54	$\cdot7054_{34}$	·04	$\cdot8508_{23}$	·54	$\cdot9382_{12}$	·04	$\cdot97932_{50}$
0·05	$0{\cdot}5199_{40}$	0·55	$0{\cdot}7088_{35}$	1·05	$0{\cdot}8531_{23}$	1·55	$0{\cdot}9394_{12}$	2·05	$0{\cdot}97982_{48}$
·06	$\cdot5239_{40}$	·56	$\cdot7123_{34}$	·06	$\cdot8554_{23}$	·56	$\cdot9406_{12}$	·06	$\cdot98030_{47}$
·07	$\cdot5279_{40}$	·57	$\cdot7157_{33}$	·07	$\cdot8577_{22}$	·57	$\cdot9418_{11}$	·07	$\cdot98077_{47}$
·08	$\cdot5319_{40}$	·58	$\cdot7190_{34}$	·08	$\cdot8599_{22}$	·58	$\cdot9429_{12}$	·08	$\cdot98124_{45}$
·09	$\cdot5359_{39}$	·59	$\cdot7224_{33}$	·09	$\cdot8621_{22}$	·59	$\cdot9441_{11}$	·09	$\cdot98169_{45}$
0·10	$0{\cdot}5398_{40}$	0·60	$0{\cdot}7257_{34}$	1·10	$0{\cdot}8643_{22}$	1·60	$0{\cdot}9452_{11}$	2·10	$0{\cdot}98214_{43}$
·11	$\cdot5438_{40}$	·61	$\cdot7291_{33}$	·11	$\cdot8665_{21}$	·61	$\cdot9463_{11}$	·11	$\cdot98257_{43}$
·12	$\cdot5478_{39}$	·62	$\cdot7324_{33}$	·12	$\cdot8686_{22}$	·62	$\cdot9474_{10}$	·12	$\cdot98300_{41}$
·13	$\cdot5517_{40}$	·63	$\cdot7357_{32}$	·13	$\cdot8708_{21}$	·63	$\cdot9484_{11}$	·13	$\cdot98341_{41}$
·14	$\cdot5557_{39}$	·64	$\cdot7389_{33}$	·14	$\cdot8729_{20}$	·64	$\cdot9495_{10}$	·14	$\cdot98382_{40}$
0·15	$0{\cdot}5596_{40}$	0·65	$0{\cdot}7422_{32}$	1·15	$0{\cdot}8749_{21}$	1·65	$0{\cdot}9505_{10}$	2·15	$0{\cdot}98422_{39}$
·16	$\cdot5636_{39}$	·66	$\cdot7454_{32}$	·16	$\cdot8770_{20}$	·66	$\cdot9515_{10}$	·16	$\cdot98461_{39}$
·17	$\cdot5675_{39}$	·67	$\cdot7486_{31}$	·17	$\cdot8790_{20}$	·67	$\cdot9525_{10}$	·17	$\cdot98500_{37}$
·18	$\cdot5714_{39}$	·68	$\cdot7517_{32}$	·18	$\cdot8810_{20}$	·68	$\cdot9535_{10}$	·18	$\cdot98537_{37}$
·19	$\cdot5753_{40}$	·69	$\cdot7549_{31}$	·19	$\cdot8830_{19}$	·69	$\cdot9545_{9}$	·19	$\cdot98574_{36}$
0·20	$0{\cdot}5793_{39}$	0·70	$0{\cdot}7580_{31}$	1·20	$0{\cdot}8849_{20}$	1·70	$0{\cdot}9554_{10}$	2·20	$0{\cdot}98610_{35}$
·21	$\cdot5832_{39}$	·71	$\cdot7611_{31}$	·21	$\cdot8869_{20}$	·71	$\cdot9564_{9}$	·21	$\cdot98645_{34}$
·22	$\cdot5871_{39}$	·72	$\cdot7642_{31}$	·22	$\cdot8888_{19}$	·72	$\cdot9573_{9}$	·22	$\cdot98679_{34}$
·23	$\cdot5910_{38}$	·73	$\cdot7673_{31}$	·23	$\cdot8907_{18}$	·73	$\cdot9582_{9}$	·23	$\cdot98713_{32}$
·24	$\cdot5948_{39}$	·74	$\cdot7704_{30}$	·24	$\cdot8925_{19}$	·74	$\cdot9591_{8}$	·24	$\cdot98745_{33}$
0·25	$0{\cdot}5987_{39}$	0·75	$0{\cdot}7734_{30}$	1·25	$0{\cdot}8944_{18}$	1·75	$0{\cdot}9599_{9}$	2·25	$0{\cdot}98778_{31}$
·26	$\cdot6026_{38}$	·76	$\cdot7764_{30}$	·26	$\cdot8962_{18}$	·76	$\cdot9608_{8}$	·26	$\cdot98809_{31}$
·27	$\cdot6064_{39}$	·77	$\cdot7794_{29}$	·27	$\cdot8980_{17}$	·77	$\cdot9616_{9}$	·27	$\cdot98840_{30}$
·28	$\cdot6103_{38}$	·78	$\cdot7823_{29}$	·28	$\cdot8997_{18}$	·78	$\cdot9625_{8}$	·28	$\cdot98870_{29}$
·29	$\cdot6141_{38}$	·79	$\cdot7852_{29}$	·29	$\cdot9015_{17}$	·79	$\cdot9633_{8}$	·29	$\cdot98899_{29}$
0·30	$0{\cdot}6179_{38}$	0·80	$0{\cdot}7881_{29}$	1·30	$0{\cdot}9032_{17}$	1·80	$0{\cdot}9641_{8}$	2·30	$0{\cdot}98928_{28}$
·31	$\cdot6217_{38}$	·81	$\cdot7910_{29}$	·31	$\cdot9049_{17}$	·81	$\cdot9649_{7}$	·31	$\cdot98956_{27}$
·32	$\cdot6255_{38}$	·82	$\cdot7939_{28}$	·32	$\cdot9066_{16}$	·82	$\cdot9656_{8}$	·32	$\cdot98983_{27}$
·33	$\cdot6293_{38}$	·83	$\cdot7967_{28}$	·33	$\cdot9082_{17}$	·83	$\cdot9664_{7}$	·33	$\cdot99010_{26}$
·34	$\cdot6331_{37}$	·84	$\cdot7995_{28}$	·34	$\cdot9099_{16}$	·84	$\cdot9671_{7}$	·34	$\cdot99036_{25}$
0·35	$0{\cdot}6368_{38}$	0·85	$0{\cdot}8023_{28}$	1·35	$0{\cdot}9115_{16}$	1·85	$0{\cdot}9678_{8}$	2·35	$0{\cdot}99061_{25}$
·36	$\cdot6406_{37}$	·86	$\cdot8051_{27}$	·36	$\cdot9131_{16}$	·86	$\cdot9686_{7}$	·36	$\cdot99086_{25}$
·37	$\cdot6443_{37}$	·87	$\cdot8078_{28}$	·37	$\cdot9147_{15}$	·87	$\cdot9693_{6}$	·37	$\cdot99111_{23}$
·38	$\cdot6480_{37}$	·88	$\cdot8106_{27}$	·38	$\cdot9162_{15}$	·88	$\cdot9699_{7}$	·38	$\cdot99134_{24}$
·39	$\cdot6517_{37}$	·89	$\cdot8133_{26}$	·39	$\cdot9177_{15}$	·89	$\cdot9706_{7}$	·39	$\cdot99158_{22}$
0·40	$0{\cdot}6554_{37}$	0·90	$0{\cdot}8159_{27}$	1·40	$0{\cdot}9192_{15}$	1·90	$0{\cdot}9713_{6}$	2·40	$0{\cdot}99180_{22}$
·41	$\cdot6591_{37}$	·91	$\cdot8186_{26}$	·41	$\cdot9207_{15}$	·91	$\cdot9719_{7}$	·41	$\cdot99202_{22}$
·42	$\cdot6628_{37}$	·92	$\cdot8212_{26}$	·42	$\cdot9222_{14}$	·92	$\cdot9726_{6}$	·42	$\cdot99224_{21}$
·43	$\cdot6664_{36}$	·93	$\cdot8238_{26}$	·43	$\cdot9236_{15}$	·93	$\cdot9732_{6}$	·43	$\cdot99245_{21}$
·44	$\cdot6700_{36}$	·94	$\cdot8264_{25}$	·44	$\cdot9251_{14}$	·94	$\cdot9738_{6}$	·44	$\cdot99266_{20}$
0·45	$0{\cdot}6736_{36}$	0·95	$0{\cdot}8289_{26}$	1·45	$0{\cdot}9265_{14}$	1·95	$0{\cdot}9744_{6}$	2·45	$0{\cdot}99286_{19}$
·46	$\cdot6772_{36}$	·96	$\cdot8315_{25}$	·46	$\cdot9279_{13}$	·96	$\cdot9750_{6}$	·46	$\cdot99305_{19}$
·47	$\cdot6808_{36}$	·97	$\cdot8340_{25}$	·47	$\cdot9292_{14}$	·97	$\cdot9756_{5}$	·47	$\cdot99324_{19}$
·48	$\cdot6844_{35}$	·98	$\cdot8365_{24}$	·48	$\cdot9306_{13}$	·98	$\cdot9761_{6}$	·48	$\cdot99343_{18}$
·49	$\cdot6879_{36}$	·99	$\cdot8389_{24}$	·49	$\cdot9319_{13}$	·99	$\cdot9767_{5}$	·49	$\cdot99361_{18}$
0·50	$0{\cdot}6915$	1·00	$0{\cdot}8413$	1·50	$0{\cdot}9332$	2·00	$0{\cdot}9772$	2·50	$0{\cdot}99379$

x	$\Phi(x)$	diff	x	$\Phi(x)$	diff	x	$\Phi(x)$	diff
2·50	0·99379	17	2·70	0·99653	11	2·90	0·99813	6
·51	·99396	17	·71	·99664	10	·91	·99819	6
·52	·99413	17	·72	·99674	9	·92	·99825	6
·53	·99430	16	·73	·99683	10	·93	·99831	5
·54	·99446	15	·74	·99693	9	·94	·99836	5
2·55	0·99461	16	2·75	0·99702	9	2·95	0·99841	5
·56	·99477	15	·76	·99711	9	·96	·99846	5
·57	·99492	14	·77	·99720	8	·97	·99851	5
·58	·99506	14	·78	·99728	8	·98	·99856	5
·59	·99520	14	·79	·99736	8	·99	·99861	4
2·60	0·99534	13	2·80	0·99744	8	3·0	0·99865	38
·61	·99547	13	·81	·99752	8	3·1	·99903	28
·62	·99560	13	·82	·99760	7	3·2	·99931	21
·63	·99573	12	·83	·99767	7	3·3	·99952	14
·64	·99585	13	·84	·99774	7	3·4	·99966	11
2·65	0·99598	11	2·85	0·99781	7	3·5	0·99977	7
·66	·99609	12	·86	·99788	7	3·6	·99984	5
·67	·99621	11	·87	·99795	6	3·7	·99989	4
·68	·99632	11	·88	·99801	6	3·8	·99993	2
·69	·99643	10	·89	·99807	6	3·9	·99995	2
2·70	0·99653		2·90	0·99813		4·0	0·99997	

x	$\phi(x)$	x	$\phi(x)$
0·0	0·3989	2·0	0·0540
0·1	·3970	2·1	·0440
0·2	·3910	2·2	·0355
0·3	·3814	2·3	·0283
0·4	·3683	2·4	·0224
0·5	0·3521	2·5	0·0175
0·6	·3332	2·6	·0136
0·7	·3123	2·7	·0104
0·8	·2897	2·8	·0079
0·9	·2661	2·9	·0060
1·0	0·2420	3·0	0·0044
1·1	·2179	3·1	·0033
1·2	·1942	3·2	·0024
1·3	·1714	3·3	·0017
1·4	·1497	3·4	·0012
1·5	0·1295	3·5	0·0009
1·6	·1109	3·6	·0006
1·7	·0940	3·7	·0004
1·8	·0790	3·8	·0003
1·9	·0656	3·9	·0002
2·0	0·0540	4·0	0·0001

The function tabulated is $\Phi(x) = \dfrac{1}{\sqrt{2\pi}} \displaystyle\int_{-\infty}^{x} e^{-\frac{1}{2}t^2}dt$. $\Phi(x)$ is the probability that a random variable, normally distributed with zero mean and unit variance, will be less than x. The last two columns give the ordinate $\phi(x) = \dfrac{1}{\sqrt{2\pi}} e^{-\frac{1}{2}x^2}$ of the normal frequency curve.

The critical table below gives on the left the range of values of x for which $\Phi(x)$ takes the value on the right, correct to the last figure given; in critical cases, take the upper of the two values of $\Phi(x)$ indicated.

x	$\Phi(x)$	x	$\Phi(x)$	x	$\Phi(x)$	x	$\Phi(x)$
3·075		3·263	0·9994	3·731	0·99990	3·916	0·99995
	0·9990		0·9995		0·99991		0·99996
3·105		3·320		3·759		3·976	
	0·9991		0·9996		0·99992		0·99997
3·138		3·389		3·791		4·055	
	0·9992		0·9997		0·99993		0·99998
3·174		3·480		3·826		4·173	
	0·9993		0·9998		0·99994		0·99999
3·215		3·615		3·867		4·417	
	0·9994		0·9999		0·99995		1·00000

Percentage points of the normal distribution

P	x	P	x	P	x	P	x	P	x	P	x
50	0·0000	5·0	1·6449	3·0	1·8808	2·0	2·0537	1·0	2·3263	5·0	1·6449
45	0·1257	4·8	1·6646	2·9	1·8957	1·9	2·0749	0·9	2·3656	1·0	2·3263
40	0·2533	4·6	1·6849	2·8	1·9110	1·8	2·0969	0·8	2·4089	0·1	3·0902
35	0·3853	4·4	1·7060	2·7	1·9268	1·7	2·1201	0·7	2·4573	0·01	3·7190
30	0·5244	4·2	1·7279	2·6	1·9431	1·6	2·1444	0·6	2·5121	$0.0^{2}1$	4·2649
25	0·6745	4·0	1·7507	2·5	1·9600	1·5	2·1701	0·5	2·5758	2·5	1·9600
20	0·8416	3·8	1·7744	2·4	1·9774	1·4	2·1973	0·4	2·6521	0·5	2·5758
15	1·0364	3·6	1·7991	2·3	1·9954	1·3	2·2262	0·3	2·7478	0·05	3·2905
10	1·2816	3·4	1·8250	2·2	2·0141	1·2	2·2571	0·2	2·8782	$0.0^{2}5$	3·8906
5	1·6449	3·2	1·8522	2·1	2·0335	1·1	2·2904	0·1	3·0902	$0.0^{3}5$	4·4172

This table gives the percentage points x where $\dfrac{P}{100} = \dfrac{1}{\sqrt{2\pi}} \displaystyle\int_{x}^{\infty} e^{-\frac{1}{2}t^2}dt$. The value x is that which is exceeded by a random variable, normally distributed with zero mean and unit variance, with probability $P/100$.

Percentage points of the t-distribution

P	25	10	5	2	1	0·2	0·1	$\dfrac{120}{\nu}$
$\nu = 1$	2·41	6·31	12·71	31·82	63·66	318·3	636·6	
2	1·60	2·92	4·30	6·96	9·92	22·33	31·60	
3	1·42	2·35	3·18	4·54	5·84	10·21	12·92	
4	1·34	2·13	2·78	3·75	4·60	7·17	8·61	
5	1·30	2·02	2·57	3·36	4·03	5·89	6·87	
6	1·27	1·94	2·45	3·14	3·71	5·21	5·96	
7	1·25	1·89	2·36	3·00	3·50	4·79	5·41	
8	1·24	1·86	2·31	2·90	3·36	4·50	5·04	
9	1·23	1·83	2·26	2·82	3·25	4·30	4·78	
10	1·22	1·81	2·23	2·76	3·17	4·14	4·59	12
12	1·21	1·78	2·18	2·68	3·05	3·93	4·32	10
15	1·20	1·75	2·13	2·60	2·95	3·73	4·07	8
20	1·18	1·72	2·09	2·53	2·85	3·55	3·85	6
24	1·18	1·71	2·06	2·49	2·80	3·47	3·75	5
30	1·17	1·70	2·04	2·46	2·75	3·39	3·65	4
40	1·17	1·68	2·02	2·42	2·70	3·31	3·55	3
60	1·16	1·67	2·00	2·39	2·66	3·23	3·46	2
120	1·16	1·66	1·98	2·36	2·62	3·16	3·37	1
∞	1·15	1·64	1·96	2·33	2·58	3·09	3·29	0

The function tabulated is t_P defined by the equation

$$\frac{P}{100} = \frac{1}{\sqrt{\nu\pi}} \frac{\Gamma(\tfrac{1}{2}\nu + \tfrac{1}{2})}{\Gamma(\tfrac{1}{2}\nu)} \int\limits_{|t| \geqslant t_P} \frac{dt}{(1 + t^2/\nu)^{\frac{1}{2}(\nu+1)}}.$$

If t is the ratio of a random variable, normally distributed with zero mean, to an independent estimate of its standard deviation based on ν degrees of freedom, $P/100$ is the probability that $|t| \geqslant t_P$.

Interpolation ν-wise should be linear in $120/\nu$.

Other percentage points may be found approximately, except when ν and P are both small, by using the fact that the variable

$$y = \pm \sinh^{-1}(\sqrt{3t^2/2\nu}),$$

where y has the same sign as t, is approximately normally distributed with zero mean and variance $3/(2\nu - 1)$.

Percentage points of the χ^2-distribution

P	99·5	99	97·5	95		10	5	2·5	1	0·5	0·1
$\nu=1$	$0\cdot0^4393$	$0\cdot0^8157$	$0\cdot0^8982$	0·00393		2·71	3·84	5·02	6·63	7·88	10·83
2	0·0100	0·0201	0·0506	0·103		4·61	5·99	7·38	9·21	10·60	13·81
3	0·0717	0·115	0·216	0·352		6·25	7·81	9·35	11·34	12·84	16·27
4	0·207	0·297	0·484	0·711		7·78	9·49	11·14	13·28	14·86	18·47
5	0·412	0·554	0·831	1·15		9·24	11·07	12·83	15·09	16·75	20·52
6	0·676	0·872	1·24	1·64		10·64	12·59	14·45	16·81	18·55	22·46
7	0·989	1·24	1·69	2·17		12·02	14·07	16·01	18·48	20·28	24·32
8	1·34	1·65	2·18	2·73		13·36	15·51	17·53	20·09	21·95	26·12
9	1·73	2·09	2·70	3·33		14·68	16·92	19·02	21·67	23·59	27·88
10	2·16	2·56	3·25	3·94		15·99	18·31	20·48	23·21	25·19	29·59
11	2·60	3·05	3·82	4·57		17·28	19·68	21·92	24·73	26·76	31·26
12	3·07	3·57	4·40	5·23		18·55	21·03	23·34	26·22	28·30	32·91
13	3·57	4·11	5·01	5·89		19·81	22·36	24·74	27·69	29·82	34·53
14	4·07	4·66	5·63	6·57		21·06	23·68	26·12	29·14	31·32	36·12
15	4·60	5·23	6·26	7·26		22·31	25·00	27·49	30·58	32·80	37·70
16	5·14	5·81	6·91	7·96		23·54	26·30	28·85	32·00	34·27	39·25
17	5·70	6·41	7·56	8·67		24·77	27·59	30·19	33·41	35·72	40·79
18	6·26	7·01	8·23	9·39		25·99	28·87	31·53	34·81	37·16	42·31
19	6·84	7·63	8·91	10·12		27·20	30·14	32·85	36·19	38·58	43·82
20	7·43	8·26	9·59	10·85		28·41	31·41	34·17	37·57	40·00	45·31
21	8·03	8·90	10·28	11·59		29·62	32·67	35·48	38·93	41·40	46·80
22	8·64	9·54	10·98	12·34		30·81	33·92	36·78	40·29	42·80	48·27
23	9·26	10·20	11·69	13·09		32·01	35·17	38·08	41·64	44·18	49·73
24	9·89	10·86	12·40	13·85		33·20	36·42	39·36	42·98	45·56	51·18
25	10·52	11·52	13·12	14·61		34·38	37·65	40·65	44·31	46·93	52·62
26	11·16	12·20	13·84	15·38		35·56	38·89	41·92	45·64	48·29	54·05
27	11·81	12·88	14·57	16·15		36·74	40·11	43·19	46·96	49·64	55·48
28	12·46	13·56	15·31	16·93		37·92	41·34	44·46	48·28	50·99	56·89
29	13·12	14·26	16·05	17·71		39·09	42·56	45·72	49·59	52·34	58·30
30	13·79	14·95	16·79	18·49		40·26	43·77	46·98	50·89	53·67	59·70
40	20·71	22·16	24·43	26·51		51·81	55·76	59·34	63·69	66·77	73·40
50	27·99	29·71	32·36	34·76		63·17	67·50	71·42	76·15	79·49	86·66
60	35·53	37·48	40·48	43·19		74·40	79·08	83·30	88·38	91·95	99·61
70	43·28	45·44	48·76	51·74		85·53	90·53	95·02	100·4	104·2	112·3
80	51·17	53·54	57·15	60·39		96·58	101·9	106·6	112·3	116·3	124·8
90	59·20	61·75	65·65	69·13		107·6	113·1	118·1	124·1	128·3	137·2
100	67·33	70·06	74·22	77·93		118·5	124·3	129·6	135·8	140·2	149·4

The function tabulated is χ_P^2 defined by the equation $\dfrac{P}{100} = \dfrac{1}{2^{\nu/2}\Gamma(\frac{1}{2}\nu)} \displaystyle\int_{\chi_P^2}^{\infty} x^{\frac{1}{2}\nu-1} e^{-x/2}\,dx.$ If x is a variable distributed as χ^2 with ν degrees of freedom, $P/100$ is the probability that $x \geq \chi_P^2$. For $\nu < 100$, linear interpolation in ν is adequate. For $\nu > 100$, $\sqrt{2\chi^2}$ is approximately normally distributed with mean $\sqrt{2\nu-1}$ and unit variance.

Properties of matter

Physical constants

Universal gas constant	$R_0 = 8{\cdot}31 \times 10^3$ J/kg mole K	Velocity of light in vacuum	$c = 3{\cdot}00 \times 10^8$ m/s
Boltzmann's constant	$k = 1{\cdot}380 \times 10^{-23}$ J/K $= 8{\cdot}62 \times 10^{-5}$ eV/K	Absolute permittivity of free space	$\epsilon_0 = 8{\cdot}85 \times 10^{-12}$ F/m
Universal gravitational constant	$G = 6{\cdot}67 \times 10^{-11}$ Nm²/kg²	Absolute permeability of free space	$\mu_0 = 4\pi \times 10^{-7}$ H/m
Mean radius of earth	$R = 6371$ km	Charge of an electron	$e = 1{\cdot}6 \times 10^{-19}$ C
Gravitational acceleration (standard gravity)	$g = 9{\cdot}81$ m/s²	Mass of an electron	$m_e = 9{\cdot}11 \times 10^{-31}$ kg
Stefan-Boltzmann constant	$\sigma = 5{\cdot}67 \times 10^{-8}$ W/m² K⁴	Charge/mass ratio of an electron	$e/m_e = 1{\cdot}76 \times 10^{11}$ C/kg
Avogadro's number	$N = 6{\cdot}02 \times 10^{26}$/kg mole	Mass of a proton	$m_p = 1{\cdot}67 \times 10^{-27}$ kg
Loschmidt's number	$2{\cdot}69 \times 10^{25}$/m³	Impedance of free space	$120\pi\Omega$
Planck's constant	$h = 6{\cdot}62 \times 10^{-34}$ J s $= 4{\cdot}14 \times 10^{-15}$ eVs	Bohr magneton	$\beta = 9{\cdot}27 \times 10^{-24}$ A m²
		Wavelength of 1 eV photon	$1{\cdot}24\,\mu m$

The periodic table

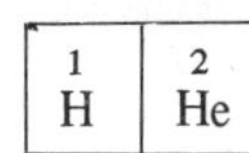

1 H	2 He

IA	IIA	IIIA	IVA	VA	VIA	VIIA	VIII	VIII	VIII	IB	IIB	IIIB	IVB	VB	VIB	VIIB	
3 Li	4 Be											5 B	6 C	7 N	8 O	9 F	10 Ne
11 Na	12 Mg											13 Al	14 Si	15 P	16 S	17 Cl	18 Ar
19 K	20 Ca	21 Sc	22 Ti	23 V	24 Cr	25 Mn	26 Fe	27 Co	28 Ni	29 Cu	30 Zn	31 Ga	32 Ge	33 As	34 Se	35 Br	36 Kr
37 Rb	38 Sr	39 Y	40 Zr	41 Nb	42 Mo	43 Tc	44 Ru	45 Rh	46 Pd	47 Ag	48 Cd	49 In	50 Sn	51 Sb	52 Te	53 I	54 Xe
55 Cs	56 Ba	57 La	72 Hf	73 Ta	74 W	75 Re	76 Os	77 Ir	78 Pt	79 Au	80 Hg	81 Tl	82 Pb	83 Bi	84 Po	85 At	86 Rn
87 Fr	88 Ra	89 Ac															

IIIA

58 Ce	59 Pr	60 Nd	61 Pm	62 Sm	63 Eu	64 Gd	65 Tb	66 Dy	67 Ho	68 Er	69 Tm	70 Yb	71 Lu
90 Th	91 Pa	92 U	93 Np	94 Pu	95 Am	96 Cm	97 Bk	98 Cf	99 Es	100 Fm	101 Md	102 No	103 Lw

Atomic properties of the elements

(for free neutral atoms in the ground state)

Z	Atomic number
AW	Atomic weight in a.m.u. ($_6C^{12} = 12{\cdot}000$)
V_i	First ionization potential in eV
K, L, M, N, O, P, Q	Principal quantum number = 1, 2, 3, 4, 5, 6, 7
s, p, d, f, g, h	Azimuthal quantum number = 0, 1, 2, 3, 4, 5

				K	L		M			N				O				
Z	Element	AW	V_i	1s	2s	2p	3s	3p	3d	4s	4p	4d	4f	5s	5p	5d	5f	5g
1	H Hydrogen	1·008	13·5	1														
2	He Helium	4·003	24·5	2														
3	Li Lithium	6·939	5·4	2	1													
4	Be Beryllium	9·012	9·3	2	2													
5	B Boron	10·811	8·3	2	2	1												
6	C Carbon	12·011	11·2	2	2	2												
7	N Nitrogen	14·007	14·5	2	2	3												
8	O Oxygen	15·999	13·6	2	2	4												
9	F Fluorine	18·998	17·3	2	2	5												
10	Ne Neon	20·183	21·5	2	2	6												
11	Na Sodium	22·990	5·1	2	2	6	1											
12	Mg Magnesium	24·312	7·6	2	2	6	2											
13	Al Aluminium	26·982	6·0	2	2	6	2	1										
14	Si Silicon	28·086	8·1	2	2	6	2	2										
15	P Phosphorus	30·974	10·9	2	2	6	2	3										
16	S Sulphur	32·064	10·3	2	2	6	2	4										
17	Cl Chlorine	35·453	13·0	2	2	6	2	5										
18	Ar Argon	39·948	15·7	2	2	6	2	6										
19	K Potassium	39·102	4·3	2	2	6	2	6		1								
20	Ca Calcium	40·080	6·1	2	2	6	2	6		2								
21	Sc Scandium	44·956	6·7	2	2	6	2	6	1	2								
22	Ti Titanium	47·900	6·8	2	2	6	2	6	2	2								
23	V Vanadium	50·942	6·7	2	2	6	2	6	3	2								
24	Cr Chromium	51·996	6·7	2	2	6	2	6	5	1								
25	Mn Manganese	54·938	7·4	2	2	6	2	6	5	2								
26	Fe Iron	55·847	7·8	2	2	6	2	6	6	2								
27	Co Cobalt	58·933	7·8	2	2	6	2	6	7	2								
28	Ni Nickel	58·710	7·6	2	2	6	2	6	8	2								
29	Cu Copper	63·540	7·7	2	2	6	2	6	10	1								
30	Zn Zinc	65·370	9·4	2	2	6	2	6	10	2								
31	Ga Gallium	69·720	6·0	2	2	6	2	6	10	2	1							
32	Ge Germanium	72·590	8·1	2	2	6	2	6	10	2	2							
33	As Arsenic	74·922	10·5	2	2	6	2	6	10	2	3							
34	Se Selenium	78·960	9·7	2	2	6	2	6	10	2	4							
35	Br Bromine	79·909	11·8	2	2	6	2	6	10	2	5							
36	Kr Krypton	83·800	13·9	2	2	6	2	6	10	2	6							
37	Rb Rubidium	85·470	4·2	2	2	6	2	6	10	2	6			1				
38	Sr Strontium	87·620	5·7	2	2	6	2	6	10	2	6			2				
39	Y Yttrium	88·905	6·5	2	2	6	2	6	10	2	6	1		2				
40	Zr Zirconium	91·220	6·9	2	2	6	2	6	10	2	6	2		2				
41	Nb Niobium	92·906		2	2	6	2	6	10	2	6	4		1				
42	Mo Molybdenum	95·940	7·4	2	2	6	2	6	10	2	6	5		1				
43	Tc Technetium			2	2	6	2	6	10	2	6	6		1				
44	Ru Ruthenium	101·070	7·7	2	2	6	2	6	10	2	6	7		1				
45	Rh Rhodium	102·905	7·7	2	2	6	2	6	10	2	6	8		1				
46	Pd Palladium	106·400	8·3	2	2	6	2	6	10	2	6	10						
47	Ag Silver	107·870	7·5	2	2	6	2	6	10	2	6	10		1				
48	Cd Cadmium	112·400	9·0	2	2	6	2	6	10	2	6	10		2				
49	In Indium	114·820	5·8	2	2	6	2	6	10	2	6	10		2	1			
50	Sn Tin	118·690	7·3	2	2	6	2	6	10	2	6	10		2	2			
51	Sb Antimony	121·750	8·5	2	2	6	2	6	10	2	6	10		2	3			
52	Te Tellurium	127·600	9·0	2	2	6	2	6	10	2	6	10		2	4			
53	I Iodine	126·904	10·6	2	2	6	2	6	10	2	6	10		2	5			
54	Xe Xenon	131·300	12·1	2	2	6	2	6	10	2	6	10		2	6			

Z	Element	AW	V_i	K	L	M	N				O				P						Q
							4s	4p	4d	4f	5s	5p	5d	5g	6s	6p	6d	6f	6g	6h	7s
55	Cs Caesium	132·905	3·9	2	8	18	2	6	10		2	6			1						
56	Ba Barium	137·340	5·2	2	8	18	2	6	10		2	6			2						
57	La Lanthanum	138·910	5·6	2	8	18	2	6	10		2	6	1		2						
58	Ce Cerium	140·120	6·5	2	8	18	2	6	10	2	2	6			2						
59	Pr Praseodymium	140·907	5·8	2	8	18	2	6	10	3	2	6			2						
60	Nd Neodymium	144·240	6·3	2	8	18	2	6	10	4	2	6			2						
61	Pm Promethium			2	8	18	2	6	10	5	2	6			2						
62	Sm Samarium	150·350	6·6	2	8	18	2	6	10	6	2	6			2						
63	Eu Europium	151·960	5·6	2	8	18	2	6	10	7	2	6			2						
64	Gd Gadolineum	157·250	6·7	2	8	18	2	6	10	7	2	6	1		2						
65	Tb Terbium	158·924	6·7	2	8	18	2	6	10	9	2	6			2						
66	Dy Dysprosium	162·500	6·8	2	8	18	2	6	10	10	2	6			2						
67	Ho Holmium	164·930		2	8	18	2	6	10	11	2	6			2						
68	Er Erbium	167·260		2	8	18	2	6	10	12	2	6			2						
69	Tm Thulium	168·934		2	8	18	2	6	10	13	2	6			2						
70	Yb Ytterbium	173·040	7·1	2	8	18	2	6	10	14	2	6			2						
71	Lu Lutetium	174·970		2	8	18	2	6	10	14	2	6	1		2						
72	Hf Hafnium	178·490		2	8	18	2	6	10	14	2	6	2		2						
73	Ta Tantalum	180·948		2	8	18	2	6	10	14	2	6	3		2						
74	W Tungsten	183·85	8·1	2	8	18	2	6	10	14	2	6	4		2						
75	Re Rhenium	186·2		2	8	18	2	6	10	14	2	6	5		2						
76	Os Osmium	190·2		2	8	18	2	6	10	14	2	6	6		2						
77	Ir Iridium	192·2		2	8	18	2	6	10	14	2	6	9								
78	Pt Platinum	195·09	8·9	2	8	18	2	6	10	14	2	6	9		1						
79	Au Gold	196·967	9·2	2	8	18	2	6	10	14	2	6	10		1						
80	Hg Mercury	200·59	10·4	2	8	18	2	6	10	14	2	6	10		2						
81	Tl Thallium	204·37	6·1	2	8	18	2	6	10	14	2	6	10		2	1					
82	Pb Lead	207·19	7·4	2	8	18	2	6	10	14	2	6	10		2	2					
83	Bi Bismuth	208·98	8·0	2	8	18	2	6	10	14	2	6	10		2	3					
84	Po Polonium			2	8	18	2	6	10	14	2	6	10		2	4					
85	At Astatine			2	8	18	2	6	10	14	2	6	10		2	5					
86	Rn Radon			2	8	18	2	6	10	14	2	6	10		2	6					
87	Fr Francium			2	8	18	2	6	10	14	2	6	10		2	6					1
88	Ra Radium			2	8	18	2	6	10	14	2	6	10		2	6					2
89	Ac Actinium			2	8	18	2	6	10	14	2	6	10		2	6	1				2
90	Th Thorium	232·038		2	8	18	2	6	10	14	2	6	10		2	6	2				2
91	Pa Protactinium			2	8	18	2	6	10	14	2	6	10		2	6	3				2
92	U Uranium	238·03		2	8	18	2	6	10	14	2	6	10		2	6	4				2
93	Np Neptunium			2	8	18	2	6	10	14	2	6	10	5	2	6					2
94	Pu Plutonium			2	8	18	2	6	10	14	2	6	10	5	2	6	1				2
95	Am Americium			2	8	18	2	6	10	14	2	6	10	6	2	6	1				2
96	Cm Curium			2	8	18	2	6	10	14	2	6	10	7	2	6	1				2
97	Bk Berkelium			2	8	18	2	6	10	14	2	6	10	8	2	6	1				2
98	Cf Californium			2	8	18	2	6	10	14	2	6	10	9	2	6	1				2
99	Es Einsteinium			2	8	18	2	6	10	14	2	6	10	10	2	6	1				2
100	Fm Fermium			2	8	18	2	6	10	14	2	6	10	11	2	6	1				2
101	Md Mendelevium			2	8	18	2	6	10	14	2	6	10	12	2	6	1				2
102	No Nobelium																				
103	Lw Lawrencium																				

Physical properties of solids

CS   Crystal structure: BCC body-centred cubic
                           FCC face-centred cubic
                           CPH close-packed hexagonal

ρ Mass density (kg/litre) $\approx$ specific gravity

t_m Melting point (°C)

t_b Boiling point (°C)

h_{if} Latent heat of fusion (kJ/kg)

h_{fg} Latent heat of vaporization (kJ/kg)

k Thermal conductivity at or near 0°C (W/m K)

c Specific heat capacity at or near 0°C (J/kg K)

α Coefficient of linear thermal expansion (10^{-6}/K)

ρ_e Electrical resistivity at 20°C (units as shown)

α_e Temperature coefficient of resistance, 0–100°C (10^{-3}/K)

ϵ_r Dielectric constant (relative permittivity) at $\lesssim$ 1 MHz, 20°C

tan δ   Loss factor at $\sim$ 1 MHz, 20°C (units of 10^{-4})

	CS	ρ	t_m	t_b	h_{if}	h_{fg}	k	c	α	$\rho_e(n\Omega\,m)$	α_e
Metallic elements											
Aluminium	FCC	2·7	660	2400	387	9460	205	880	23	27	4·2
Copper	FCC	8·96	1083	2580	205	5230	390	380	17	16·8	4·3
Gold	FCC	19·3	1063	2660	66	1750	310	145	14	23	3·9
Iron	BCC/FCC	7·9	1535	2900	270	6600	76	437	12	97	6·5
Lead	FCC	11·3	327	1750	24	850	35	126	29	206	4·3
Nickel	FCC	8·9	1453	2820	305	5850	91	444	13	68	6·8
Platinum	FCC	21·5	1769	3800	113	2400	69	125	9	106	3·9
Silver	FCC	10·5	961	2180	105	2330	418	232	19	16	4·1
Tantalum	BCC	16·6	3000	5300	160		54	140	6	135	~3·5
Tin	Diamond/ tetragonal	7·3	232	~2500	59	2400	64	224	23	120	~4·5
Titanium	CPH/BCC	4·5	1680	3300	435		17	500	9	550	~3·5
Tungsten	BCC	19·3	3380	~6000	185		190	130	4·5	55	4·6
Zinc	CPH	7·1	420	907	110	1750	113	384	31	59	4·2

	ρ	t_m	k	c	α	$\rho_e(n\Omega\,m)$	α_e
Alloys							
Brass (65/35)	8·45	927	120	370	20	69	1·6
Constantan (60/40)	8·9	1320	22	410		490	~0·02
Dural (4·4% Cu)	2·8	640	150	900	23	~52	~2·3
Manganin (84% Cu)	8·5		22	405		440	~0
Nichrome (80/20)	8·36		13	430	12·5	1030	0·18
Phosphor-Bronze	8·92	~1050	~75	380	18	115	3·5
Steel (mild)	~7·85		~50	~450	~11	~120	~3·0

	ρ	t_m	k	c	α	$\rho_e(M\Omega\,m)$	ϵ_r	$\tan\delta$
Non-metals								
Alumina	3·9	2050	21	1050	8	10^3–10^6	4·5–8·4	2–100
Brick	1·4–2·2		0·4–0·8	800	3–9	1–2		
Concrete	2·4		1·0–1·5	1100	10–14			
Dry ground	~1·6					0·01–0·1		
Glass	2·4–3·5	~1100	0·4–1·1	500–800	3–10	$5\cdot10^3$–10^6	5–8	13–100
Granite	2·7		2–4	800	6–9		7–9	
Ice[†]	0·92	0	2·3	2100				
Mica	2·8		~0·5	840		10^5–10^9	5–7	1–2
Nylon 6	1·14	200–220	0·25–0·33	1600	80–130	10^4–10^7	3–7	200–1300
Paper (dry)	~1·0		0·06			10^4	1·9–2·9	20–45
Perspex	1·2	85–115*	0·19–0·23	1450	50–80		2·5–3·5	160–300
Polystyrene	1·06	80–105*	0·08–0·2	1300	60–80	10^{10}	2·4–3·5	<20
Polythene	0·93	65–130*	0·25–0·5	2200	110–220	10^5	2·3	2–5
PTFE	2·2		0·23–0·27	1050	90–130	10^9	2·1	< 3
PVC (plasticized)	1·7	70–80*	0·16–0·19		50–250	10^4–10^7	4–6	600
Porcelain	2·4	1550	0·8–1·85	1100	2·2	10^4–10^7	5·5–7	60–100
Quartz (crystal)	2·65		5–9	730	7·5–13·7	10^6–$2\cdot10^8$	4·5–5	

* Softening temperature.

[†] h_{if} = 333.

	ρ	t_{m}	k	c	α	ρ_{e} (MΩ m)	ϵ_{r}	tan δ
Non-metals–continued								
Quartz (fused)	2·2		1·3	840	0·5	10^{10}	3·8	2
Rubber (natural, vulcanized)	1·1–1·2	125	∼0·15		∼200	∼10^7	2–3·5	280
Sandstone	2·4		1·1–2·3	900	5–12		10	
Timber (along grain)	0·4–0·8		∼0·15	∼1600	3–5		2–9	350–600

Mechanical properties of solids

E Young's modulus (kN/mm²)	ν Poisson's ratio
G Shear modulus (kN/mm²)	σ_{y} Proof or yield stress (N/mm²)
K Bulk modulus (kN/mm²)	σ_{f} Ultimate (failure) stress (N/mm²)

Values of σ_{y} and σ_{f}, particularly, usually depend strongly on the preparation and condition of a material. The ranges given are typical but not necessarily exhaustive, and unless otherwise stated those for metals refer to drawn or wrought, rather than cast, material of commercial purity.

	E	G	K	ν	σ_{y}	σ_{f}
Metallic elements						
Aluminium	70	26	75	0·34	30–140	60–160
Copper	124	46	130	0·35	47–320	200–350
Gold	80	28	167	0·42	0–210	110–230
Iron (wrought)	195	76		0·29	160	350
Iron (cast)	115	45		0·25		140–320
Lead	16	6		0·44		15–18
Nickel	205	79	176	0·31	140–660	480–730
Platinum	168	61	240	0·38	15–180	125–200
Silver	76	28	100	0·37	55–300	140–380
Tantalum	186					340–930
Tin	47	17	52	0·36	9–14	15–200
Titanium	110	41	110	0·34	200–500	250–700
Tungsten	360	140				1000–4000
Zinc	97	36	100	0·35		110–200
Alloys						
Brass (65/35)	105	38	115	0·35	62–430	330–530
Constantan (60/40)	163	61	157	0·33	200–440	400–570
Dural (4·4% Cu)	70	27	70	0·33	125–450	230–500
Manganin (84% Cu)	124	47				465
Mumetal (77% Ni)	220					540–910
Nichrome (80/20)	186					170–900
Phosphor-Bronze	100			0·38	110–670	330–750
Steel: mild	210	81	170	0·30	240	480
Steel: high-yield structural	210	81	170	0·30	450	600
Steel: ultra high strength					1600	2000

	E	ν	σ_f (tension)	σ_f (compression)
Non-metals				
Alumina	200–400	0·24	140–200	1000–2500
Brick (grade A)	10–50			69–140
Concrete (28-day)	10–17	0·1–0·21		27–55
Glass	50–80	0·2–0·27	30–90	
Granite	40–70			90–235
Nylon 6	1–2·5		70–85	50–100
Perspex	2·7–3·5		50–75	80–140
Polystyrene	2·5–4·0		35–60	80–110
Polythene	0·1–1·0		7–38	15–20
PTFE	0·4–0·6		17–28	5–12
PVC (plasticized)	∼0·3		14–40	75–100
Rubber (natural, vulcanized)	∼0·001–1	0·46–0·49	14–40	
Sandstone	14–55			30–135
Timber (along grain)	8–13		20–110	50–100

Work functions

ϕ_t and ϕ_p are the least energies, in electron volts, to extract an electron from the element by thermionic and photoelectric emission respectively.

Element	ϕ_t	ϕ_p	Element	ϕ_t	ϕ_p	Element	ϕ_t	ϕ_p
Aluminium		4·1	Gold	4·4	4·8	Sodium		2·3
Barium	2·5	2·5	Iron	4·5	4·6	Tantalum	4·2	4·1
Bismuth		4·3	Mercury		4·5	Thorium	3·4	3·5
Caesium	1·8	1·9	Molybdenum	4·2	4·2	Titanium	4·0	4·2
Calcium		2·7	Nickel	5·0	5·0	Tungsten	4·5	4·5
Carbon	4·6	4·8	Platinum	5·3	6·3	Zinc		4·2
Chromium	4·6	4·4	Potassium		2·2			
Copper	4·4	4·3	Silver	4·1	4·7			

Properties of semiconductors

(at room temperature)

AW	Atomic or molecular weight		E_g	Energy gap (eV)
N	Number density of atoms or molecules ($m^{-3} \times 10^{-28}$)		N_c, N_v	Effective density of states in conduction and valence bands ($m^{-3} \times 10^{-25}$)
a	Lattice constant (Å)		N_i	Intrinsic carrier concentration (m^{-3})
ρ	Mass density (kg/litre)		μ_e	Mobility of electrons (m^2/Vs $\times 10^{-2}$)
m_e^*/m_0	Effective mass of electron relative to free rest mass m_0		μ_h	Mobility of holes (m^2/Vs $\times 10^{-2}$)
m_h^*/m_0	Effective mass of hole (light, heavy and split-off) relative to m_0		ϵ_r	Dielectric constant
			E_B	Breakdown field (V/μm)

	AW	N	a	ρ	m_e^*/m_0	m_h^*/m_0	E_g	N_c	N_v	N_i	μ_e	μ_h	ϵ_r	E_B
Germanium Ge	72·60	4·42	5·66	5·32	0·12	0·04 0·28 0·08	0·67	1·04	0·60	$2\cdot4.10^{19}$	38	18	16·3	~8
Silicon Si	28·09	5·00	5·43	2·33	0·26	0·16 0·50 0·24	1·11	2·8	1·04	$1\cdot45.10^{16}$	13	5	11·7	~30
Gallium Arsenide Ga As	144·63	2·21	5·65	5·32	0·067	0·65	1·40	0·047	0·70	$\sim9.10^{12}$	85	4	12	~35
Indium Antimonide In Sb	236·58				0·013	0·18	0·17				700	10		
Cadmium Sulphide Cd S	144·48			4·82	0·27	0·07	2·5						3·4	0·18

Properties of ferromagnetic materials

Soft materials

B_{sat} Saturated flux density (T)
$(\mu_r)_{max}$ Maximum relative permeability
ρ_e Resistivity (nΩm)
w Hysteresis loss per cycle (mJ/kg)

	B_{sat}	$(\mu_r)_{max}$	ρ_e	w
Iron	2·15	5 000	100	30
Iron (4% Si)	1·97	7 000	600	20
Grain-oriented silicon-iron	2·00	30 000	550	5
Permalloy (78% Ni)	1·08	10^5	160	0·5
Supermalloy (79% Ni, 5% Mo)	0·79	10^6	600	0·1
Mumetal (77% Ni, 5% Cu)	0·65	10^5	620	
Ferrite MnZn $(Fe_2O_3)_2$		2500	2.10^8	1·0

Hard materials

B_r Permanent flux density (T)
H_c Coercive force (kA/m)
$(BH)_{max}$ is in units of kJ/m^3

	B_r	H_c	$(BH)_{max}$
Steel (0·9% C, 1% Mn)	0·90	4·0	0·8
Alnico 5 (14% Ni, 24% Co, 8% Al)	1·25	46	20
Ferroxdur, BaO $(Fe_2O_3)_6$	0·35	160	12
Fe–Co powder	0·90	82	40
Alnico 9	1·05	130	100

Superconducting materials

The critical field H_c for a superconducting material at an absolute temperature T depends on the critical field at absolute zero, H_0, according to the relation

$$H_c = H_0 \{1 - (T/T_c)^2\}$$

where T_c is the critical temperature. Values of H_0 and T_c for some superconductors are given below.

	T_c(K)	H_0(kA/m)
Aluminium	1·19	8
Gallium	1·09	4
Indium	3·41	23
Lead	7·18	65
Mercury, α	4·15	33

	T_c(K)	H_0(kA/m)
Mercury, β	3·95	27
Niobium	9·46	156
Tantalum	4·48	67
Thorium	1·37	13
Tin	3·72	25
Vanadium	5·30	105
Zinc	0·92	4
Nb_3Sn	18·5	
NbN	16	
$Nb_3(Al_{0·8}Ge_{0·2})$	20	
V_3Si	17	
V_3Ga	16·8	

Properties of liquids

ρ Mass density (kg/litre) $\approx$ specific gravity
K Bulk modulus (kN/mm^2)
t_f, t_b Freezing, boiling points (°C)
h_{fg} Latent heat of boiling (kJ/kg) at $\sim$ 1 atm
$k(t)$ Thermal conductivity (at t°C) (W/m K)

c Specific heat capacity at 20°C (J/kg K)
β Coefficient of volume thermal expansion (10^{-3}/K)
μ Dynamic viscosity at 20°C (mNs/m^2)
ρ_e Electrical resistivity at or near 20°C (Ωm)
ϵ_r Dielectric constant (relative permittivity) at 1 MHz, 20°C

	ρ	K	t_f	t_b	h_{fg}	$k(t)$	c	β	μ	ρ_e	ϵ_r
Mercury	13·5	25	−39	357	292	8 (20)	139	0·18	1·55	$9·6.10^{-7}$	
Sodium			98	883		85 (130)				10^{-7}*	
Water	1·00	2·3	0	100	2260	0·61 (20)	4180	0·21	1·00	5000†	81
Sea-water	1·03	2·3	−2·5	103			3930			0·2–0·3	
Mineral oil	$\sim$0·9	1–2				0·13 (100)	1700	0·7–0·9		$3·10^{10}$–10^{15}	$\sim$2·2
Carbon tetrachloride, CCl_4	1·6	1·1	−23	77	215	0·11 (20)	840	1·22	0·97		2·2
Acetone, C_3H_5OH	0·79	1·2	−95	57	560	0·16 (20)	2210	1·43	0·32		21
Ethyl alcohol, C_2H_5OH	0·79	1·3	−120	78	850	0·17 (20)	2500	1·08	1·20	3000	26

* At 100°C

† Distilled water

Thermodynamic properties of fluids

a Velocity of sound (m/s)
c_p Specific heat capacity at constant pressure (kJ/kg K)
c_v Specific heat capacity at constant volume (kJ/kg K)
h Specific enthalpy (kJ/kg)
k Thermal conductivity (W/m K)
p Absolute pressure (bar, or as indicated)
Pr Prandtl number, $c_p \mu / k$
s Specific entropy (kJ/kg K)
t Temperature (°C)
T Absolute temperature (K)

u Specific internal energy (kJ/kg)
v Specific volume (m³/kg, or as indicated)
z Altitude above sea-level (km)
γ Ratio of specific heats c_p / c_v
λ Mean free path (μm)
μ Dynamic viscosity (mNs/m², or cP)
ν Kinematic viscosity (mm²/s, or cSt)
ρ Mass density (kg/m³)

The subscripts s, f, g, fg indicate respectively saturation, liquid, vapour and liquid-vapour phase change.

Saturated water and steam, to 100°C

t_s	p	v_f *	v_g *	u_f	u_{fg}	u_g
0.01	0.006112	1.0002	206163	0.0	2375.6	2375.6
1	0.006566	1.0001	192607	4.2	2372.7	2376.9
2	0.007055	1.0001	179923	8.4	2369.9	2378.3
3	0.007575	1.0001	168169	12.6	2367.1	2379.7
4	0.008129	1.0000	157272	16.8	2364.3	2381.1
5	0.008718	1.0000	147163	21.0	2361.4	2382.4
6	0.009345	1.0000	137780	25.2	2358.6	2383.8
7	0.010012	1.0001	129064	29.4	2355.8	2385.2
8	0.010720	1.0001	120966	33.6	2352.9	2386.5
9	0.011472	1.0002	113435	37.8	2350.2	2388.0
10	0.012270	1.0003	106430	42.0	2347.3	2389.3
12	0.014014	1.0004	93835	50.4	2341.7	2392.1
14	0.015973	1.0007	82900	58.8	2336.0	2394.8
16	0.018168	1.0010	73384	67.1	2330.5	2397.6
18	0.020624	1.0013	65087	75.5	2324.8	2400.3
20	0.023366	1.0017	57838	83.9	2319.2	2403.1
22	0.026422	1.0022	51492	92.2	2313.6	2405.7
24	0.029821	1.0026	45926	100.6	2307.9	2408.5
26	0.033597	1.0032	41034	108.9	2302.3	2411.2
28	0.037782	1.0037	36728	117.3	2296.6	2413.9
30	0.042415	1.0043	32929	125.7	2291.0	2416.7
32	0.047534	1.0049	29572	134.0	2285.4	2419.4
34	0.053180	1.0056	26601	142.4	2279.7	2422.1
36	0.059400	1.0063	23967	150.7	2274.1	2424.8
38	0.066240	1.0070	21627	159.1	2268.4	2427.5
40	0.073750	1.0078	19546	167.5	2262.8	2430.2
42	0.081985	1.0086	17692	175.8	2257.1	2432.9
44	0.091001	1.0094	16036	184.2	2251.4	2435.6
46	0.100860	1.0103	14557	192.5	2245.8	2438.3
48	0.111620	1.0112	13233	200.9	2240.0	2440.9
50	0.123350	1.0121	12046	209.3	2234.3	2443.6
52	0.136130	1.0131	10980	217.6	2228.6	2446.2
54	0.150020	1.0140	10022	226.0	2222.9	2448.8
56	0.165110	1.0150	9159	234.4	2217.1	2451.5
58	0.181470	1.0161	8381	242.7	2211.4	2454.1
60	0.199200	1.0171	7679	251.1	2205.7	2456.7
62	0.218380	1.0182	7044	259.5	2199.9	2459.4
64	0.239120	1.0193	6469	267.8	2194.1	2461.9
66	0.261500	1.0205	5948	276.2	2188.4	2464.6
68	0.285630	1.0217	5476	284.6	2182.5	2467.1
70	0.311620	1.0228	5046	293.0	2176.7	2469.6
72	0.339580	1.0241	4656	301.4	2170.8	2472.2
74	0.369640	1.0253	4300	309.7	2165.1	2474.8
76	0.401910	1.0266	3976	318.1	2159.3	2477.3
78	0.436520	1.0279	3680	326.5	2153.3	2479.8
80	0.473600	1.0292	3409	334.9	2147.5	2482.3
82	0.513290	1.0305	3162	343.2	2141.6	2484.8
84	0.555730	1.0319	2935	351.6	2135.7	2487.3
86	0.601080	1.0333	2727	360.0	2129.6	2489.7
88	0.649480	1.0347	2537	368.4	2123.7	2492.2
90	0.701090	1.0361	2361	376.8	2117.7	2494.6
92	0.756080	1.0376	2200	385.3	2111.7	2497.0
94	0.814610	1.0391	2052	393.7	2105.7	2499.5
96	0.876860	1.0406	1915	402.1	2099.6	2501.8
98	0.943010	1.0421	1789	410.5	2093.7	2504.2
100	1.013250	1.0437	1673	419.0	2087.5	2506.5

* In this Table specific volumes are in dm³/kg (l/kg).

h_f	h_{fg}	h_g	s_f	s_{fg}	s_g	t_s
0·0	2501·6	2501·6	0·0000	9·1575	9·1575	0·01
4·2	2499·2	2503·4	0·0153	9·1158	9·1311	1
8·4	2496·8	2505·2	0·0306	9·0741	9·1047	2
12·6	2494·5	2507·1	0·0459	9·0326	9·0785	3
16·8	2492·1	2508·9	0·0611	8·9915	9·0526	4
21·0	2489·7	2510·7	0·0762	8·9507	9·0269	5
25·2	2487·4	2512·6	0·0913	8·9102	9·0015	6
29·4	2485·0	2514·4	0·1063	8·8699	8·9762	7
33·6	2482·6	2516·2	0·1213	8·8300	8·9513	8
37·8	2480·3	2518·1	0·1362	8·7903	8·9265	9
42·0	2477·9	2519·9	0·1510	8·7510	8·9020	10
50·4	2473·2	2523·6	0·1805	8·6731	8·8536	12
58·8	2468·4	2527·2	0·2098	8·5962	8·8060	14
67·1	2463·8	2530·9	0·2388	8·5205	8·7593	16
75·5	2459·0	2534·5	0·2677	8·4458	8·7135	18
83·9	2454·3	2538·2	0·2963	8·3721	8·6684	20
92·2	2449·6	2541·8	0·3247	8·2994	8·6241	22
100·6	2444·9	2545·5	0·3530	8·2276	8·5806	24
108·9	2440·2	2549·1	0·3810	8·1569	8·5379	26
117·3	2435·4	2552·7	0·4088	8·0871	8·4959	28
125·7	2430·7	2556·4	0·4365	8·0181	8·4546	30
134·0	2426·0	2560·0	0·4640	7·9500	8·4140	32
142·4	2421·2	2563·6	0·4913	7·8827	8·3740	34
150·7	2416·5	2567·2	0·5184	7·8164	8·3348	36
159·1	2411·7	2570·8	0·5453	7·7509	8·2962	38
167·5	2406·9	2574·4	0·5721	7·6862	8·2583	40
175·8	2402·1	2577·9	0·5987	7·6222	8·2209	42
184·2	2397·3	2581·5	0·6252	7·5590	8·1842	44
192·5	2392·6	2585·1	0·6514	7·4967	8·1481	46
200·9	2387·7	2588·6	0·6776	7·4349	8·1125	48
209·3	2382·9	2592·2	0·7035	7·3741	8·0776	50
217·6	2378·1	2595·7	0·7293	7·3139	8·0432	52
226·0	2373·2	2599·2	0·7550	7·2543	8·0093	54
234·4	2368·3	2602·7	0·7804	7·1955	7·9759	56
242·7	2363·5	2606·2	0·8058	7·1373	7·9431	58
251·1	2358·6	2609·7	0·8310	7·0798	7·9108	60
259·5	2353·7	2613·2	0·8560	7·0230	7·8790	62
267·8	2348·8	2616·6	0·8809	6·9668	7·8477	64
276·2	2343·9	2620·1	0·9057	6·9111	7·8168	66
284·6	2338·9	2623·5	0·9303	6·8561	7·7864	68
293·0	2333·9	2626·9	0·9548	6·8017	7·7565	70
301·4	2328·9	2630·3	0·9792	6·7478	7·7270	72
309·7	2324·0	2633·7	1·0034	6·6945	7·6979	74
318·1	2319·0	2637·1	1·0275	6·6418	7·6693	76
326·5	2313·9	2640·4	1·0514	6·5896	7·6410	78
334·9	2308·9	2643·8	1·0753	6·5379	7·6132	80
343·3	2303·8	2647·1	1·0990	6·4868	7·5858	82
351·7	2298·7	2650·4	1·1225	6·4363	7·5588	84
360·1	2293·5	2653·6	1·1460	6·3861	7·5321	86
368·5	2288·4	2656·9	1·1693	6·3365	7·5058	88
376·9	2283·2	2660·1	1·1925	6·2874	7·4799	90
385·4	2278·0	2663·4	1·2156	6·2387	7·4543	92
393·8	2272·8	2666·6	1·2386	6·1905	7·4291	94
402·2	2267·5	2669·7	1·2615	6·1427	7·4042	96
410·6	2262·3	2672·9	1·2842	6·0954	7·3796	98
419·1	2256·9	2676·0	1·3069	6·0485	7·3554	100

Saturated water and steam, to 221 bar

p	t_s	v_f^*	v_g^*	u_f	u_{fg}	u_g
0·00611	0·01	1·0002	206162·9	0·0	2375·6	2375·6
0·008	3·77	1·0000	159668·5	15·9	2364·9	2380·8
0·01	6·98	1·0001	129210·7	29·3	2355·9	2385·2
0·02	17·51	1·0012	67011·6	73·5	2326·1	2399·6
0·03	24·10	1·0027	45670·0	101·0	2307·6	2408·6
0·04	28·98	1·0040	34803·3	121·4	2293·9	2415·3
0·05	32·90	1·0052	28194·5	137·8	2282·8	2420·6
0·06	36·18	1·0064	23740·6	151·5	2273·6	2425·1
0·07	39·03	1·0074	20530·4	163·4	2265·5	2428·9
0·08	41·54	1·0084	18103·8	173·9	2258·4	2432·3
0·09	43·79	1·0094	16203·4	183·3	2252·0	2435·3
0·10	45·83	1·0102	14673·7	191·8	2246·3	2438·1
0·11	47·71	1·0111	13415·2	199·7	2240·8	2440·5
0·12	49·45	1·0119	12361·0	206·9	2236·0	2442·9
0·13	51·06	1·0126	11464·9	213·7	2231·3	2445·0
0·14	52·58	1·0133	10693·4	220·0	2227·0	2447·0
0·15	54·00	1·0140	10022·1	226·0	2222·9	2448·9
0·16	55·34	1·0147	9432·4	231·6	2219·1	2450·7
0·17	56·62	1·0154	8910·3	236·9	2215·4	2452·3
0·18	57·83	1·0160	8444·6	242·0	2211·9	2453·9
0·19	58·98	1·0166	8026·6	246·8	2208·6	2455·4
0·20	60·09	1·0172	7649·2	251·5	2205·4	2456·9
0·22	62·16	1·0183	6994·6	260·1	2199·5	2459·6
0·24	64·08	1·0194	6446·2	268·2	2193·9	2462·1
0·26	65·87	1·0204	5979·9	275·7	2188·7	2464·4
0·28	67·55	1·0214	5578·4	282·7	2183·8	2466·5
0·30	69·13	1·0223	5229·0	289·3	2179·3	2468·5
0·32	70·62	1·0232	4922·0	295·6	2174·9	2470·5
0·34	72·03	1·0241	4650·1	301·5	2170·8	2472·3
0·36	73·38	1·0249	4407·6	307·1	2167·0	2474·0
0·38	74·66	1·0257	4189·8	312·5	2163·1	2475·6
0·40	75·89	1·0265	3993·2	317·7	2159·5	2477·2
0·42	77·06	1·0273	3814·8	322·6	2156·1	2478·7
0·44	78·19	1·0280	3652·2	327·3	2152·7	2480·0
0·46	79·28	1·0287	3503·2	331·9	2149·6	2481·5
0·48	80·33	1·0294	3366·3	336·3	2146·5	2482·7
0·50	81·35	1·0301	3240·1	340·5	2143·4	2484·0
0·52	82·33	1·0308	3123·3	344·6	2140·5	2485·2
0·54	83·28	1·0314	3014·8	348·6	2137·8	2486·4
0·56	84·19	1·0320	2913·9	352·4	2135·1	2487·5
0·58	85·09	1·0327	2819·7	356·2	2132·3	2488·6
0·60	85·95	1·0333	2731·7	359·8	2129·9	2489·7
0·62	86·80	1·0339	2649·1	363·4	2127·2	2490·7
0·64	87·62	1·0344	2571·5	366·8	2124·9	2491·7
0·66	88·42	1·0350	2498·5	370·2	2122·5	2492·7
0·68	89·20	1·0356	2429·7	373·5	2120·1	2493·6
0·70	89·96	1·0361	2364·7	376·7	2117·8	2494·6
0·72	90·70	1·0367	2303·1	379·8	2115·7	2495·5
0·74	91·43	1·0372	2244·8	382·9	2113·4	2496·3
0·76	92·14	1·0377	2189·5	385·8	2111·4	2497·2
0·78	92·83	1·0382	2136·9	388·8	2109·2	2498·0
0·80	93·51	1·0387	2086·9	391·6	2107·2	2498·8

* In this Table specific volumes are in dm^3/kg (l/kg).

h_f	h_{fg}	h_g	s_f	s_{fg}	s_g	p
0·0	2501·6	2501·6	0·0000	9·1575	9·1575	0·00611
15·9	2492·6	2508·5	0·0576	9·0008	9·0584	0·008
29·3	2485·1	2514·4	0·1060	8·8707	8·9767	0·01
73·5	2460·1	2533·6	0·2606	8·4640	8·7246	0·02
101·0	2444·6	2545·6	0·3543	8·2242	8·5785	0·03
121·4	2433·1	2554·5	0·4225	8·0530	8·4755	0·04
137·8	2423·8	2561·6	0·4763	7·9197	8·3960	0·05
151·5	2416·0	2567·5	0·5209	7·8103	8·3312	0·06
163·4	2409·2	2572·6	0·5591	7·7176	8·2767	0·07
173·9	2403·2	2577·1	0·5926	7·6369	8·2295	0·08
183·3	2397·8	2581·1	0·6224	7·5657	8·1881	0·09
191·8	2393·0	2584·8	0·6493	7·5018	8·1511	0·10
199·7	2388·4	2588·1	0·7738	7·3438	8·1176	0·11
206·9	2384·3	2591·2	0·6964	7·3908	8·0872	0·12
213·7	2380·3	2594·0	0·7172	7·3420	8·0592	0·13
220·0	2376·7	2596·7	0·7367	7·2966	8·0333	0·14
226·0	2373·2	2599·2	0·7549	7·2544	8·0093	0·15
231·6	2370·0	2601·6	0·7721	7·2147	7·9868	0·16
236·9	2366·9	2603·8	0·7883	7·1775	7·9658	0·17
242·0	2363·9	2605·9	0·8036	7·1423	7·9459	0·18
246·8	2361·1	2607·9	0·8182	7·1090	7·9272	0·19
251·5	2358·4	2609·9	0·8321	7·0773	7·9094	0·20
260·1	2353·4	2613·5	0·8581	7·0183	7·8764	0·22
268·2	2348·6	2616·8	0·8820	6·9644	7·8464	0·24
275·7	2344·2	2619·9	0·9041	6·9147	7·8188	0·26
282·7	2340·0	2622·7	0·9248	6·8684	7·7932	0·28
289·3	2336·1	2625·4	0·9441	6·8254	7·7695	0·30
295·6	2332·4	2628·0	0·9623	6·7850	7·7473	0·32
301·5	2328·9	2630·4	0·9795	6·7470	7·7265	0·34
307·1	2325·6	2632·7	0·9958	6·7111	7·7069	0·36
312·5	2322·3	2634·8	1·0113	6·6771	7·6884	0·38
317·7	2319·2	2636·9	1·0261	6·6448	7·6709	0·40
322·6	2316·3	2638·9	1·0402	6·6140	7·6542	0·42
327·3	2313·4	2640·7	1·0538	6·5845	7·6383	0·44
331·9	2310·7	2642·6	1·0667	6·5564	7·6231	0·46
336·3	2308·0	2644·3	1·0792	6·5294	7·6086	0·48
340·6	2305·4	2646·0	1·0912	6·5035	7·5947	0·50
344·7	2302·9	2647·6	1·1028	6·4786	7·5814	0·52
348·7	2300·5	2649·2	1·1140	6·4545	7·5685	0·54
352·5	2298·2	2650·7	1·1248	6·4313	7·5561	0·56
356·3	2295·8	2652·1	1·1353	6·4089	7·5442	0·58
359·9	2293·7	2653·6	1·1455	6·3872	7·5327	0·60
363·5	2291·4	2654·9	1·1553	6·3663	7·5216	0·62
366·9	2289·4	2656·3	1·1649	6·3459	7·5108	0·64
370·3	2287·3	2657·6	1·1742	6·3261	7·5003	0·66
373·6	2285·2	2658·8	1·1832	6·3070	7·4902	0·68
376·8	2283·3	2660·1	1·1921	6·2883	7·4804	0·70
379·9	2281·4	2661·3	1·2007	6·2701	7·4708	0·72
383·0	2279·4	2662·4	1·2090	6·2526	7·4616	0·74
385·9	2277·7	2663·6	1·2172	6·2353	7·4525	0·76
388·9	2275·8	2664·7	1·2252	6·2185	7·4437	0·78
391·7	2274·1	2665·8	1·2330	6·2022	7·4352	0·80

Saturated water and steam, to 221 bar

p	t_s	v_f^*	v_g^*	u_f	u_{fg}	u_g
0.82	94.18	1.0392	2039.2	394.4	2105.2	2499.6
0.84	94.83	1.0397	1993.8	397.2	2103.2	2500.4
0.86	95.47	1.0402	1950.4	399.9	2101.3	2501.2
0.88	96.10	1.0407	1908.9	402.5	2099.4	2501.9
0.90	96.71	1.0412	1869.1	405.1	2097.6	2502.7
0.92	97.32	1.0416	1831.1	407.7	2095.6	2503.3
0.94	97.91	1.0421	1794.6	410.2	2093.8	2504.0
0.96	98.50	1.0425	1759.6	412.6	2092.2	2504.8
0.98	99.07	1.0430	1726.0	415.0	2090.5	2505.5
1.00	99.63	1.0434	1693.7	417.4	2088.6	2506.0
1.10	102.32	1.0455	1549.2	428.7	2080.5	2509.2
1.20	104.81	1.0476	1428.1	439.3	2072.8	2512.0
1.30	107.13	1.0495	1325.0	449.1	2065.7	2514.8
1.40	109.32	1.0513	1236.3	458.3	2059.0	2517.2
1.50	111.37	1.0530	1159.0	466.9	2052.6	2519.6
1.60	113.32	1.0547	1091.1	475.2	2046.4	2521.6
1.70	115.17	1.0563	1030.9	483.0	2040.7	2523.7
1.80	116.93	1.0579	977.18	490.5	2035.1	2525.6
1.90	118.62	1.0594	928.95	497.7	2029.8	2527.5
2.00	120.23	1.0608	885.40	504.5	2024.7	2529.2
2.20	123.27	1.0636	809.80	517.4	2015.1	2532.4
2.40	126.09	1.0663	746.41	529.3	2006.0	2535.4
2.60	128.73	1.0688	692.47	540.6	1997.5	2538.2
2.80	131.21	1.0712	646.00	551.2	1989.4	2540.6
3.00	133.54	1.0735	605.53	561.1	1982.0	2543.0
3.20	135.76	1.0757	569.95	570.6	1974.7	2545.2
3.40	137.86	1.0779	538.43	579.5	1967.7	2547.2
3.60	139.87	1.0799	510.29	588.1	1961.1	2549.2
3.80	141.79	1.0819	485.02	596.4	1954.6	2551.0
4.00	143.63	1.0839	462.20	604.3	1948.5	2552.7
4.20	145.39	1.0858	441.47	611.8	1942.5	2554.4
4.40	147.09	1.0876	422.57	619.1	1936.8	2556.0
4.60	148.73	1.0894	405.26	626.2	1931.3	2557.5
4.80	150.31	1.0911	389.34	633.0	1925.8	2558.8
5.00	151.85	1.0928	374.66	639.6	1920.6	2560.2
5.20	153.33	1.0945	361.06	645.9	1915.6	2561.5
5.40	154.77	1.0961	348.44	652.2	1910.5	2562.7
5.60	156.16	1.0977	336.69	658.2	1905.8	2564.0
5.80	157.52	1.0993	325.72	664.1	1901.0	2565.1
6.00	158.84	1.1009	315.46	669.7	1896.5	2566.2
6.20	160.12	1.1024	305.84	675.3	1892.0	2567.3
6.40	161.38	1.1039	296.80	680.8	1887.5	2568.2
6.60	162.60	1.1053	288.29	686.1	1883.2	2569.2
6.80	163.79	1.1068	280.26	691.2	1879.0	2570.2
7.00	164.96	1.1082	272.68	696.3	1874.8	2571.1
7.20	166.10	1.1096	265.50	701.2	1870.8	2572.0
7.40	167.21	1.1110	258.70	706.1	1866.8	2572.9
7.60	168.30	1.1123	252.24	710.9	1862.8	2573.7
7.80	169.37	1.1137	246.10	715.4	1859.0	2574.4
8.00	170.41	1.1150	240.26	720.0	1855.3	2575.3

* In this Table specific volumes are in dm^3/kg (l/kg).

h_f	h_{fg}	h_g	s_f	s_{fg}	s_g	p
394·5	2272·3	2666·8	1·2407	6·1861	7·4268	0·82
397·3	2270·6	2667·9	1·2481	6·1706	7·4187	0·84
400·0	2268·9	2668·9	1·2554	6·1553	7·4107	0·86
402·6	2267·3	2669·9	1·2626	6·1404	7·4030	0·88
405·2	2265·7	2670·9	1·2696	6·1258	7·3954	0·90
407·8	2264·0	2671·8	1·2765	6·1114	7·3879	0·92
410·3	2262·4	2672·7	1·2832	6·0975	7·3807	0·94
412·7	2261·0	2673·7	1·2898	6·0838	7·3736	0·96
415·1	2259·5	2674·6	1·2963	6·0703	7·3666	0·98
417·5	2257·9	2675·4	1·3027	6·0571	7·3598	1·00
428·8	2250·8	2679·6	1·3330	5·9947	7·3277	1·10
439·4	2244·0	2683·4	1·3609	5·9375	7·2984	1·20
449·2	2237·8	2687·0	1·3868	5·8847	7·2715	1·30
458·4	2231·9	2690·3	1·4109	5·8356	7·2465	1·40
467·1	2226·3	2693·4	1·4336	5·7898	7·2234	1·50
475·4	2220·8	2696·2	1·4550	5·7467	7·2017	1·60
483·2	2215·8	2699·0	1·4752	5·7061	7·1813	1·70
490·7	2210·8	2701·5	1·4944	5·6678	7·1622	1·80
497·9	2206·1	2704·0	1·5127	5·6313	7·1440	1·90
504·7	2201·6	2706·3	1·5301	5·5967	7·1268	2·00
517·6	2193·0	2710·6	1·5628	5·5321	7·0949	2·20
529·6	2184·9	2714·5	1·5929	5·4728	7·0657	2·40
540·9	2177·3	2718·2	1·6209	5·4180	7·0389	2·60
551·5	2170·0	2721·5	1·6471	5·3669	7·0140	2·80
561·4	2163·3	2724·7	1·6717	5·3192	6·9909	3·00
570·9	2156·7	2727·6	1·6948	5·2744	6·9692	3·20
579·9	2150·4	2730·3	1·7168	5·2321	6·9489	3·40
588·5	2144·4	2732·9	1·7376	5·1921	6·9297	3·60
596·8	2138·5	2735·3	1·7575	5·1540	6·9115	3·80
604·7	2132·9	2737·6	1·7764	5·1179	6·8943	4·00
612·3	2127·5	2739·8	1·7946	5·0833	6·8779	4·20
619·6	2122·3	2741·9	1·8120	5·0502	6·8622	4·40
626·7	2117·2	2743·9	1·8287	5·0186	6·8473	4·60
633·5	2112·2	2745·7	1·8448	4·9881	6·8329	4·80
640·1	2107·4	2747·5	1·8604	4·9588	6·8192	5·00
646·5	2102·8	2749·3	1·8754	4·9305	6·8059	5·20
652·8	2098·1	2750·9	1·8899	4·9033	6·7932	5·40
658·8	2093·7	2752·5	1·9040	4·8769	6·7809	5·60
664·7	2089·3	2754·0	1·9176	4·8514	6·7690	5·80
670·4	2085·1	2755·5	1·9308	4·8267	6·7575	6·00
676·0	2080·9	2756·9	1·9437	4·8027	6·7464	6·20
681·5	2076·7	2758·2	1·9562	4·7794	6·7356	6·40
686·8	2072·7	2759·5	1·9684	4·7568	6·7252	6·60
692·0	2068·8	2760·8	1·9803	4·7347	6·7150	6·80
697·1	2064·9	2762·0	1·9918	4·7134	6·7052	7·00
702·0	2061·2	2763·2	2·0031	4·6925	6·6956	7·20
706·9	2057·4	2764·3	2·0141	4·6721	6·6862	7·40
711·7	2053·7	2765·4	2·0249	4·6522	6·6771	7·60
716·3	2050·1	2766·4	2·0354	4·6329	6·6683	7·80
720·9	2046·6	2767·5	2·0457	4·6139	6·6596	8·00

Saturated water and steam, to 221 bar

p	t_s	v_f^*	v_g^*	u_f	u_{fg}	u_g
8·20	171·44	1·1163	234·69	724·5	1851·6	2576·1
8·40	172·45	1·1176	229·38	729·0	1847·8	2576·7
8·60	173·43	1·1188	224·31	733·2	1844·3	2577·5
8·80	174·40	1·1201	219·46	737·5	1840·7	2578·2
9·00	175·36	1·1213	214·82	741·6	1837·2	2578·8
9·20	176·29	1·1226	210·37	745·8	1833·7	2579·5
9·40	177·21	1·1238	206·10	749·7	1830·3	2580·1
9·60	178·12	1·1250	202·01	753·7	1827·0	2580·7
9·80	179·01	1·1262	198·08	757·6	1823·7	2581·3
10·00	179·88	1·1274	194·30	761·5	1820·4	2581·9
11·00	184·06	1·1331	177·39	779·9	1804·7	2584·6
12·00	187·96	1·1386	163·21	797·0	1789·8	2586·8
13·00	191·60	1·1438	151·14	813·2	1775·7	2588·9
14·00	195·04	1·1489	140·73	828·5	1762·3	2590·8
15·00	198·28	1·1538	131·67	842·9	1749·5	2592·4
16·00	201·37	1·1586	123·70	856·6	1737·1	2593·8
17·00	204·30	1·1633	116·64	869·8	1725·3	2595·1
18·00	207·11	1·1678	110·33	882·4	1713·8	2596·2
19·00	209·79	1·1723	104·67	894·6	1702·7	2597·2
20·00	212·37	1·1766	99·55	906·2	1691·9	2598·1
21·00	214·85	1·1809	94·90	917·4	1681·5	2598·9
22·00	217·24	1·1850	90·66	928·3	1671·3	2599·6
23·00	219·55	1·1891	86·78	938·9	1661·3	2600·2
24·00	221·78	1·1932	83·21	949·0	1651·7	2600·7
25·00	223·94	1·1972	79·92	958·9	1642·2	2601·1
26·00	226·03	1·2011	76·87	968·6	1633·0	2601·6
27·00	228·06	1·2050	74·03	977·9	1623·9	2601·8
28·00	230·04	1·2088	71·40	987·1	1615·0	2602·1
29·00	231·96	1·2126	68·94	996·0	1606·3	2602·3
30·00	233·84	1·2163	66·63	1004·7	1597·8	2602·4
31·00	235·66	1·2200	64·47	1013·2	1589·2	2602·4
32·00	237·44	1·2237	62·44	1021·5	1581·0	2602·5
33·00	239·18	1·2273	60·53	1029·6	1572·9	2602·5
34·00	240·88	1·2310	58·73	1037·6	1564·8	2602·4
35·00	242·54	1·2345	57·03	1045·4	1557·0	2602·4
36·00	244·16	1·2381	55·42	1053·0	1549·2	2602·2
37·00	245·75	1·2416	53·89	1060·6	1541·4	2602·0
38·00	247·31	1·2451	52·44	1068·0	1533·9	2601·8
39·00	248·84	1·2486	51·06	1075·2	1526·4	2601·7
40·00	250·33	1·2521	49·75	1082·4	1518·9	2601·3
42·00	253·24	1·2589	47·31	1096·3	1504·4	2600·7
44·00	256·05	1·2657	45·08	1109·8	1490·1	2600·0
46·00	258·76	1·2725	43·04	1122·9	1476·1	2599·0
48·00	261·38	1·2792	41·16	1135·7	1462·5	2598·1
50·00	263·92	1·2858	39·43	1148·1	1449·0	2597·1
52·00	266·38	1·2925	37·82	1160·2	1435·8	2595·9
54·00	268·77	1·2990	36·33	1172·0	1422·6	2594·6
56·00	271·09	1·3056	34·94	1183·5	1409·8	2593·3
58·00	273·36	1·3122	33·65	1194·8	1397·1	2591·9
60·00	275·56	1·3187	32·43	1205·8	1384·6	2590·4

* In this Table specific volumes are in dm^3/kg (l/kg).

h_f	h_{fg}	h_g	s_f	s_{fg}	s_g	p
725·4	2043·1	2768·5	2·0558	4·5953	6·6511	8·20
729·9	2039·5	2769·4	2·0657	4·5772	6·6429	8·40
734·2	2036·2	2770·4	2·0753	4·5595	6·6348	8·60
738·5	2032·8	2771·3	2·0848	4·5421	6·6269	8·80
742·6	2029·5	2772·1	2·0941	4·5251	6·6192	9·00
746·8	2026·2	2773·0	2·1033	4·5083	6·6116	9·20
750·8	2023·0	2773·8	2·1122	4·4920	6·6042	9·40
754·8	2019·8	2774·6	2·1210	4·4759	6·5969	9·60
758·7	2016·7	2775·4	2·1297	4·4601	6·5898	9·80
762·6	2013·6	2776·2	2·1382	4·4446	6·5828	10·00
746·8	2026·2	2773·0	2·1033	4·5083	6·6116	9·20
781·1	1998·6	2779·7	2·1786	4·3712	6·5498	11·00
798·4	1984·3	2782·7	2·2160	4·3034	6·5194	12·00
814·7	1970·7	2785·4	2·2509	4·2404	6·4913	13·00
830·1	1957·7	2787·8	2·2836	4·1815	6·4651	14·00
844·6	1945·3	2789·9	2·3144	4·1262	6·4406	15·00
858·5	1933·2	2791·7	2·3436	4·0740	6·4176	16·00
871·8	1921·6	2793·4	2·3712	4·0246	6·3958	17·00
884·5	1910·3	2794·8	2·3976	3·9775	6·3751	18·00
896·8	1899·3	2796·1	2·4227	3·9328	6·3555	19·00
908·6	1888·6	2797·2	2·4468	3·8899	6·3367	20·00
919·9	1878·3	2798·2	2·4699	3·8488	6·3187	21·00
930·9	1868·2	2799·1	2·4921	3·8094	6·3015	22·00
941·6	1858·2	2799·8	2·5136	3·7714	6·2850	23·00
951·9	1848·5	2800·4	2·5342	3·7348	6·2690	24·00
961·9	1839·0	2800·9	2·5542	3·6995	6·2537	25·00
971·7	1829·7	2801·4	2·5736	3·6652	6·2388	26·00
981·2	1820·5	2801·7	2·5923	3·6321	6·2244	27·00
990·5	1811·5	2802·0	2·6105	3·6000	6·2105	28·00
999·5	1802·7	2802·2	2·6282	3·5687	6·1969	29·00
1008·3	1794·0	2802·3	2·6455	3·5383	6·1838	30·00
1017·0	1785·3	2802·3	2·6622	3·5088	6·1710	31·00
1025·4	1776·9	2802·3	2·6785	3·4800	6·1585	32·00
1033·7	1768·6	2802·3	2·6945	3·4518	6·1463	33·00
1041·8	1760·3	2802·1	2·7100	3·4245	6·1345	34·00
1049·7	1752·3	2802·0	2·7252	3·3977	6·1229	35·00
1057·5	1744·2	2801·7	2·7401	3·3714	6·1115	36·00
1065·2	1736·2	2801·4	2·7547	3·3457	6·1004	37·00
1072·7	1728·4	2801·1	2·7689	3·3207	6·0896	38·00
1080·1	1720·7	2800·8	2·7828	3·2961	6·0789	39·00
1087·4	1712·9	2800·3	2·7965	3·2720	6·0685	40·00
1101·6	1697·8	2799·4	2·8231	3·2251	6·0482	42·00
1115·4	1682·9	2798·3	2·8487	3·1799	6·0286	44·00
1128·8	1668·2	2797·0	2·8735	3·1362	6·0097	46·00
1141·8	1653·9	2795·7	2·8974	3·0939	5·9913	48·00
1154·5	1639·7	2794·2	2·9207	3·0528	5·9735	50·00
1166·9	1625·7	2792·6	2·9432	3·0129	5·9561	52·00
1179·0	1611·8	2790·8	2·9651	2·9741	5·9392	54·00
1190·8	1598·2	2789·0	2·9864	2·9363	5·9227	56·00
1202·4	1584·6	2787·0	3·0071	2·8994	5·9065	58·00
1213·7	1571·3	2785·0	3·0274	2·8632	5·8906	60·00

Saturated water and steam, to 221 bar

p	t_s	v_f^{*}	v_g^{*}	u_f	u_fg	u_g
62·00	277·71	1·3252	31·29	1216·7	1372·2	2588·9
64·00	279·80	1·3318	30·23	1227·3	1359·9	2587·2
66·00	281·85	1·3383	29·22	1237·7	1347·8	2585·5
68·00	283·85	1·3448	28·27	1248·0	1335·7	2583·7
70·00	285·80	1·3514	27·37	1258·0	1323·8	2581·8
72·00	287·71	1·3579	26·52	1267·9	1312·1	2580·0
74·00	289·59	1·3645	25·71	1277·7	1300·2	2577·9
76·00	291·42	1·3711	24·94	1287·3	1288·6	2575·9
78·00	293·22	1·3777	24·22	1296·8	1277·1	2573·8
80·00	294·98	1·3843	23·52	1306·1	1265·6	2571·7
82·00	296·71	1·3909	22·86	1315·3	1254·3	2569·5
84·00	298·40	1·3976	22·23	1324·4	1242·9	2567·3
86·00	300·07	1·4043	21·62	1333·3	1231·6	2564·9
88·00	301·71	1·4111	21·05	1342·3	1220·3	2562·6
90·00	303·31	1·4179	20·49	1351·0	1209·1	2560·2
92·00	304·89	1·4247	19·96	1359·7	1198·0	2557·6
94·00	306·45	1·4316	19·45	1368·2	1186·9	2555·1
96·00	307·98	1·4385	18·96	1376·8	1175·9	2552·6
98·00	309·48	1·4455	18·49	1385·2	1164·7	2550·0
100·00	310·96	1·4526	18·04	1393·6	1153·7	2547·3
102·00	312·42	1·4597	17·60	1401·8	1142·8	2544·6
104·00	313·86	1·4668	17·18	1409·9	1131·9	2541·9
106·00	315·27	1·4740	16·78	1418·1	1121·0	2539·1
108·00	316·67	1·4813	16·39	1426·2	1109·9	2536·1
110·00	318·04	1·4887	16·01	1434·2	1099·0	2533·2
112·00	319·40	1·4961	15·64	1442·1	1088·1	2530·2
114·00	320·73	1·5037	15·29	1450·1	1077·2	2527·2
116·00	322·05	1·5113	14·94	1457·9	1066·3	2524·2
118·00	323·35	1·5189	14·61	1465·7	1055·3	2521·0
120·00	324·64	1·5267	14·29	1473·4	1044·4	2517·8
122·00	325·90	1·5346	13·97	1481·1	1033·4	2514·5
124·00	327·15	1·5425	13·67	1488·8	1022·4	2511·1
126·00	328·39	1·5506	13·37	1496·4	1011·4	2507·7
128·00	329·61	1·5588	13·08	1503·9	1000·2	2504·2
130·00	330·81	1·5671	12·80	1511·5	989·1	2500·6
132·00	332·00	1·5755	12·53	1519·1	977·9	2497·0
134·00	333·18	1·5841	12·26	1526·6	966·6	2493·2
136·00	334·34	1·5927	12·00	1534·0	955·4	2489·4
138·00	335·49	1·6015	11·75	1541·5	944·0	2485·5
140·00	336·63	1·6105	11·50	1549·0	932·5	2481·4
142·00	337·75	1·6196	11·26	1556·4	920·9	2477·4
144·00	338·86	1·6289	11·02	1563·8	909·4	2473·2
146·00	339·96	1·6383	10·79	1571·3	897·6	2468·9
148·00	341·04	1·6480	10·56	1578·6	885·8	2464·5
150·00	342·12	1·6578	10·34	1586·0	873·9	2460·0
152·00	343·18	1·6678	10·13	1593·4	861·9	2455·4
154·00	344·23	1·6780	9·92	1600·9	849·8	2450·7
156·00	345·27	1·6885	9·71	1608·3	837·6	2445·8
158·00	346·30	1·6992	9·51	1615·7	825·3	2441·0
160·00	347·32	1·7102	9·31	1623·0	812·9	2435·9

* In this Table specific volumes are in dm^3/kg (l/kg).

h_f	h_{fg}	h_g	s_f	s_{fg}	s_g	p
1224.9	1558.0	2782.9	3.0472	2.8281	5.8753	62.00
1235.8	1544.8	2780.6	3.0665	2.7936	5.8601	64.00
1246.5	1531.8	2778.3	3.0854	2.7598	5.8452	66.00
1257.1	1518.8	2775.9	3.1039	2.7266	5.8305	68.00
1267.5	1505.9	2773.4	3.1220	2.6941	5.8161	70.00
1277.7	1493.2	2770.9	3.1398	2.6621	5.8019	72.00
1287.8	1480.4	2768.2	3.1572	2.6307	5.7879	74.00
1297.7	1467.8	2765.5	3.1743	2.5998	5.7741	76.00
1307.5	1455.2	2762.7	3.1912	2.5693	5.7605	78.00
1317.2	1442.7	2759.9	3.2077	2.5393	5.7470	80.00
1326.7	1430.3	2757.0	3.2240	2.5097	5.7337	82.00
1336.1	1417.9	2754.0	3.2400	2.4806	5.7206	84.00
1345.4	1405.5	2750.9	3.2558	2.4518	5.7076	86.00
1354.7	1393.1	2747.8	3.2714	2.4233	5.6947	88.00
1363.8	1380.8	2744.6	3.2867	2.3953	5.6820	90.00
1372.8	1368.5	2741.3	3.3019	2.3674	5.6693	92.00
1381.7	1356.3	2738.0	3.3168	2.3400	5.6568	94.00
1390.6	1344.1	2734.7	3.3316	2.3128	5.6444	96.00
1399.4	1331.8	2731.2	3.3462	2.2858	5.6320	98.00
1408.1	1319.6	2727.7	3.3606	2.2592	5.6198	100.00
1416.7	1307.5	2724.2	3.3748	2.2328	5.6076	102.00
1425.2	1295.4	2720.6	3.3889	2.2066	5.5955	104.00
1433.7	1283.2	2716.9	3.4029	2.1806	5.5835	106.00
1442.2	1270.9	2713.1	3.4167	2.1548	5.5715	108.00
1450.6	1258.7	2709.3	3.4304	2.1292	5.5596	110.00
1458.9	1246.5	2705.4	3.4439	2.1038	5.5477	112.00
1467.2	1234.3	2701.5	3.4574	2.0784	5.5358	114.00
1475.4	1222.1	2697.5	3.4707	2.0533	5.5240	116.00
1483.6	1209.8	2693.4	3.4840	2.0281	5.5121	118.00
1491.7	1197.5	2689.2	3.4971	2.0032	5.5003	120.00
1499.8	1185.1	2684.9	3.5101	1.9784	5.4885	122.00
1507.9	1172.7	2680.6	3.5231	1.9535	5.4766	124.00
1515.9	1160.3	2676.2	3.5359	1.9289	5.4648	126.00
1523.9	1147.7	2671.6	3.5487	1.9042	5.4529	128.00
1531.9	1135.1	2667.0	3.5614	1.8795	5.4409	130.00
1539.9	1122.4	2662.3	3.5741	1.8549	5.4290	132.00
1547.8	1109.7	2657.5	3.5867	1.8302	5.4169	134.00
1555.7	1096.9	2652.6	3.5992	1.8056	5.4048	136.00
1563.6	1084.0	2647.6	3.6116	1.7811	5.3927	138.00
1571.5	1070.9	2642.4	3.6241	1.7563	5.3804	140.00
1579.4	1057.8	2637.2	3.6365	1.7316	5.3681	142.00
1587.3	1044.6	2631.9	3.6488	1.7069	5.3557	144.00
1595.2	1031.2	2626.4	3.6611	1.6821	5.3432	146.00
1603.0	1017.8	2620.8	3.6734	1.6573	5.3307	148.00
1610.9	1004.2	2615.1	3.6857	1.6323	5.3180	150.00
1618.8	990.5	2609.3	3.6979	1.6074	5.3053	152.00
1626.7	976.7	2603.4	3.7102	1.5822	5.2924	154.00
1634.6	962.7	2597.3	3.7224	1.5571	5.2795	156.00
1642.5	948.7	2591.2	3.7347	1.5317	5.2664	158.00
1650.4	934.5	2584.9	3.7470	1.5063	5.2533	160.00

Saturated water and steam, to 221 bar

p	t_s	v_f^{*}	v_g^{*}	u_f	u_fg	u_g
162.00	348.32	1.7214	9.12	1630.5	800.4	2430.9
164.00	349.32	1.7330	8.93	1638.0	787.7	2425.7
166.00	350.31	1.7446	8.74	1645.4	775.1	2420.5
168.00	351.29	1.7569	8.55	1653.4	761.6	2415.0
170.00	352.26	1.7695	8.37	1661.5	747.8	2409.3
172.00	353.22	1.7825	8.19	1669.6	733.9	2403.5
174.00	354.16	1.7961	8.01	1677.7	719.9	2397.6
176.00	355.11	1.8101	7.84	1685.7	705.8	2391.5
178.00	356.04	1.8247	7.67	1693.7	691.6	2385.3
180.00	356.96	1.8399	7.50	1701.7	677.3	2378.9
182.00	357.87	1.8557	7.33	1709.7	662.6	2372.3
184.00	358.78	1.8722	7.16	1717.8	647.7	2365.5
186.00	359.67	1.8894	7.00	1725.8	632.7	2358.5
188.00	360.56	1.9074	6.84	1733.9	617.3	2351.3
190.00	361.44	1.9262	6.68	1742.1	601.6	2343.7
192.00	362.31	1.9460	6.52	1750.5	585.3	2335.8
194.00	363.17	1.9669	6.36	1758.9	568.8	2327.7
196.00	364.03	1.9889	6.20	1767.7	551.4	2319.2
198.00	364.87	2.0124	6.04	1776.7	533.5	2310.2
200.00	365.71	2.0374	5.87	1785.9	514.9	2300.7
202.00	366.54	2.0643	5.71	1795.5	495.2	2290.7
204.00	367.37	2.0935	5.55	1805.6	474.4	2280.0
206.00	368.18	2.1256	5.38	1816.2	452.2	2268.4
208.00	368.99	2.1612	5.20	1827.6	428.2	2255.9
210.00	369.79	2.2018	5.02	1840.1	402.0	2242.0
212.00	370.58	2.2489	4.83	1853.9	372.6	2226.5
214.00	371.37	2.3059	4.62	1869.7	338.8	2208.4
216.00	372.14	2.3788	4.39	1888.4	298.4	2186.8
218.00	372.92	2.4819	4.12	1912.8	245.8	2158.6
220.00	373.68	2.6675	3.73	1951.6	162.8	2114.4
221.20	374.15	3.1700	3.17	2037.3	0.0	2037.3

* In this Table specific volumes are in $\mathrm{dm^3/kg}$ (l/kg).

h_f	h_{fg}	h_g	s_f	s_{fg}	s_g	p
1658.4	920.2	2578.6	3.7592	1.4809	5.2401	162.00
1666.4	905.7	2572.1	3.7716	1.4552	5.2268	164.00
1674.4	891.2	2565.6	3.7841	1.4292	5.2133	166.00
1682.9	875.8	2558.7	3.7973	1.4022	5.1995	168.00
1691.6	860.0	2551.6	3.8106	1.3750	5.1856	170.00
1700.3	844.1	2544.4	3.8239	1.3475	5.1714	172.00
1709.0	828.1	2537.1	3.8372	1.3199	5.1571	174.00
1717.6	811.9	2529.5	3.8502	1.2923	5.1425	176.00
1726.2	795.6	2521.8	3.8635	1.2643	5.1278	178.00
1734.8	779.1	2513.9	3.8766	1.2361	5.1127	180.00
1743.5	762.2	2505.7	3.8897	1.2078	5.0975	182.00
1752.2	745.1	2497.3	3.9029	1.1790	5.0819	184.00
1760.9	727.8	2488.7	3.9161	1.1499	5.0660	186.00
1769.8	710.0	2479.8	3.9295	1.1202	5.0497	188.00
1778.7	691.8	2470.5	3.9430	1.0900	5.0330	190.00
1787.9	673.0	2460.9	3.9568	1.0590	5.0158	192.00
1797.1	653.9	2451.0	3.9708	1.0273	4.9981	194.00
1806.7	633.9	2440.6	3.9851	0.9947	4.9798	196.00
1816.5	613.2	2429.7	3.9998	0.9610	4.9608	198.00
1826.6	591.6	2418.2	4.0151	0.9259	4.9410	200.00
1837.2	568.9	2406.1	4.0310	0.8891	4.9201	202.00
1848.3	544.8	2393.1	4.0476	0.8506	4.8982	204.00
1860.0	519.2	2379.2	4.0653	0.8095	4.8748	206.00
1872.6	491.5	2364.1	4.0843	0.7653	4.8496	208.00
1886.3	461.2	2347.5	4.1049	0.7173	4.8222	210.00
1901.6	427.3	2328.9	4.1279	0.6638	4.7917	212.00
1919.0	388.4	2307.4	4.1543	0.6027	4.7570	214.00
1939.8	341.9	2281.7	4.1858	0.5299	4.7157	216.00
1966.9	281.5	2248.4	4.2271	0.4357	4.6628	218.00
2010.3	186.3	2196.6	4.2934	0.2880	4.5814	220.00
2107.4	0.0	2107.4	4.4429	0.0000	4.4429	221.20

Superheated steam, to 220 bar and 800°C

p (t_s)	t:	t_s	50	100	150	200	250
0	u		2571·7	2642·5	2714·6	2787·8	2862·4
	h		2594·8	2688·7	2783·8	2880·1	2977·8
0·02	v^*	67012	74524	86080	97628	109171	120711
(17·50)	u	2399·6	2445·4	2516·3	2588·4	2661·7	2736·3
	h	2533·6	2594·4	2688·5	2783·7	2880·0	2977·7
	s	8·7246	8·9226	9·1934	9·4327	9·6479	9·8441
0·06	v^*	23741	24812	28676	32532	36383	40232
(36·20)	u	2425·1	2444·6	2515·9	2588·2	2661·5	2736·2
	h	2567·5	2593·5	2688·0	2783·4	2879·8	2977·6
	s	8·3312	8·4135	8·6854	8·9251	9·1406	9·3369
0·10	v^*	14674	14869	17195	19512	21825	24136
(45·80)	u	2438·1	2444·0	2515·6	2588·0	2661·4	2736·0
	h	2584·8	2592·7	2687·5	2783·1	2879·6	2977·4
	s	8·1511	8·1757	8·4486	8·6888	8·9045	9·1010
0·50	v^*	3240·1		3418·1	3889·3	4356·0	4820·5
(81·30)	u	2484·0		2511·7	2585·6	2659·9	2735·1
	h	2646·0		2682·6	2780·1	2877·7	2976·1
	s	7·5947		7·6953	7·9406	8·1587	8·3564
1·00	v^*	1693·7		1695·5	1936·3	2172·3	2406·1
(99·60)	u	2506·0		2506·7	2582·7	2658·2	2733·9
	h	2675·4		2676·2	2776·3	2875·4	2974·5
	s	7·3598		7·3618	7·6137	7·8349	8·0342
1·50	v^*	1159·0			1285·2	1444·4	1601·3
(111·4)	u	2519·6			2579·7	2656·2	2732·7
	h	2693·4			2772·5	2872·9	2972·9
	s	7·2234			7·4194	7·6439	7·8447
2	v^*	885·40			959·54	1080·4	1198·9
(120·2)	u	2529·2			2576·6	2654·4	2731·4
	h	2706·3			2768·5	2870·5	2971·2
	s	7·1268			7·2794	7·5072	7·7096
3	v^*	605·53			633·74	716·35	796·44
(133·5)	u	2543·0			2570·3	2650·6	2729·0
	h	2724·7			2760·4	2865·5	2967·9
	s	6·9909			7·0771	7·3119	7·5176
4	v^*	462·20			470·66	534·26	595·19
(143·6)	u	2552·7			2563·7	2646·7	2726·4
	h	2737·6			2752·0	2860·4	2964·5
	s	6·8943			6·9285	7·1708	7·3800
5	v^*	374·66				424·96	474·43
(151·8)	u	2560·2				2642·6	2723·9
	h	2747·5				2855·1	2961·1
	s	6·8192				7·0592	7·2721
6	v^*	315·46				352·04	393·91
(158·8)	u	2566·2				2638·5	2721·3
	h	2755·5				2849·7	2957·6
	s	6·7575				6·9662	7·1829
7	v^*	272·68				299·92	336·37
(165·0)	u	2571·1				2634·3	2718·5
	h	2762·0				2844·2	2954·0
	s	6·7052				6·8859	7·1066

* In this Table specific volumes are in dm^3/kg (l/kg).

300	400	500	600	700	800	p (t_s)
						0
2938.4	3095.1	3258.4	3428.7	3605.8	3789.5	
3076.9	3279.7	3489.2	3705.6	3928.9	4158.7	
132251	155329	178405	201482	224558	247634	0.02
2812.3	2969.0	3132.4	3302.6	3479.7	3663.4	(17.50)
3076.8	3279.7	3489.2	3705.6	3928.8	4158.7	
10.025	10.351	10.641	10.904	11.146	11.371	
44079	51773	59467	67159	74852	82544	0.06
2812.2	2969.0	3132.4	3302.6	3479.7	3663.4	(36.20)
3076.7	3279.6	3489.2	3705.6	3928.8	4158.7	
9.5179	9.8441	10.134	10.397	10.639	10.864	
26445	31062	35679	40295	44910	49526	0.10
2812.2	2969.0	3132.3	3302.6	3479.7	3663.4	(45.80)
3076.6	3279.6	3489.1	3705.5	3928.8	4158.7	
9.2820	9.6083	9.8984	10.162	10.404	10.628	
5283.9	6209.1	7133.5	8057.4	8981.0	9904.4	0.50
2811.5	2968.5	3132.0	3302.3	3479.5	3663.3	(81.30)
3075.7	3279.0	3488.7	3705.2	3928.6	4158.5	
8.5380	8.8649	9.1552	9.4185	9.6606	9.8855	
2638.7	3102.5	3565.3	4027.7	4489.8	4951.7	1.00
2810.6	2968.0	3131.6	3302.0	3479.2	3663.1	(99.60)
3074.5	3278.2	3488.1	3704.8	3928.2	4158.3	
8.2166	8.5442	8.8348	9.0982	9.3405	9.5654	
1757.0	2066.9	2375.9	2684.5	2992.7	3300.8	1.50
2809.8	2967.5	3131.2	3301.7	3479.0	3662.9	(111.4)
3073.3	3277.5	3487.6	3704.4	3927.9	4158.0	
8.0280	8.3562	8.6472	8.9108	9.1531	9.3781	
1316.2	1549.2	1781.2	2012.9	2244.2	2475.4	2
2808.9	2966.9	3130.8	3301.4	3478.8	3662.7	(120.2)
3072.1	3276.7	3487.0	3704.0	3927.6	4157.8	
7.8937	8.2226	8.5139	8.7776	9.0201	9.2452	
875.29	1031.4	1186.5	1341.2	1495.7	1649.9	3
2807.1	2965.8	3130.1	3300.8	3478.3	3662.3	(133.5)
3069.7	3275.2	3486.0	3703.2	3927.0	4157.3	
7.7034	8.0338	8.3257	8.5898	8.8325	9.0577	
654.85	772.50	889.19	1005.4	1121.4	1237.2	4
2805.3	2964.6	3129.2	3300.1	3477.8	3662.0	(143.6)
3067.2	3273.6	3484.9	3702.3	3926.4	4156.9	
7.5675	7.8994	8.1919	8.4563	8.6992	8.9246	
522.58	617.16	710.78	803.95	896.85	989.56	5
2803.5	2963.5	3128.4	3299.5	3477.4	3661.6	(151.8)
3064.8	3272.1	3483.8	3701.5	3925.8	4156.4	
7.4614	7.7948	8.0879	8.3526	8.5957	8.8213	
434.39	513.61	591.84	669.63	747.14	824.47	6
2801.7	2962.4	3127.6	3298.9	3476.8	3661.2	(158.8)
3062.3	3270.6	3482.7	3700.7	3925.1	4155.9	
7.3740	7.7090	8.0027	8.2678	8.5111	8.7368	
371.39	439.64	506.89	573.68	640.21	706.55	7
2799.8	2961.3	3126.8	3298.3	3476.4	3660.9	(165.0)
3059.8	3269.0	3481.6	3699.9	3924.5	4155.5	
7.2997	7.6362	7.9305	8.1959	8.4395	8.6653	

Superheated steam, to 220 bar and 800°C

p (t_s)	t:	t_s	50	100	150	200	250
8 (170·4)	v^*	240·26				260·79	293·21
	u	2575·3				2630·0	2715·8
	h	2767·5				2838·6	2950·4
	s	6·6596				6·8148	7·0397
9 (175·4)	v^*	214·82				230·32	259·63
	u	2578·8				2625·4	2713·1
	h	2772·1				2832·7	2946·8
	s	6·6192				6·7508	6·9800
10 (179·9)	v^*	194·30				205·92	232·75
	u	2581·9				2620·9	2710·3
	h	2776·2				2826·8	2943·0
	s	6·5828				6·6922	6·9259
15 (198·3)	v^*	131·67				132·38	151·99
	u	2592·4				2596·1	2695·5
	h	2789·9				2794·7	2923·5
	s	6·4406				6·4508	6·7099
20 (212·4)	v^*	99·549					111·45
	u	2598·1					2679·5
	h	2797·2					2902·4
	s	6·3367					6·5454
30 (233·8)	v^*	66·632					70·551
	u	2602·4					2643·1
	h	2802·3					2854·8
	s	6·1838					6·2857
40 (250·3)	v^*	49·749					
	u	2601·3					
	h	2800·3					
	s	6·0685					
50 (263·9)	v^*	39·425					
	u	2597·1					
	h	2794·2					
	s	5·9735					
60 (275·6)	v^*	32·433					
	u	2590·4					
	h	2785·0					
	s	5·8907					
70 (285·8)	v^*	27·368					
	u	2581·8					
	h	2773·4					
	s	5·8161					
80 (295·0)	v^*	23·521					
	u	2571·7					
	h	2759·9					
	s	5·7470					
90 (303·3)	v^*	20·493					
	u	2560·2					
	h	2744·6					
	s	5·6820					

* In this Table specific volumes are in dm^3/kg (l/kg).

300	400	500	600	700	800	p (t_s)
324·14	384·16	443·17	501·72	560·01	618·11	8
2798·0	2960·2	3126·0	3297·7	3475·9	3660·5	(170·4)
3057·3	3267·5	3480·5	3699·1	3923·9	4155·0	
7·2348	7·5729	7·8678	8·1336	8·3773	8·6033	
287·39	341·01	393·61	445·76	497·63	549·33	9
2796·0	2959·1	3125·2	3297·0	3475·4	3660·1	(175·4)
3054·7	3266·0	3479·4	3698·2	3923·3	4154·5	
7·1771	7·5169	7·8124	8·0785	8·3225	8·5486	
257·98	306·49	353·96	400·98	447·73	494·30	10
2794·1	2957·9	3124·3	3296·4	3475·0	3659·8	(179·9)
3052·1	3264·4	3478·3	3697·4	3922·7	4154·1	
7·1251	7·4665	7·7627	8·0292	8·2734	8·4997	
169·70	202·92	235·03	266·66	298·03	329·21	15
2784·4	2952·2	3120·3	3293·3	3472·6	3657·9	(198·3)
3038·9	3256·6	3472·8	3693·3	3919·6	4151·7	
6·9207	7·2709	7·5703	7·8385	8·0838	8·3108	
125·50	151·13	175·55	199·50	223·17	246·66	20
2774·0	2946·4	3116·2	3290·2	3470·2	3656·1	(212·4)
3025·0	3248·7	3467·3	3689·2	3916·5	4149·4	
6·7696	7·1296	7·4323	7·7022	7·9485	8·1763	
81·159	99·310	116·08	132·34	148·32	164·12	30
2751·6	2934·6	3108·0	3284·0	3465·3	3652·3	(233·8)
2995·1	3232·5	3456·2	3681·0	3910·3	4144·7	
6·5422	6·9246	7·2345	7·5079	7·7564	7·9857	
58·833	73·376	86·341	98·763	110·90	122·85	40
2726·7	2922·2	3099·6	3277·7	3460·5	3648·6	(250·3)
2962·0	3215·7	3445·0	3672·8	3904·1	4140·0	
6·3642	6·7733	7·0909	7·3680	7·6187	7·8495	
45·301	57·791	68·494	78·616	88·446	98·093	50
2699·0	2909·3	3091·2	3271·4	3455·7	3644·8	(263·9)
2925·5	3198·3	3433·7	3664·5	3897·9	4135·3	
6·2105	6·6508	6·9770	7·2578	7·5108	7·7431	
36·145	47·379	56·592	65·184	73·478	81·587	60
2668·1	2895·8	3082·6	3265·1	3450·8	3641·2	(275·6)
2885·0	3180·1	3422·2	3656·2	3891·7	4130·7	
6·0692	6·5462	6·8818	7·1664	7·4217	7·6554	
29·457	39·922	48·086	55·590	62·787	69·798	70
2633·2	2881·7	3074·0	3258·8	3445·9	3637·4	(285·8)
2839·4	3161·2	3410·6	3647·9	3885·4	4126·0	
5·9327	6·4536	6·7993	7·0880	7·3456	7·5808	
24·264	34·310	41·704	48·394	54·770	60·956	80
2592·7	2867·1	3065·2	3252·3	3441·0	3633·7	(295·0)
2736·8	3141·6	3398·8	3639·5	3879·2	4121·3	
5·7942	6·3694	6·7262	7·0191	7·2790	7·5158	
	29·929	36·737	42·798	48·534	54·030	90
	2851·8	3056·2	3245·9	3436·2	3630·0	(303·3)
	3121·2	3386·8	3631·1	3873·0	4116·7	
	6·2915	6·6600	6·9574	7·2196	7·4579	

Superheated steam, to 220 bar and 800°C

p (t_S)	t:	t_S	400	500	600	700	800
100	v^*	18·041	26·408	32·760	38·320	43·546	48·580
(311·0)	u	2547·3	2835·8	3047·0	3239·5	3431·3	3626·2
	h	2727·7	3099·9	3374·6	3622·7	3866·8	4112·0
	s	5·6198	6·2182	6·5994	6·9013	7·1660	7·4058
110	v^*	16·007	23·512	29·503	34·656	39·466	44·081
(318·0)	u	2533·2	2819·2	3037·7	3233·0	3426·4	3622·4
	h	2709·3	3077·8	3362·2	3614·2	3860·5	4107·3
	s	5·5596	6·1483	6·5432	6·8499	7·1170	7·3584
120	v^*	14·285	21·084	26·786	31·603	36·066	40·332
(324·6)	u	2517·8	2801·8	3028·2	3226·5	3421·5	3618·7
	h	2689·2	3054·8	3349·6	3605·7	3854·3	4102·7
	s	5·5003	6·0810	6·4906	6·8022	7·0718	7·3147
130	v^*	12·800	19·015	24·485	29·019	33·189	37·160
(330·8)	u	2500·6	2783·5	3018·5	3219·9	3416·5	3614·9
	h	2667·0	3030·7	3336·8	3597·1	3848·0	4098·0
	s	5·4409	6·0155	6·4409	6·7577	7·0298	7·2743
140	v^*	11·498	17·227	22·509	26·804	30·723	34·441
(336·6)	u	2481·4	2764·4	3008·7	3213·2	3411·6	3611·1
	h	2642·4	3005·6	3323·8	3588·5	3841·7	4093·3
	s	5·3804	5·9513	6·3937	6·7159	6·9906	7·2367
150	v^*	10·343	15·661	20·795	24·834	28·587	32·086
(342·1)	u	2460·0	2744·2	2998·7	3206·5	3406·6	3607·3
	h	2615·1	2979·1	3310·6	3579·8	3835·4	4088·6
	s	5·3180	5·8876	6·3487	6·6764	6·9536	7·2013
160	v^*	9·3099	14·275	19·293	23·204	26·717	30·025
(347·3)	u	2435·9	2722·9	2988·4	3199·7	3401·6	3603·6
	h	2584·9	2951·3	3297·1	3571·0	3829·1	4084·0
	s	5·2533	5·8240	6·3054	6·6389	6·9188	7·1681
170	v^*	8·3721	13·034	17·966	21·721	25·068	28·207
(352·3)	u	2409·3	2700·1	2978·1	3192·9	3396·6	3599·8
	h	2551·6	2921·7	3283·5	3562·2	3822·8	4079·3
	s	5·1856	5·7599	6·2636	6·6031	6·8857	7·1366
180	v^*	7·4973	11·913	16·785	20·403	23·603	26·591
(357·0)	u	2378·9	2675·9	2967·5	3186·1	3391·6	3596·0
	h	2513·9	2890·3	3269·6	3553·4	3816·5	4074·6
	s	5·1127	5·6947	6·2232	6·5688	6·8542	7·1067
190	v^*	6·6759	10·889	15·726	19·223	22·291	25·146
(361·4)	u	2343·7	2649·8	2956·6	3179·3	3386·7	3592·2
	h	2470·5	2856·7	3255·4	3544·5	3810·2	4070·0
	s	5·0330	5·6278	6·1839	6·5360	6·8241	7·0783
200	v^*	5·8745	9·9470	14·771	18·161	21·111	23·845
(365·7)	u	2300·7	2621·6	2945·7	3172·3	3381·6	3588·4
	h	2418·2	2820·5	3241·1	3535·5	3803·8	4065·3
	s	4·9410	5·5585	6·1456	6·5043	6·7953	7·0511
210	v^*	5·0225	9·0714	13·907	17·201	20·044	22·669
(369·8)	u	2242·0	2590·8	2934·5	3165·3	3376·6	3584·6
	h	2347·5	2781·3	3226·5	3526·5	3797·5	4060·6
	s	4·8222	5·4863	6·1082	6·4737	6·7677	7·0251
220	v^*	3·7347	8·2510	13·119	16·327	19·074	21·599
(373·7)	u	2114·4	2557·3	2923·1	3158·2	3371·5	3580·7
	h	2196·6	2738·8	3211·7	3517·4	3791·1	4055·9
	s	4·5814	5·4102	6·0716	6·4441	6·7410	7·0001

* In this Table specific volumes are in dm³/kg (l/kg).

Supercritical steam, to 1000 bar and 800°C

p	t:	400	500	600	700	800
240	v^*	6·7392	11·737	14·798	17·377	19·729
	h	2641·2	3181·4	3499·1	3778·3	4046·6
	s	5·2430	6·0003	6·3876	6·6905	6·9529
260	v^*	5·2812	10·565	13·505	15·941	18·147
	h	2511·7	3150·2	3480·6	3765·5	4037·2
	s	5·0326	5·9311	6·3340	6·6431	6·9090
280	v^*	3·8219	9·5566	12·397	14·712	16·792
	h	2330·7	3118·1	3461·9	3752·6	4027·8
	s	4·7503	5·8636	6·2829	6·5934	6·8677
300	v^*	2·8306	8·6808	11·436	13·647	15·619
	h	2161·8	3085·0	3443·0	3739·7	4018·5
	s	4·4896	5·7972	6·2340	6·5560	6·8288
350	v^*	2·1108	6·9253	9·5194	11·520	13·275
	h	1993·1	2998·3	3395·1	3707·3	3995·1
	s	4·2214	5·6349	6·1194	6·4584	6·7400
400	v^*	1·9091	5·6156	8·0884	9·9302	11,521
	h	1934·1	2906·8	3346·4	3674·8	3971·7
	s	4·1190	5·4762	6·0135	6·3701	6·6606
450	v^*	1·8013	4·6249	6·9842	8·6988	10·160
	h	1900·6	2813·5	3297·4	3642·4	3948·4
	s	4·0554	5·3226	5·9143	6·2890	6·5885
500	v^*	1·7291	3·8822	6·1113	7·7197	9·0759
	h	1877·7	2723·0	3248·3	3610·2	3925·3
	s	4·0083	5·1782	5·8207	6·2138	6·5222
600	v^*	1·6324	2·9515	4·8350	6·2690	7·4603
	h	1847·3	2570·6	3151·6	3547·0	3879·6
	s	3·9383	4·9374	5·6477	6·0775	6·4031
700	v^*	1·5671	2·4668	3·9719	5·2566	6·3208
	h	1827·8	2467·1	3060·4	3486·3	3835·3
	s	3·8855	4·7688	5·4931	5·9562	6·2979
800	v^*	1·5180	2·1881	3·3792	4·5193	5·4805
	h	1814·2	2397·4	2980·3	3428·7	3792·8
	s	3·8425	4·6488	5·3595	5·8470	6·2034
900	v^*	1·4788	2·0129	2·9668	3·9642	4·8407
	h	1804·6	2349·9	2913·5	3374·6	3752·4
	s	3·8059	4·5602	5·2468	5·7479	6·1179
1000	v^*	1·4464	1·8934	2·6681	3·5356	4·3411
	h	1797·6	2316·1	2857·5	3324·4	3714·3
	s	3·7738	4·4913	5·1505	5·6579	6·0397

* In this Table specific volumes are in dm^3/kg (l/kg).

Saturated water and steam

t_s	v_f	v_g	c_{pf}	c_{pg}	k_f	k_g	μ_f	μ_g	Pr_f	Pr_g	t_s
0.01	0.00100	206.2	4.217	1.854	0.569	0.0173	1.755	0.0088	13.02	0.942	0.01 *
10	0.00100	106.4	4.193	1.860	0.587	0.0185	1.301	0.0091	9.29	0.915	10
20	0.00100	57.8	4.182	1.866	0.603	0.0191	1.002	0.0094	6.95	0.918	20
30	0.00100	32.9	4.179	1.875	0.618	0.0198	0.797	0.0097	5.39	0.923	30
40	0.00101	19.5	4.179	1.885	0.632	0.0204	0.651	0.0101	4.31	0.930	40
50	0.00101	12.05	4.181	1.899	0.643	0.0210	0.544	0.0104	3.53	0.939	50
60	0.00102	7.68	4.185	1.915	0.653	0.0217	0.462	0.0107	2.96	0.947	60
70	0.00102	5.05	4.190	1.936	0.662	0.0224	0.400	0.0111	2.53	0.956	70
80	0.00103	3.41	4.197	1.962	0.670	0.0231	0.350	0.0114	2.19	0.966	80
90	0.00104	2.36	4.205	1.992	0.676	0.0240	0.311	0.0117	1.93	0.976	90
100	0.00104	1.673	4.216	2.028	0.681	0.0249	0.278	0.0121	1.723	0.986	100
125	0.00107	0.770	4.254	2.147	0.687	0.0272	0.219	0.0133	1.358	1.047	125
150	0.00109	0.392	4.310	2.314	0.687	0.0300	0.180	0.0144	1.133	1.110	150
175	0.00112	0.217	4.389	2.542	0.679	0.0334	0.153	0.0156	0.990	1.185	175
200	0.00116	0.127	4.497	2.843	0.665	0.0375	0.133	0.0167	0.902	1.270	200
225	0.00120	0.0783	4.648	3.238	0.644	0.0427	0.1182	0.0179	0.853	1.36	225
250	0.00125	0.0500	4.867	3.772	0.616	0.0495	0.1065	0.0191	0.841	1.45	250
275	0.00132	0.0327	5.202	4.561	0.582	0.0587	0.0972	0.0202	0.869	1.56	275
300	0.00140	0.0216	5.762	5.863	0.541	0.0719	0.0897	0.0214	0.955	1.74	300
325	0.00153	0.0142	6.861	8.440	0.493	0.0929	0.0790	0.0230	1.100	2.09	325
350	0.00174	0.00880	10.10	17.15	0.437	0.1343	0.0648	0.0258	1.50	3.29	350
360	0.00190	0.00694	14.6	25.1	0.400	0.168	0.0582	0.0275	2.11	3.89	360
374.15	0.00317	0.00317	∞	∞	0.24	0.24	0.045	0.045	∞	∞	374.15 †

* Triple point.

† Critical point.

Ammonia, NH_3

t_s	p*	v_f	v_g	h_f	h_g	s_f	s_g	By 50 K		By 100 K	
				Saturated				h	s	h	s
−40	0.0718	0.00145	1.552	zero	1390	zero	5.963	1499	6.387	1606	6.736
−35	0.0932	0.00146	1.216	22.3	1398	0.095	5.872	1508	6.292	1616	6.639
−30	0.1196	0.00148	0.963	44.7	1406	0.188	5.785	1517	6.203	1626	6.547
−25	0.1516	0.00149	0.772	67.2	1413	0.279	5.703	1526	6.119	1636	6.461
−20	0.190	0.00150	0.624	89.8	1420	0.368	5.624	1535	6.039	1646	6.379
−15	0.236	0.00152	0.509	112.3	1426	0.457	5.549	1543	5.963	1656	6.301
−10	0.291	0.00153	0.418	135.4	1433	0.544	5.477	1552	5.891	1665	6.227
−5	0.355	0.00155	0.347	158.2	1439	0.630	5.407	1560	5.822	1675	6.157
0	0.429	0.00157	0.289	181.2	1444	0.715	5.340	1568	5.756	1684	6.090
5	0.516	0.00158	0.243	204.5	1450	0.799	5.276	1576	5.694	1693	6.027
10	0.615	0.00160	0.206	227.7	1454	0.881	5.214	1583	5.634	1702	5.967
15	0.728	0.00162	0.175	251.4	1459	0.963	5.154	1590	5.576	1711	5.909
20	0.857	0.00164	0.149	275.2	1463	1.044	5.095	1597	5.521	1719	5.853
25	1.001	0.00166	0.128	298.9	1466	1.124	5.039	1604	5.468	1728	5.800
30	1.167	0.00168	0.111	323.1	1469	1.204	4.984	1610	5.417	1736	5.750
35	1.350	0.00170	0.096	347.5	1471	1.282	4.930	1616	5.368	1744	5.702
40	1.554	0.00173	0.083	371.5	1473	1.360	4.877	1622	5.321	1752	5.655
45	1.782	0.00175	0.073	396.8	1474	1.437	4.825	1628	5.275	1760	5.610
50	2.033	0.00178	0.063	421.9	1475	1.515	4.773	1633	5.230	1767	5.567

Freezing point at 1 atm $= -77.7\ °C$
Critical point: Celsius temp. $= 132.4\ °C$
pressure $= 11.30\ MN/m^2\ (113.0\ bar)$

* Pressures in this Table are in MN/m^2 (10 bar).

Dichlorofluoromethane (Freon-12) CCl_2F_2

t_s	p^*	Saturated						Superheated By 20 K		By 40 K	
		v_f	v_g	h_f	h_g	s_f	s_g	h	s	h	s
−40	0.0641	0.00066	0.2421	zero	169.6	zero	0.7274	180.8	0.7737	192.4	0.8178
−35	0.0806	0.00067	0.1955	4.4	171.9	0.0187	0.7220	183.3	0.7681	195.1	0.8120
−30	0.1003	0.00067	0.1595	8.9	174.2	0.0371	0.7171	185.8	0.7631	197.8	0.8068
−25	0.1236	0.00068	0.1313	13.3	176.5	0.0552	0.7127	188.3	0.7586	200.4	0.8021
−20	0.1508	0.00069	0.1089	17.8	178.7	0.0731	0.7088	190.8	0.7546	203.1	0.7979
−15	0.1825	0.00069	0.0911	22.3	181.0	0.0906	0.7052	193.2	0.7510	205.7	0.7942
−10	0.219	0.00070	0.0767	26.9	183.2	0.1080	0.7020	195.7	0.7477	208.3	0.7909
−5	0.261	0.00071	0.0650	31.4	185.4	0.1251	0.6991	198.1	0.7449	210.9	0.7879
0	0.308	0.00072	0.0554	36.1	187.5	0.1420	0.6966	200.5	0.7423	213.5	0.7853
5	0.362	0.00072	0.0475	40.7	189.7	0.1587	0.6942	202.9	0.7401	216.1	0.7830
10	0.423	0.00073	0.0409	45.4	191.7	0.1752	0.6921	205.2	0.7381	218.6	0.7810
15	0.491	0.00074	0.0354	50.1	193.8	0.1915	0.6902	207.5	0.7363	221.2	0.7792
20	0.567	0.00075	0.0308	54.9	195.8	0.2078	0.6885	209.8	0.7348	223.7	0.7777
25	0.651	0.00076	0.0269	59.7	197.7	0.2239	0.6869	212.1	0.7334	226.1	0.7763
30	0.745	0.00077	0.0235	64.6	199.6	0.2399	0.6854	214.3	0.7321	228.6	0.7751
35	0.847	0.00079	0.0206	69.5	201.5	0.2559	0.6839	216.4	0.7310	231.0	0.7741
40	0.960	0.00080	0.0182	74.6	203.2	0.2718	0.6825	218.5	0.7300	233.4	0.7732
45	1.084	0.00081	0.0160	79.7	204.9	0.2877	0.6812	220.6	0.7291	235.7	0.7724
50	1.219	0.00083	0.0142	84.9	206.5	0.3037	0.6797	222.6	0.7282	238.0	0.7718

Freezing point at 1 atm   = − 155.0 °C
Critical point: Celsius temp. =   112.0 °C
pressure   =   4.115 MN/m² (41.15 bar)

* Pressures in this table are in MN/m² (10 bar).

Carbon dioxide, CO_2

t_s	p^*	Saturated						Superheated By 30 K		By 60 K	
		v_f	v_g	h_f	h_g	s_f	s_g	h	s	h	s
−40	1.005	0.00090	0.0382	zero	321.1	zero	1.377	355.4	1.507	383.0	1.611
−35	1.20	0.00091	0.0320	9.7	322.2	0.039	1.352	356.9	1.485	385.6	1.588
−30	1.43	0.00093	0.0270	19.5	323.1	0.079	1.328	358.7	1.464	388.0	1.566
−25	1.68	0.00095	0.0229	29.5	323.7	0.119	1.304	360.4	1.442	390.3	1.545
−20	1.97	0.00097	0.0195	39.7	323.7	0.158	1.280	361.8	1.421	392.5	1.525
−15	2.29	0.00099	0.0166	50.2	323.2	0.198	1.256	363.0	1.401	394.5	1.505
−10	2.65	0.00102	0.0142	60.9	322.3	0.238	1.231	363.9	1.381	396.2	1.486
−5	3.04	0.00105	0.0122	72.0	320.5	0.278	1.205	364.6	1.361	397.8	1.467
0	3.48	0.00108	0.0104	83.7	318.1	0.320	1.178	364.9	1.342	399.3	1.449
5	3.97	0.00111	0.00879	96.0	312.9	0.364	1.143	364.9	1.322	400.4	1.431
10	4.50	0.00116	0.00743	109.1	307.2	0.407	1.107	364.7	1.302	401.4	1.414
15	5.08	0.00121	0.00623	123.3	301.0	0.454	1.071	364.0	1.282	402.2	1.396
20	5.73	0.00129	0.00516	139.1	292.3	0.506	1.028	362.9	1.261	402.7	1.379
25	6.44	0.00140	0.00413	159.7	279.9	0.573	0.976	361.5	1.241	403.0	1.362
30	7.21	0.00169	0.00294	191.2	253.1	0.682	0.886	359.6	1.220	402.9	1.345
† 31.05	7.38	0.00214	0.00214	223.0	223.0	0.780	0.780	359.1	1.216	402.9	1.341

Freezing point at 1 atm = − 56.6 °C

* Pressures in this Table are in MN/m² (10 bar).
† Critical point.

Air at atmospheric pressure

t	v	c_p	k	μ	Pr	t
−100	0.488	1.01	0.016	0.012	0.75	−100
0	0.773	1.01	0.024	0.017	0.72	0
100	1.057	1.02	0.032	0.022	0.70	100
200	1.341	1.03	0.039	0.026	0.69	200
300	1.624	1.05	0.045	0.030	0.69	300
400	1.908	1.07	0.051	0.033	0.70	400
500	2.191	1.10	0.056	0.036	0.70	500
600	2.473	1.12	0.061	0.039	0.71	600
700	2.756	1.14	0.066	0.042	0.72	700
800	3.039	1.16	0.071	0.044	0.73	800

This Table may be used with reasonable accuracy for values of c_p, k, μ and Pr of N_2, O_2 and CO.

International Standard Atmosphere

ρ_0 = density at sea level = $1 \cdot 225 \ \text{kg/m}^3$

z	p	T	ρ/ρ_0	v	k	λ	a
−2·5	1·352	304·4	1·263	12·07	0·0266	0·053	349·8
−2·0	1·278	301·2	1·207	12·53	0·0264	0·055	347·9
−1·5	1·207	297·9	1·152	13·01	0·0261	0·058	346·0
−1·0	1·139	294·7	1·100	13·52	0·0259	0·060	344·1
−0·5	1·075	291·4	1·049	14·05	0·0256	0·063	342·2
0	1·013	288·2	1·000	14·61	0·0253	0·066	340·3
0·5	0·955	284·9	0·953	15·20	0·0251	0·070	338·4
1·0	0·899	281·7	0·907	15·81	0·0248	0·073	336·4
1·5	0·846	278·4	0·864	16·46	0·0246	0·077	334·5
2·0	0·795	275·2	0·822	17·15	0·0243	0·081	332·5
2·5	0·747	271·9	0·781	17·87	0·0241	0·085	330·6
3·0	0·701	268·7	0·742	18·63	0·0238	0·089	328·6
3·5	0·658	265·4	0·705	19·43	0·0235	0·094	326·6
4·0	0·617	262·2	0·669	20·28	0·0233	0·099	324·6
4·5	0·578	258·9	0·634	21·17	0·0230	0·105	322·6
5·0	0·540	255·7	0·601	22·11	0·0228	0·110	320·5
6·0	0·472	249·2	0·539	24·16	0·0222	0·123	316·5
7·0	0·411	242·7	0·482	26·46	0·0217	0·138	312·3
8·0	0·357	236·2	0·429	29·04	0·0212	0·155	308·1
9·0	0·308	229·7	0·381	31·96	0·0206	0·174	303·8
10·0	0·265	223·3	0·338	35·25	0·0201	0·196	299·5
11	0·227	216·8	0·298	38·99	0·0195	0·223	295·2
12	0·194	216·7	0·255	45·57	0·0195	0·260	295·1
13	0·166	216·7	0·218	53·33	0·0195	0·305	295·1
14	0·142	216·7	0·186	62·39	0·0195	0·357	295·1
15	0·121	216·7	0·159	73·00	0·0195	0·417	295·1
16	0·104	216·7	0·136	85·40	0·0195	0·488	295·1
17	0·088	216·7	0·116	99·90	0·0195	0·571	295·1
18	0·076	216·7	0·099	116·9	0·0195	0·668	295·1
19	0·065	216·7	0·085	136·7	0·0195	0·781	295·1
20	0·055	216·7	0·073	159·9	0·0195	0·914	295·1
21	0·047	217·6	0·062	188·4	0·0196	1·073	295·7
22	0·040	218·6	0·053	222·0	0·0197	1·260	296·4
23	0·035	219·6	0·045	261·4	0·0198	1·477	297·0
24	0·030	220·6	0·038	307·4	0·0199	1·731	297·7
25	0·025	221·6	0·033	361·4	0·0199	2·027	298·4
26	0·022	222·5	0·028	424·4	0·0200	2·372	299·1
27	0·019	223·5	0·024	498·1	0·0201	2·773	299·7
28	0·016	224·5	0·020	584·1	0·0202	3·240	300·4
29	0·014	225·5	0·018	684·4	0·0203	3·783	301·0
30	0·012	226·5	0·015	801·3	0·0203	4·413	301·7

At sea level the normal composition of clean dry air, by volume and (weight) percent, is:

Nitrogen	78·08 (75·52)	Carbon dioxide	0·03 (0·05)	Methane	0·0002 (0·0001)	Nitrous oxide	5×10^{-5} (8×10^{-5})
Oxygen	20·95 (23·14)	Neon	0·002 (0·001)	Krypton	0·0001 (0·0003)	Xenon	9×10^{-6} (4×10^{-5})
Argon	0·93 (1.28)	Helium	0·0005 (7×10^{-5})	Hydrogen	5×10^{-5} (3×10^{-6})		

The approximate composition may be taken to be: nitrogen/oxygen 79/21 (77/23).
The average molecular weight of dry air is 28·96, and its gas constant $R = 0 \cdot 2871 \ \text{kJ/kg K}$.

Properties of gases

ρ	Mass density at s.t.p.* (kg/m^3)
t_f, t_b	Freezing, boiling points at 1 atm (°C)
t_c, p_c, ρ_c	Critical temperature, pressure, density (°C, bar, kg/m^3)
k	Thermal conductivity at s.t.p. (mW/m K)
c_p	Specific heat capacity at constant pressure, for s.t.p. (kJ/kg K)
γ	Ratio of specific heat capacities
β	Coefficient of volume expansion at constant pressure, for 1 atm, 0–100°C (10^{-3}/K)
μ	Dynamic viscosity at 1 atm, 20°C (μNs/m^2)
λ	Mean free path at s.t.p. (nm)
ϵ_r	Dielectric constant at s.t.p. 10 GHz

* Standard temperature and pressure: 0°C, 1 atm.

	ρ	t_f	t_b	t_c	p_c	ρ_c	k	c_p	γ	β	μ	λ	ϵ_r
Air	1·293			−141	37·7		24·1	1·004	1·40	3·67	18·1		1·00058
Oxygen	1·429	−219	−183	−119	50·4	430	24·4	0·915	1·40	~3·7	20·0	63	1·00053
Nitrogen	1·250	−210	−196	−147	33·9	311	24·3	1·039	1·40	3·67	17·4	59	1·00059
Hydrogen	0·090	−259	−253	−240	13·0	31	168·4	14·2	1·41	3·659*	8·8	111	1·00026
Carbon Dioxide	1·977	−57		31	74·0	460	14·5	0·819	1·30	3·72	14·6	40	1·00099
Helium	0·178		−269	−268	2·3	69	141·5	5·19	1·63	3·658*	19·4	174	1·00070
Neon	0·900	−249	−246	−229	27·2	484	46·5		1·64	3·660*	31·0	124	1·00127
Argon	1·784	−189	−186	−122	48·6	531	16·2	0·520	1·67		22·2	63	1·00056

* At 1·3 atm.

Thermochemical data for equilibrium reactions

Stoichiometric equations

$\sum \nu_i A_i = 0$, where ν_i is the *stoichiometric coefficient* of the substance whose *chemical symbol* is A_i.

(1) $-2H + H_2 = 0$

(2) $-2N + N_2 = 0$

(3) $-2O + O_2 = 0$

(4) $-2NO + N_2 + O_2 = 0$

(5) $-H_2 - \frac{1}{2}O_2 + H_2O = 0$

(6) $-\frac{1}{2}H_2 - OH + H_2O = 0$

(7) $-CO - \frac{1}{2}O_2 + CO_2 = 0$

(8) $-CO - H_2O + CO_2 + H_2 = 0$

(9) $-\frac{1}{2}N_2 - \frac{3}{2}H_2 + NH_3 = 0$

Standard enthalpy of reaction

$\Delta H_T^\circ = \sum_i \nu_i [\bar{h}_i]_T^\circ$ where $[h_i]_T^\circ$ = enthalpy per kmol of substance A_i, at 1 atm pressure and absolute temperature T.

Warning: This table lists *absolute* temperatures.

Temp. K	Reaction number				ΔH_T°/MJ					Temp. K
	1	2	3	4	5	6	7	8	9	
200	−434.7	−944.1	−496.9	−180.4	−240.9	−280.2	−282.1	−41.21	−43.71	200
298	−436.0	−945.3	−498.4	−180.6	−241.8	−281.3	−283.0	−41.17	−45.90	298
400	−437.3	−946.6	−499.8	−180.7	−242.8	−282.4	−283.5	−40.63	−48.04	400
600	−439.7	−948.9	−502.1	−180.7	−244.8	−284.1	−283.6	−38.88	−51.39	600
800	−442.1	−951.1	−503.9	−180.8	−246.5	−285.5	−283.3	−36.82	−53.66	800
1000	−444.5	−953.0	−505.4	−180.9	−247.9	−286.6	−282.6	−34.74	−55.07	1000

Standard enthalpy of reaction—*continued*

Temp. K	1	2	3	4	5 $\Delta H_T^\circ/\text{MJ}$	6	7	8	9	Temp. K
1200	−446.7	−954.7	−506.7	−180.9	−249.0	−287.4	−281.8	−32.79	−55.83	1200
1400	−448.7	−956.1	−507.8	−181.0	−249.9	−287.9	−280.9	−30.98	−56.07	1400
1600	−450.6	−957.5	−508.9	−181.0	−250.6	−288.4	−279.9	−29.29	−55.99	1600
1800	−452.3	−958.7	−509.8	−181.0	−251.2	−288.6	−278.9	−27.71	−55.66	1800
2000	−453.8	−959.9	−510.6	−181.0	−251.7	−288.8	−277.9	−26.22	−55.19	2000
2200	−455.2	−961.0	−511.4	−180.8	−252.1	−288.9	−276.8	−24.79	−54.61	2200
2400	−456.4	−962.1	−512.0	−180.7	−252.4	−289.0	−275.8	−23.41	−53.92	2400
2600	−457.6	−963.1	−512.5	−180.4	−252.7	−289.0	−274.8	−22.07	−53.12	2600
2800	−458.6	−964.1	−513.0	−180.1	−253.0	−288.9	−273.7	−20.77	−52.22	2800
3000	−459.6	−965.0	−513.4	−179.7	−253.3	−288.9	−272.7	−19.49	−51.20	3000
3200	−460.4	−966.0	−513.8	−179.3	−253.5	−288.8	−271.7	−18.19	−50.10	3200
3400	−461.2	−967.0	−514.1	−178.7	−253.8	−288.7	−270.7	−16.91	−48.94	3400
3600	−461.9	−968.1	−514.4	−178.2	−254.1	−288.6	−269.8	−15.62	−47.75	3600
3800	−462.5	−969.2	−514.6	−177.6	−254.5	−288.5	−268.8	−14.33	−46.49	3800
4000	−463.0	−970.4	−514.8	−176.9	−254.8	−288.4	−267.8	−13.00	−45.19	4000
4500	−464.0	−973.8	−515.3	−175.2	−255.9	−288.1	−265.5	− 9.57	−41.68	4500
5000	−464.6	−977.9	−515.9	−173.2	−257.2	−288.0	−263.1	− 5.95	−37.79	5000
5500	−464.8	−982.9	−516.5	−171.1	−258.6	−287.9	−260.7	− 2.10	−33.56	5500
6000	−464.7	−989.0	−517.2	−169.0	−260.3	−287.9	−258.2	2.01	−28.98	6000

Equilibrium constants

$$\log_{10} K_p = \sum_i \nu_i \log_{10} p_i', \quad \text{where } p_i' = \text{partial pressure of substance } A_i, \text{ in atmospheres (atm)}.$$

Warning: This table lists *absolute* temperatures.

Temp. K	1	2	3	4	5 $\log_{10} K_p$	6	7	8	9	Temp. K
200	108.644	240.810	123.984	45.858	60.792	70.267	69.356	8.564	6.708	200
298	71.224	159.600	81.208	30.342	40.048	46.137	45.066	5.018	2.869	298
400	51.752	117.408	58.946	22.284	29.240	33.567	32.431	3.191	0.778	400
600	32.672	76.162	37.148	14.420	18.633	21.242	20.087	1.454	−1.380	600
800	23.078	55.488	26.202	10.486	13.289	15.044	13.916	0.627	−2.523	800
1000	17.292	43.056	19.614	8.124	10.062	11.309	10.221	0.159	−3.233	1000
1200	13.414	34.754	15.208	6.550	7.899	8.811	7.764	−0.135	−3.716	1200
1400	10.630	28.812	12.054	5.424	6.347	7.021	6.014	−0.333	−4.064	1400
1600	8.532	24.350	9.684	4.580	5.180	5.677	4.706	−0.474	−4.325	1600
1800	6.896	20.874	7.836	3.924	4.270	4.631	3.693	−0.577	−4.528	1800
2000	5.580	18.092	6.356	3.398	3.540	3.793	2.884	−0.656	−4.689	2000
2200	4.502	15.810	5.142	2.968	2.942	3.107	2.226	−0.716	−4.819	2200
2400	3.600	13.908	4.130	2.610	2.443	2.535	1.679	−0.764	−4.927	2400
2600	2.834	12.298	3.272	2.308	2.021	2.052	1.219	−0.802	−5.016	2600
2800	2.178	10.914	2.536	2.050	1.658	1.637	0.825	−0.833	−5.092	2800
3000	1.606	9.716	1.898	1.826	1.343	1.278	0.485	−0.858	−5.156	3000
3200	1.106	8.664	1.340	1.630	1.067	0.963	0.189	−0.878	−5.211	3200
3400	0.664	7.736	0.846	1.458	0.824	0.687	−0.071	−0.895	−5.259	3400
3600	0.270	6.910	0.408	1.306	0.607	0.440	−0.302	−0.909	−5.300	3600
3800	−0.084	6.172	0.014	1.170	0.413	0.220	−0.508	−0.921	−5.336	3800
4000	−0.402	5.504	−0.340	1.048	0.238	0.022	−0.692	−0.930	−5.368	4000
4500	−1.074	4.094	−1.086	0.794	−0.133	−0.397	−1.079	−0.946	−5.431	4500
5000	−1.612	2.962	−1.686	0.592	−0.430	−0.731	−1.386	−0.956	−5.477	5000
5500	−2.054	2.032	−2.176	0.428	−0.675	−1.004	−1.635	−0.960	−5.511	5500
6000	−2.422	1.250	−2.584	0.294	−0.880	−1.232	−1.841	−0.961	−5.536	6000

Standard free enthalpy of reaction

At a given temperature, the standard free enthalpy of reaction ΔG_T° (or *standard Gibbs function change*) may be calculated from the listed value of $\log_{10} K_p$ by the following equation:

$$\Delta G_T^\circ = -R_0 T \log_e K_p$$
$$= -0 \cdot 01914 \, T \log_{10} K_p \ \text{MJ}.$$

Thermodynamics and fluid mechanics

In the following, T is absolute temperature, Q heat input per unit mass, W work output per unit mass and V velocity; other symbols are as in Thermodynamic Properties (page 47) or are defined below.

Thermodynamic relations

Basic relations

Enthalpy: $h = u + pv$

Helmholtz function: $f = u - Ts$

Gibbs function: $g = h - Ts$

$\mathrm{d}h = T\,\mathrm{d}s + v\,\mathrm{d}p$

$\mathrm{d}f = -p\,\mathrm{d}v - s\,\mathrm{d}T$

$\mathrm{d}g = v\,\mathrm{d}p - s\,\mathrm{d}T$

Maxwell's relations

$$\left(\frac{\partial T}{\partial v}\right)_s = -\left(\frac{\partial p}{\partial s}\right)_v$$

$$\left(\frac{\partial T}{\partial p}\right)_s = \left(\frac{\partial v}{\partial s}\right)_p$$

$$\left(\frac{\partial p}{\partial T}\right)_v = \left(\frac{\partial s}{\partial v}\right)_T$$

$$\left(\frac{\partial v}{\partial T}\right)_p = -\left(\frac{\partial s}{\partial p}\right)_T.$$

Specific heats

$$c_\mathrm{p} = \left(\frac{\partial h}{\partial T}\right)_p \; ; \; c_\mathrm{v} = \left(\frac{\partial u}{\partial T}\right)_v$$

Coefficients

Volume expansion $\beta = \dfrac{1}{v}\left(\dfrac{\partial v}{\partial T}\right)_p$

Compressibility $\kappa = -\dfrac{1}{v}\left(\dfrac{\partial v}{\partial p}\right)_T$

Joule-Thomson $\mu = \left(\dfrac{\partial T}{\partial p}\right)_h$

(for a perfect gas $\mu = 0$)

$$c_\mathrm{p} - c_\mathrm{v} = \beta^2 \frac{Tv}{\kappa}$$

(for a perfect gas $c_\mathrm{p} - c_\mathrm{v} = R$).

Equations of state

Perfect gas: $pv = RT = \dfrac{R_0 T}{M}$

for a gas of molecular weight M with gas constant R.

Van der Waals' gas: $(p + a/v^2)(v - b) = RT$

where a, b are constants of the gas.

Process relations

Reversible polytropic, closed system

$pv^n = \text{constant}$

$$W = \frac{p_2 v_2 - p_1 v_1}{1 - n} \quad (n \neq 1)$$

For a perfect gas,

$$W = \frac{R}{1 - n}\,(T_2 - T_1)$$

$$Q = \left(c_\mathrm{v} + \frac{R}{1 - n}\right)(T_2 - T_1)$$

$$\frac{T_2}{T_1} = \left(\frac{p_2}{p_1}\right)^{(n-1)/n}$$

and, if adiabatic ($Q = 0$), $n = \gamma$.

Reversible isothermal, closed system

$$Q = T(s_2 - s_1)$$

(for a perfect gas, $pv = \text{constant}$ and

$$Q = W = RT \ln (v_2/v_1)).$$

Steady flow

In terms of enthalpy, the energy equation relating any two sections of a steady flow at which properties are uniform is

$$Q - W = h_2 - h_1 + \tfrac{1}{2}(V^2_2 - V^2_1) + g(z_2 - z_1)$$

where g is the acceleration of gravity and W is shaft work.

Nozzle flow:

For an expansion in which pv^n is constant the critical pressure ratio is

$$\frac{p_2}{p_1} = \left(\frac{2}{n+1}\right)^{n/(n-1)}$$

which gives the mass flux

$$\frac{m}{A_2} = \sqrt{\left\{n\left(\frac{2}{n+1}\right)^{(n+1)/(n-1)}\frac{p_1}{v_1}\right\}} \; .$$

Pressure and area are related by

$$\frac{dA}{dp} = vA\left(\frac{1}{V^2} - \frac{1}{a^2}\right)$$

where a is the sonic velocity, given by

$$a^2 = -v^2\left(\frac{\partial p}{\partial v}\right)_s = \left(\frac{\partial p}{\partial \rho}\right)_s$$

For *any process* the entropy change of a perfect gas is

$$s_2 - s_1 = c_v \ln\frac{p_2}{p_1} + c_p \ln\frac{v_2}{v_1}$$

$$= c_v \ln\frac{T_2}{T_1} + R \ln\frac{v_2}{v_1}$$

$$= c_p \ln\frac{T_2}{T_1} - R \ln\frac{p_2}{p_1}$$

Equations for fluid flow

Continuity

$$\frac{\partial \rho}{\partial t} = -\operatorname{div}(\rho\mathbf{V})$$

(for an incompressible fluid, div $\mathbf{V} = 0$)

Momentum

$$\mathbf{F} = \frac{\partial}{\partial t}\int_\tau \rho\mathbf{V}\,d\tau + \int_A \rho\mathbf{V}(\mathbf{V}.d\mathbf{A})$$

where $\mathbf{F}$ is the net force on a control volume τ bounded by a surface A. For steady flow

$$\mathbf{F} = \int_A \rho\mathbf{V}(\mathbf{V}.d\mathbf{A})$$

Energy

$$\dot{q} + \operatorname{div}(k\operatorname{grad} T) = \rho\left(\frac{\partial u}{\partial t} + \mathbf{V}.\operatorname{grad} u\right) + p\operatorname{div}\mathbf{V}$$

at any point; terms for dissipation and radiation may be included in $\dot{q}$, the rate of heat production per unit volume. The integrated form for a control volume τ bounded by a surface A is

$$\dot{q} - \dot{w} = \frac{\partial}{\partial t}\int_\tau (u + \tfrac{1}{2}V^2 + gz)\rho\,d\tau +$$

$$\int_A \rho(u + \tfrac{1}{2}V^2 + gz + \tfrac{p}{\rho})\mathbf{V}.d\mathbf{A}$$

where $\dot{q}$, $\dot{w}$ are now the total rates of heat input and work output for the volume. For steady flow,

$$\dot{q} - \dot{w} = \int \rho\left(u + \tfrac{1}{2}V^2 + gz + \tfrac{p}{\rho}\right)\mathbf{V}.d\mathbf{A}$$

The Navier-Stokes equations

In Cartesian coordinates

$$\rho\left(\frac{\partial V_x}{\partial t} + V_x\frac{\partial V_x}{\partial x} + V_y\frac{\partial V_x}{\partial y} + V_z\frac{\partial V_x}{\partial z}\right)$$

$$= F_x + \frac{\partial \sigma_x}{\partial x} + \frac{\partial \tau_{yx}}{\partial y} + \frac{\partial \tau_{zx}}{\partial z}$$

etc., where F_x is a body force per unit volume and

$$\sigma_x = -p + 2\mu\frac{\partial V_x}{\partial x} - \tfrac{2}{3}\mu\left(\frac{\partial V_x}{\partial x} + \frac{\partial V_y}{\partial y} + \frac{\partial V_z}{\partial z}\right)$$

etc.

$$\tau_{xy} = \tau_{yx} = \mu\left(\frac{\partial V_x}{\partial y} + \frac{\partial V_y}{\partial x}\right)$$

etc. These give, for constant viscosity,

$$\frac{DV_x}{Dt} = -\frac{1}{\rho}\frac{\partial p}{\partial x} + \nu\left(\frac{\partial^2 V_x}{\partial x^2} + \frac{\partial^2 V_x}{\partial y^2} + \frac{\partial^2 V_x}{\partial z^2}\right)$$

$$+ \tfrac{1}{3}\nu\frac{\partial \theta}{\partial x} + \frac{F_x}{\rho}$$

etc. or

$$\frac{D\mathbf{V}}{Dt} = -\frac{1}{\rho}\nabla p + \nu\nabla^2\mathbf{V} + \tfrac{1}{3}\nu\nabla\theta + \frac{\mathbf{F}}{\rho}$$

where

$$\frac{D}{Dt} \equiv \frac{\partial}{\partial t} + \mathbf{V}.\nabla$$

and

$$\theta = \frac{\partial V_x}{\partial x} + \frac{\partial V_y}{\partial y} + \frac{\partial V_z}{\partial z} = \nabla.\mathbf{V}$$

Stream function and velocity potential

For two-dimensional incompressible flow the stream function is a scalar field ψ such that, in Cartesian coordinates,

$$V_x = -\frac{\partial \psi}{\partial y}, \ V_y = \frac{\partial \psi}{\partial x}.$$

If in addition the flow is irrotational, the velocity potential is a scalar field ϕ such that

$$V_x = -\frac{\partial \phi}{\partial x}, \ V_y = -\frac{\partial \phi}{\partial y}.$$

ϕ and ψ are both harmonic functions satisfying Laplace's equation; the complex potential $\phi + j\psi$ satisfies the Cauchy-Riemann conditions.

Dimensionless groups

General

h heat transfer coefficient
L characteristic dimension

Froude number $\qquad$ $\mathrm{Fr} = \dfrac{V}{\sqrt{Lg}}$

Grashof number $\qquad$ $\mathrm{Gr} = \dfrac{\beta g \rho^2 L^3 \Delta T}{\mu^2}$

Mach number $\qquad$ $M = \dfrac{V}{a}$

Nusselt number $\qquad$ $\mathrm{Nu} = \dfrac{hL}{k}$

Prandtl number $\qquad$ $\mathrm{Pr} = \dfrac{c_\mathrm{p}\mu}{k}$

Reynolds number $\qquad$ $\mathrm{Re} = \dfrac{VL\rho}{\mu}$

Stanton number $\qquad$ $\mathrm{St} = \dfrac{h}{\rho V c_\mathrm{p}}$

Hydraulic machines

P power
N speed
D diameter
Q discharge
H head

Power number $\dfrac{P}{\rho N^3 D^5}$

Discharge number $\dfrac{Q}{ND^3}$

Head number $\dfrac{gH}{N^2 D^2}$

Reynolds number $\dfrac{\rho ND^2}{\mu}$

Addison's shape parameter:

Turbine $\dfrac{N\sqrt{(P/\rho)}}{(gH)^{5/4}}$

Pump $\dfrac{N\sqrt{Q}}{(gH)^{3/4}}$

Convective heat transfer: empirical formulae

Natural convection

$Nu = C(GrPr)^n$ where C and n are as follows.

	(GrPr)	C	n
Vertical plates	10^4 to 10^9	0·59	0·25
and cylinders	10^9 to 10^{12}	0·13	0·33
Horizontal pipes	10^3 to 10^9	0·53	0·25
Horizontal plates:			
Heated facing up or	10^5 to 2×10^7	0·54	0·25
cooled facing down	2×10^7 to 3×10^{10}	0·14	0·33
Heated facing down			
or cooled facing up	3×10^5 to 3×10^{10}	0·27	0·25

Forced convection (average values)

Laminar flow:

Over flat plate $\quad\quad Nu = 0.664 \, (Re)^{1/2} \, (Pr)^{1/3}$

Fully-developed pipe flow $\quad Nu = \dfrac{0.0668(D/L)\,RePr}{1 + 0.04\,[D/L(RePr)]^{2/3}} + 3.65$.

(D = diameter, L = length, Re based on diameter)

Turbulent flow:

Over flat plate $\quad Nu = 0.036 \, Pr^{1/3} \, Re^{0.8}$

In pipe $\quad\quad\quad Nu = 0.023 \, Pr^{0.4} \, Re^{0.8}$.

(Re based on diameter)

In the above, properties should be evaluated at the mean film temperature and free-stream velocity, except that bulk properties should be used for forced convection in pipe flow.

Black-body radiation

The power radiated by a black body in all directions over a solid angle 2π, per unit surface area and per unit frequency interval is

$$E_\nu = \frac{2\pi h \nu^3/c^2}{e^{h\nu/kT} - 1}$$

in the region of the frequency ν, where h is Planck's constant, k Boltzmann's constant, and c the velocity of light; alternatively, the power per unit wavelength interval in the region of the wavelength λ is

$$E_\lambda = \frac{2\pi h c^2/\lambda^5}{e^{hc/\lambda kT} - 1}$$

The total power per unit area is

$$E = \int_0^\infty E_\nu \, d\nu = \int_0^\infty E_\lambda \, d\lambda = \sigma T^4$$

in which σ is the Stefan-Boltzmann constant. The wavelength λ_m at which E_λ is a maximum is given by the *Wien displacement law:*

$$\lambda_m T = 0.0029 \text{ m K}$$

Generalized compressibility chart

The compressibility factor Z as a function of reduced pressure and temperature p_R and T_R is to a good approximation the same for all gases; the function is plotted here with p_R as independent variable and T_R as parameter. Reduced pressure and temperature are the ratios of actual values to the critical values p_c, T_c.

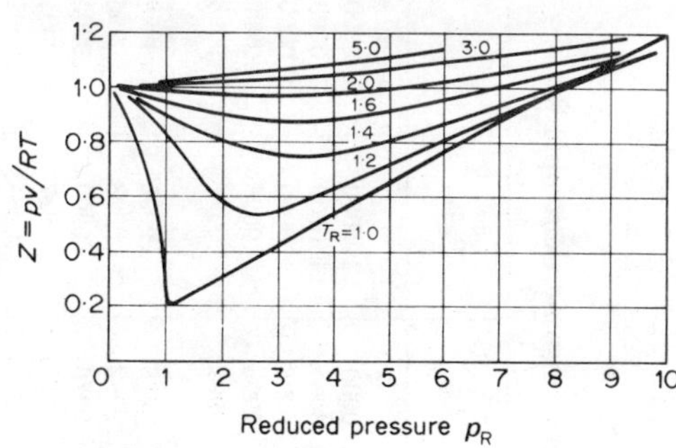

Tables for compressible flow of a perfect gas

M	Mach number
a	Speed of sound
A	Cross-sectional area of duct
θ	Prandtl-Meyer angle
I	Impulse function
L_{max}	Maximum length for choked flow in a duct of diameter D with friction coefficient f

Asterisk (*) denotes reference value when $M = 1$

Subscript *or* superscript 0 denotes stagnation condition

Subscripts 1, 2 denote conditions upstream, downstream of shock

Subscript n denotes normal to shock

Table

Flow parameters versus M for subsonic isentropic flow, $\gamma = 1{\cdot}4$

M	p/p_0	ρ/ρ_0	T/T_0	a/a_0	$A*/A$
.00	1.0000	1.0000	1.0000	1.0000	.00000
.01	.9999	1.0000	1.0000	1.0000	.01728
.02	.9997	.9998	.9999	1.0000	.03455
.03	.9994	.9996	.9998	.9999	.05181
.04	.9989	.9992	.9997	.9998	.06905
.05	.9983	.9988	.9995	.9998	.08627
.06	.9975	.9982	.9993	.9996	.1035
.07	.9966	.9976	.9990	.9995	.1206
.08	.9955	.9968	.9987	.9994	.1377
.09	.9944	.9960	.9984	.9992	.1548
.10	.9930	.9950	.9980	.9990	.1718
.11	.9916	.9940	.9976	.9988	.1887
.12	.9900	.9928	.9971	.9986	.2056
.13	.9883	.9916	.9966	.9983	.2224
.14	.9864	.9903	.9961	.9980	.2391
.15	.9844	.9888	.9955	.9978	.2557
.16	.9823	.9873	.9949	.9974	.2723
.17	.9800	.9857	.9943	.9971	.2887
.18	.9776	.9840	.9936	.9968	.3051
.19	.9751	.9822	.9928	.9964	.3213
.20	.9725	.9803	.9921	.9960	.3374
.21	.9697	.9783	.9913	.9956	.3534
.22	.9668	.9762	.9904	.9952	.3693
.23	.9638	.9740	.9895	.9948	.3851
.24	.9607	.9718	.9886	.9943	.4007
.25	.9575	.9694	.9877	.9938	.4162
.26	.9541	.9670	.9867	.9933	.4315
.27	.9506	.9645	.9856	.9928	.4467
.28	.9470	.9619	.9846	.9923	.4618
.29	.9433	.9592	.9835	.9917	.4767
.30	.9395	.9564	.9823	.9911	.4914
.31	.9355	.9535	.9811	.9905	.5059
.32	.9315	.9506	.9799	.9899	.5203
.33	.9274	.9476	.9787	.9893	.5345
.34	.9231	.9445	.9774	.9886	.5486

M	p/p_0	ρ/ρ_0	T/T_0	a/a_0	$A*/A$
.35	.9188	.9413	.9761	.9880	.5624
.36	.9143	.9380	.9747	.9873	.5761
.37	.9098	.9347	.9733	.9866	.5896
.38	.9052	.9313	.9719	.9859	.6029
.39	.9004	.9278	.9705	.9851	.6160
.40	.8956	.9243	.9690	.9844	.6289
.41	.8907	.9207	.9675	.9836	.6416
.42	.8857	.9170	.9659	.9828	.6541
.43	.8807	.9132	.9643	.9820	.6663
.44	.8755	.9094	.9627	.9812	.6784
.45	.8703	.9055	.9611	.9803	.6903
.46	.8650	.9016	.9594	.9795	.7019
.47	.8596	.8976	.9577	.9786	.7134
.48	.8541	.8935	.9560	.9777	.7246
.49	.8486	.8894	.9542	.9768	.7356
.50	.8430	.8852	.9524	.9759	.7464
.51	.8374	.8809	.9506	.9750	.7569
.52	.8317	.8766	.9487	.9740	.7672
.53	.8259	.8723	.9468	.9730	.7773
.54	.8201	.8679	.9449	.9721	.7872
.55	.8142	.8634	.9430	.9711	.7968
.56	.8082	.8589	.9410	.9701	.8063
.57	.8022	.8544	.9390	.9690	.8155
.58	.7962	.8498	.9370	.9680	.8244
.59	.7901	.8451	.9349	.9669	.8331
.60	.7840	.8405	.9328	.9658	.8416
.61	.7778	.8357	.9307	.9647	.8499
.62	.7716	.8310	.9286	.9636	.8579
.63	.7654	.8262	.9265	.9625	.8657
.64	.7591	.8213	.9243	.9614	.8732
.65	.7528	.8164	.9221	.9603	.8806
.66	.7465	.8115	.9199	.9591	.8877
.67	.7401	.8066	.9176	.9579	.8945
.68	.7338	.8016	.9153	.9567	.9012
.69	.7274	.7966	.9131	.9555	.9076

Table 1 (Continued)

M	p/p_0	ρ/ρ_0	T/T_0	a/a_0	A^*/A
.70	.7209	.7916	.9107	.9543	.9138
.71	.7145	.7865	.9084	.9531	.9197
.72	.7080	.7814	.9061	.9519	.9254
.73	.7016	.7763	.9037	.9506	.9309
.74	.6951	.7712	.9013	.9494	.9362
.75	.6886	.7660	.8989	.9481	.9412
.76	.6821	.7609	.8964	.9468	.9461
.77	.6756	.7557	.8940	.9455	.9507
.78	.6690	.7505	.8915	.9442	.9551
.79	.6625	.7452	.8890	.9429	.9592
.80	.6560	.7400	.8865	.9416	.9632
.81	.6495	.7347	.8840	.9402	.9669
.82	.6430	.7295	.8815	.9389	.9704
.83	.6365	.7242	.8789	.9375	.9737
.84	.6300	.7189	.8763	.9361	.9769
.85	.6235	.7136	.8737	.9347	.9797
.86	.6170	.7083	.8711	.9333	.9824
.87	.6106	.7030	.8685	.9319	.9849
.88	.6041	.6977	.8659	.9305	.9872
.89	.5977	.6924	.8632	.9291	.9893
.90	.5913	.6870	.8606	.9277	.9912
.91	.5849	.6817	.8579	.9262	.9929
.92	.5785	.6764	.8552	.9248	.9944
.93	.5721	.6711	.8525	.9233	.9958
.94	.5658	.6658	.8498	.9218	.9969
.95	.5595	.6604	.8471	.9204	.9979
.96	.5532	.6551	.8444	.9189	.9986
.97	.5469	.6498	.8416	.9174	.9992
.98	.5407	.6445	.8389	.9159	.9997
.99	.5345	.6392	.8361	.9144	.9999
1.00	.5283	.6339	.8333	.9129	1.0000

Table 2

Flow parameters versus M for supersonic isentropic flow, $\gamma = 1\cdot4$

M	$\dfrac{p}{p_0}$	$\dfrac{\rho}{\rho_0}$	$\dfrac{T}{T_0}$	$\dfrac{a}{a_0}$	$\dfrac{A^*}{A}$	$\dfrac{\frac{\rho}{2}V^2}{p_0}$	θ
1.00	.5283	.6339	.8333	.9129	1.0000	.3698	0
1.01	.5221	.6287	.8306	.9113	.9999	.3728	.04473
1.02	.5160	.6234	.8278	.9098	.9997	.3758	.1257
1.03	.5099	.6181	.8250	.9083	.9993	.3787	.2294
1.04	.5039	.6129	.8222	.9067	.9987	.3815	.3510
1.05	.4979	.6077	.8193	.9052	.9980	.3842	.4874
1.06	.4919	.6024	.8165	.9036	.9971	.3869	.6367
1.07	.4860	.5972	.8137	.9020	.9961	.3895	.7973
1.08	.4800	.5920	.8108	.9005	.9949	.3919	.9680
1.09	.4742	.5869	.8080	.8989	.9936	.3944	1.148
1.10	.4684	.5817	.8052	.8973	.9921	.3967	1.336
1.11	.4626	.5766	.8023	.8957	.9905	.3990	1.532
1.12	.4568	.5714	.7994	.8941	.9888	.4011	1.735
1.13	.4511	.5663	.7966	.8925	.9870	.4032	1.944
1.14	.4455	.5612	.7937	.8909	.9850	.4052	2.160
1.15	.4398	.5562	.7908	.8893	.9828	.4072	2.381
1.16	.4343	.5511	.7879	.8877	.9806	.4090	2.607
1.17	.4287	.5461	.7851	.8860	.9782	.4108	2.839
1.18	.4232	.5411	.7822	.8844	.9758	.4125	3.074
1.19	.4178	.5361	.7793	.8828	.9732	.4141	3.314
1.20	.4124	.5311	.7764	.8811	.9705	.4157	3.558
1.21	.4070	.5262	.7735	.8795	.9676	.4171	3.806
1.22	.4017	.5213	.7706	.8778	.9647	.4185	4.057
1.23	.3964	.5164	.7677	.8762	.9617	.4198	4.312
1.24	.3912	.5115	.7648	.8745	.9586	.4211	4.569
1.25	.3861	.5067	.7619	.8729	.9553	.4223	4.830
1.26	.3809	.5019	.7590	.8712	.9520	.4233	5.093
1.27	.3759	.4971	.7561	.8695	.9486	.4244	5.359
1.28	.3708	.4923	.7532	.8679	.9451	.4253	5.627
1.29	.3658	.4876	.7503	.8662	.9415	.4262	5.898
1.30	.3609	.4829	.7474	.8645	.9378	.4270	6.170
1.31	.3560	.4782	.7445	.8628	.9341	.4277	6.445
1.32	.3512	.4736	.7416	.8611	.9302	.4283	6.721
1.33	.3464	.4690	.7387	.8595	.9263	.4289	7.000
1.34	.3417	.4644	.7358	.8578	.9223	.4294	7.279

Table 2 (Continued)

M	$\dfrac{p}{p_0}$	$\dfrac{\rho}{\rho_0}$	$\dfrac{T}{T_0}$	$\dfrac{a}{a_0}$	$\dfrac{A^*}{A}$	$\dfrac{\frac{\rho}{2}V^2}{p_0}$	θ
1.35	.3370	.4598	.7329	.8561	.9182	.4299	7.561
1.36	.3323	.4553	.7300	.8544	.9141	.4303	7.844
1.37	.3277	.4508	.7271	.8527	.9099	.4306	8.128
1.38	.3232	.4463	.7242	.8510	.9056	.4308	8.413
1.39	.3187	.4418	.7213	.8493	.9013	.4310	8.699
1.40	.3142	.4374	.7184	.8476	.8969	.4311	8.987
1.41	.3098	.4330	.7155	.8459	.8925	.4312	9.276
1.42	.3055	.4287	.7126	.8442	.8880	.4312	9.565
1.43	.3012	.4244	.7097	.8425	.8834	.4311	9.855
1.44	.2969	.4201	.7069	.8407	.8788	.4310	10.15
1.45	.2927	.4158	.7040	.8390	.8742	.4308	10.44
1.46	.2886	.4116	.7011	.8373	.8695	.4306	10.73
1.47	.2845	.4074	.6982	.8356	.8647	.4303	11.02
1.48	.2804	.4032	.6954	.8339	.8599	.4299	11.32
1.49	.2764	.3991	.6925	.8322	.8551	.4295	11.61
1.50	.2724	.3950	.6897	.8305	.8502	.4290	11.91
1.51	.2685	.3909	.6868	.8287	.8453	.4285	12.20
1.52	.2646	.3869	.6840	.8270	.8404	.4279	12.49
1.53	.2608	.3829	.6811	.8253	.8354	.4273	12.79
1.54	.2570	.3789	.6783	.8236	.8304	.4266	13.09
1.55	.2533	.3750	.6754	.8219	.8254	.4259	13.38
1.56	.2496	.3710	.6726	.8201	.8203	.4252	13.68
1.57	.2459	.3672	.6698	.8184	.8152	.4243	13.97
1.58	.2423	.3633	.6670	.8167	.8101	.4235	14.27
1.59	.2388	.3595	.6642	.8150	.8050	.4226	14.56
1.60	.2353	.3557	.6614	.8133	.7998	.4216	14.86
1.61	.2318	.3520	.6586	.8115	.7947	.4206	15.16
1.62	.2284	.3483	.6558	.8098	.7895	.4196	15.45
1.63	.2250	.3446	.6530	.8081	.7843	.4185	15.75
1.64	.2217	.3409	.6502	.8064	.7791	.4174	16.04
1.65	.2184	.3373	.6475	.8046	.7739	.4162	16.34
1.66	.2151	.3337	.6447	.8029	.7686	.4150	16.63
1.67	.2119	.3302	.6419	.8012	.7634	.4138	16.93
1.68	.2088	.3266	.6392	.7995	.7581	.4125	17.22
1.69	.2057	.3232	.6364	.7978	.7529	.4112	17.52
1.70	.2026	.3197	.6337	.7961	.7476	.4098	17.81
1.71	.1996	.3163	.6310	.7943	.7423	.4086	18.10
1.72	.1966	.3129	.6283	.7926	.7371	.4071	18.40
1.73	.1936	.3095	.6256	.7909	.7318	.4056	18.69
1.74	.1907	.3062	.6229	.7892	.7265	.4041	18.98
1.75	.1878	.3029	.6202	.7875	.7212	.4026	19.27
1.76	.1850	.2996	.6175	.7858	.7160	.4011	19.56
1.77	.1822	.2964	.6148	.7841	.7107	.3996	19.86
1.78	.1794	.2932	.6121	.7824	.7054	.3980	20.15
1.79	.1767	.2900	.6095	.7807	.7002	.3964	20.44
1.80	.1740	.2868	.6068	.7790	.6949	.3947	20.73
1.81	.1714	.2837	.6041	.7773	.6897	.3931	21.01
1.82	.1688	.2806	.6015	.7756	.6845	.3914	21.30
1.83	.1662	.2776	.5989	.7739	.6792	.3897	21.59
1.84	.1637	.2745	.5963	.7722	.6740	.3879	21.88
1.85	.1612	.2715	.5936	.7705	.6688	.3862	22.16
1.86	.1587	.2686	.5910	.7688	.6636	.3844	22.45
1.87	.1563	.2656	.5884	.7671	.6584	.3826	22.73
1.88	.1539	.2627	.5859	.7654	.6533	.3808	23.02
1.89	.1516	.2598	.5833	.7637	.6481	.3790	23.30
1.90	.1492	.2570	.5807	.7620	.6430	.3771	23.59
1.91	.1470	.2542	.5782	.7604	.6379	.3753	23.87
1.92	.1447	.2514	.5756	.7587	.6328	.3734	24.15
1.93	.1425	.2486	.5731	.7570	.6277	.3715	24.43
1.94	.1403	.2459	.5705	.7553	.6226	.3696	24.71
1.95	.1381	.2432	.5680	.7537	.6175	.3677	24.99
1.96	.1360	.2405	.5655	.7520	.6125	.3657	25.27
1.97	.1339	.2378	.5630	.7503	.6075	.3638	25.55
1.98	.1318	.2352	.5605	.7487	.6025	.3618	25.83
1.99	.1298	.2326	.5580	.7470	.5975	.3598	26.10
2.00	.1278	.2300	.5556	.7454	.5926	.3579	26.38
2.01	.1258	.2275	.5531	.7437	.5877	.3559	26.66
2.02	.1239	.2250	.5506	.7420	.5828	.3539	26.93
2.03	.1220	.2225	.5482	.7404	.5779	.3518	27.20
2.04	.1201	.2200	.5458	.7388	.5730	.3498	27.48

Table 2 (Continued)

M	$\dfrac{p}{p_0}$	$\dfrac{\rho}{\rho_0}$	$\dfrac{T}{T_0}$	$\dfrac{a}{a_0}$	$\dfrac{A^*}{A}$	$\dfrac{\frac{\rho}{2}V^2}{p_0}$	θ
2.05	.1182	.2176	.5433	.7371	.5682	.3478	27.75
2.06	.1164	.2152	.5409	.7355	.5634	.3458	28.02
2.07	.1146	.2128	.5385	.7338	.5586	.3437	28.29
2.08	.1128	.2104	.5361	.7322	.5538	.3417	28.56
2.09	.1111	.2081	.5337	.7306	.5491	.3396	28.83
2.10	.1094	.2058	.5313	.7289	.5444	.3376	29.10
2.11	.1077	.2035	.5290	.7273	.5397	.3355	29.36
2.12	.1060	.2013	.5266	.7257	.5350	.3334	29.63
2.13	.1043	.1990	.5243	.7241	.5304	.3314	29.90
2.14	.1027	.1968	.5219	.7225	.5258	.3293	30.16
2.15	.1011	.1946	.5196	.7208	.5212	.3272	30.43
2.16	.09956	.1925	.5173	.7192	.5167	.3252	30.69
2.17	.09802	.1903	.5150	.7176	.5122	.3231	30.95
2.18	.09650	.1882	.5127	.7160	.5077	.3210	31.21
2.19	.09500	.1861	.5104	.7144	.5032	.3189	31.47
2.20	.09352	.1841	.5081	.7128	.4988	.3169	31.73
2.21	.09207	.1820	.5059	.7112	.4944	.3148	31.99
2.22	.09064	.1800	.5036	.7097	.4900	.3127	32.25
2.23	.08923	.1780	.5014	.7081	.4856	.3106	32.51
2.24	.08785	.1760	.4991	.7065	.4813	.3085	32.76
2.25	.08648	.1740	.4969	.7049	.4770	.3065	33.02
2.26	.08514	.1721	.4947	.7033	.4727	.3044	33.27
2.27	.08382	.1702	.4925	.7018	.4685	.3023	33.53
2.28	.08252	.1683	.4903	.7002	.4643	.3003	33.78
2.29	.08123	.1664	.4881	.6986	.4601	.2982	34.03
2.30	.07997	.1646	.4859	.6971	.4560	.2961	34.28
2.31	.07873	.1628	.4837	.6955	.4519	.2941	34.53
2.32	.07751	.1609	.4816	.6940	.4478	.2920	34.78
2.33	.07631	.1592	.4794	.6924	.4437	.2900	35.03
2.34	.07512	.1574	.4773	.6909	.4397	.2879	35.28
2.35	.07396	.1556	.4752	.6893	.4357	.2859	35.53
2.36	.07281	.1539	.4731	.6878	.4317	.2839	35.77
2.37	.07168	.1522	.4709	.6863	.4278	.2818	36.02
2.38	.07057	.1505	.4688	.6847	.4239	.2798	36.26
2.39	.06948	.1488	.4668	.6832	.4200	.2778	36.50
2.40	.06840	.1472	.4647	.6817	.4161	.2758	36.75
2.41	.06734	.1456	.4626	.6802	.4123	.2738	36.99
2.42	.06630	.1439	.4606	.6786	.4085	.2718	37.23
2.43	.06527	.1424	.4585	.6771	.4048	.2698	37.47
2.44	.06426	.1408	.4565	.6756	.4010	.2678	37.71
2.45	.06327	.1392	.4544	.6741	.3973	.2658	37.95
2.46	.06229	.1377	.4524	.6726	.3937	.2639	38.18
2.47	.06133	.1362	.4504	.6711	.3900	.2619	38.42
2.48	.06038	.1347	.4484	.6696	.3864	.2599	38.66
2.49	.05945	.1332	.4464	.6681	.3828	.2580	38.89
2.50	.05853	.1317	.4444	.6667	.3793	.2561	39.12
2.51	.05762	.1302	.4425	.6652	.3757	.2541	39.36
2.52	.05674	.1288	.4405	.6637	.3722	.2522	39.59
2.53	.05586	.1274	.4386	.6622	.3688	.2503	39.82
2.54	.05500	.1260	.4366	.6608	.3653	.2484	40.05
2.55	.05415	.1246	.4347	.6593	.3619	.2465	40.28
2.56	.05332	.1232	.4328	.6579	.3585	.2446	40.51
2.57	.05250	.1218	.4309	.6564	.3552	.2427	40.75
2.58	.05169	.1205	.4289	.6549	.3519	.2409	40.96
2.59	.05090	.1192	.4271	.6535	.3486	.2390	41.19
2.60	.05012	.1179	.4252	.6521	.3453	.2371	41.41
2.61	.04935	.1166	.4233	.6506	.3421	.2353	41.64
2.62	.04859	.1153	.4214	.6492	.3389	.2335	41.86
2.63	.04784	.1140	.4196	.6477	.3357	.2317	42.09
2.64	.04711	.1128	.4177	.6463	.3325	.2298	42.31
2.65	.04639	.1115	.4159	.6449	.3294	.2280	42.53
2.66	.04568	.1103	.4141	.6435	.3263	.2262	42.75
2.67	.04498	.1091	.4122	.6421	.3232	.2245	42.97
2.68	.04429	.1079	.4104	.6406	.3202	.2227	43.19
2.69	.04362	.1067	.4086	.6392	.3172	.2209	43.40
2.70	.04295	.1056	.4068	.6378	.3142	.2192	43.62
2.71	.04229	.1044	.4051	.6364	.3112	.2174	43.84
2.72	.04165	.1033	.4033	.6350	.3083	.2157	44.05
2.73	.04102	.1022	.4015	.6337	.3054	.2140	44.27
2.74	.04039	.1010	.3998	.6323	.3025	.2123	44.48

Table 2 (Continued)

M	$\dfrac{p}{p_0}$	$\dfrac{\rho}{\rho_0}$	$\dfrac{T}{T_0}$	$\dfrac{a}{a_0}$	$\dfrac{A^*}{A}$	$\dfrac{\frac{\rho}{2}V^2}{p_0}$	θ
2.75	.03978	.09994	.3980	.6309	.2996	.2106	44.69
2.76	.03917	.09885	.3963	.6295	.2968	.2089	44.91
2.77	.03858	.09778	.3945	.6281	.2940	.2072	45.12
2.78	.03799	.09671	.3928	.6268	.2912	.2055	45.33
2.79	.03742	.09566	.3911	.6254	.2884	.2039	45.54
2.80	.03685	.09463	.3894	.6240	.2857	.2022	45.75
2.81	.03629	.09360	.3877	.6227	.2830	.2006	45.95
2.82	.03574	.09259	.3860	.6213	.2803	.1990	46.16
2.83	.03520	.09158	.3844	.6200	.2777	.1973	46.37
2.84	.03467	.09059	.3827	.6186	.2750	.1957	46.57
2.85	.03415	.08962	.3810	.6173	.2724	.1941	46.78
2.86	.03363	.08865	.3794	.6159	.2698	.1926	46.98
2.87	.03312	.08769	.3777	.6146	.2673	.1910	47.19
2.88	.03263	.08675	.3761	.6133	.2648	.1894	47.39
2.89	.03213	.08581	.3745	.6119	.2622	.1879	47.59
2.90	.03165	.08489	.3729	.6106	.2598	.1863	47.79
2.91	.03118	.08398	.3712	.6093	.2573	.1848	47.99
2.92	.03071	.08307	.3696	.6080	.2549	.1833	48.19
2.93	.03025	.08218	.3681	.6067	.2524	.1818	48.39
2.94	.02980	.08130	.3665	.6054	.2500	.1803	48.59
2.95	.02935	.08043	.3649	.6041	.2477	.1788	48.78
2.96	.02891	.07957	.3633	.6028	.2453	.1773	48.98
2.97	.02848	.07872	.3618	.6015	.2430	.1758	49.18
2.98	.02805	.07788	.3602	.6002	.2407	.1744	49.37
2.99	.02764	.97705	.3587	.5989	.2384	.1729	49.56
3.00	.02722	.07623	.3571	.5976	.2362	.1715	49.76
3.01	.02682	.07541	.3556	.5963	.2339	.1701	49.95
3.02	.02642	.07461	.3541	.5951	.2317	.1687	50.14
3.03	.02603	.07382	.3526	.5938	.2295	.1673	50.33
3.04	.02564	.07303	.3511	.5925	.2273	.1659	50.52
3.05	.02526	.07226	.3496	.5913	.2252	.1645	50.71
3.06	.02489	.07149	.3481	.5900	.2230	.1631	50.90
3.07	.02452	.07074	.3466	.5887	.2209	.1618	51.09
3.08	.02416	.06999	.3452	.5875	.2188	.1604	51.28
3.09	.02380	.06925	.3437	.5862	.2168	.1591	51.46

M	$\dfrac{p}{p_0}$	$\dfrac{\rho}{\rho_0}$	$\dfrac{T}{T_0}$	$\dfrac{a}{a_0}$	$\dfrac{A^*}{A}$	$\dfrac{\frac{\rho}{2}V^2}{p_0}$	θ
3.10	.02345	.06852	.3422	.5850	.2147	.1577	51.65
3.11	.02310	.06779	.3408	.5838	.2127	.1564	51.84
3.12	.02276	.06708	.3393	.5825	.2107	.1551	52.02
3.13	.02243	.06637	.3379	.5813	.2087	.1538	52.20
3.14	.02210	.06568	.3365	.5801	.2067	.1525	52.39
3.15	.02177	.06499	.3351	.5788	.2048	.1512	52.57
3.16	.02146	.06430	.3337	.5776	.2028	.1500	52.75
3.17	.02114	.06363	.3323	.5764	.2009	.1487	52.93
3.18	.02083	.06296	.3309	.5752	.1990	.1475	53.11
3.19	.02053	.06231	.3295	.5740	.1971	.1462	53.29
3.20	.02023	.06165	.3281	.5728	.1953	.1450	53.47
3.21	.01993	.06101	.3267	.5716	.1934	.1438	53.65
3.22	.01964	.06037	.3253	.5704	.1916	.1426	53.83
3.23	.01936	.05975	.3240	.5692	.1898	.1414	54.00
3.24	.01908	.05912	.3226	.5680	.1880	.1402	54.18
3.25	.01880	.05851	.3213	.5668	.1863	.1390	54.35
3.26	.01853	.05790	.3199	.5656	.1845	.1378	54.53
3.27	.01826	.05730	.3186	.5645	.1828	.1367	54.71
3.28	.01799	.05671	.3173	.5633	.1810	.1355	54.88
3.29	.01773	.05612	.3160	.5621	.1793	.1344	55.05
3.30	.01748	.05554	.3147	.5609	.1777	.1332	55.22
3.31	.01722	.05497	.3134	.5598	.1760	.1321	55.39
3.32	.01698	.05440	.3121	.5586	.1743	.1310	55.56
3.33	.01673	.05384	.3108	.5575	.1727	.1299	55.73
3.34	.01649	.05329	.3095	.5563	.1711	.1288	55.90
3.35	.01625	.05274	.3082	.5552	.1695	.1277	56.07
3.36	.01602	.05220	.3069	.5540	.1679	.1266	56.24
3.37	.01579	.05166	.3057	.5529	.1663	.1255	56.41
3.38	.01557	.05113	.3044	.5517	.1648	.1245	56.58
3.39	.01534	.05061	.3032	.5506	.1632	.1234	56.75
3.40	.01513	.05009	.3019	.5495	.1617	.1224	56.91
3.41	.01491	.04958	.3007	.5484	.1602	.1214	57.07
3.42	.01470	.04908	.2995	.5472	.1587	.1203	57.24
3.43	.01449	.04858	.2982	.5461	.1572	.1193	57.40
3.44	.01428	.04808	.2970	.5450	.1558	.1183	57.56

Table 2 (Continued)

M	$\dfrac{p}{p_0}$	$\dfrac{\rho}{\rho_0}$	$\dfrac{T}{T_0}$	$\dfrac{a}{a_0}$	$\dfrac{A^*}{A}$	$\dfrac{\rho}{2}V^2 \big/ p_0$	θ
3.45	.01408	.04759	.2958	.5439	.1543	.1173	57.73
3.46	.01388	.04711	.2946	.5428	.1529	.1163	57.89
3.47	.01368	.04663	.2934	.5417	.1515	.1153	58.05
3.48	.01349	.04616	.2922	.5406	.1501	.1144	58.21
3.49	.01330	.04569	.2910	.5395	.1487	.1134	58.37
3.50	.01311	.04523	.2899	.5384	.1473	.1124	58.53
3.60	.01138	.04089	.2784	.5276	.1342	.1033	60.09
3.70	9.903×10^{-3}	.03702	.2675	.5172	.1224	.09490	61.60
3.80	8.629×10^{-3}	.03355	.2572	.5072	.1117	.08722	63.04
3.90	7.532×10^{-3}	.03044	.2474	.4974	.1021	.08019	64.44
4.00	6.586×10^{-3}	.02766	.2381	.4880	.09329	.07376	65.78
4.10	5.769×10^{-3}	.02516	.2293	.4788	.08536	.06788	67.08
4.20	5.062×10^{-3}	.02292	.2208	.4699	.07818	.06251	68.33
4.30	4.449×10^{-3}	.02090	.2129	.4614	.07166	.05759	69.54
4.40	3.918×10^{-3}	.01909	.2053	.4531	.06575	.05309	70.71
4.50	3.455×10^{-3}	.01745	.1980	.4450	.06038	.04898	71.83
4.60	3.053×10^{-3}	.01597	.1911	.4372	.05550	.04521	72.92
4.70	2.701×10^{-3}	.01464	.1846	.4296	.05107	.04177	73.97
4.80	2.394×10^{-3}	.01343	.1783	.4223	.04703	.03861	74.99
4.90	2.126×10^{-3}	.01233	.1724	.4152	.04335	.03572	75.97
5.00	1.890×10^{-3}	.01134	.1667	.4082	.04000	.03308	76.92
6.00	6.334×10^{-4}	5.194×10^{-3}	.1220	.3492	.01880	.01596	84.96
7.00	2.416×10^{-4}	2.609×10^{-3}	.09259	.3043	9.602×10^{-3}	8.285×10^{-3}	90.97
8.00	1.024×10^{-4}	1.414×10^{-3}	.07246	.2692	5.260×10^{-3}	4.589×10^{-3}	95.62
9.00	4.739×10^{-5}	8.150×10^{-4}	.05814	.2411	3.056×10^{-3}	2.687×10^{-3}	99.32
10.00	2.356×10^{-5}	4.948×10^{-4}	.04762	.2182	1.866×10^{-3}	1.649×10^{-3}	102.3
100.00	2.790×10^{-12}	5.583×10^{-9}	4.998×10^{-4}	.02236	2.157×10^{-8}	1.953×10^{-8}	127.6
∞	0	0	0	0	0	0	130.5

Table 3

Parameters for shock flow, $\gamma = 1\cdot4$

M_{1n}	p_2/p_1	ρ_2/ρ_1	T_2/T_1	a_2/a_1	$p_2{}^0/p_1{}^0$	M_2 for Normal Shocks Only
1.00	1.000	1.000	1.000	1.000	1.0000	1.0000
1.01	1.023	1.017	1.007	1.003	1.0000	.9901
1.02	1.047	1.033	1.013	1.007	1.0000	.9805
1.03	1.071	1.050	1.020	1.010	1.0000	.9712
1.04	1.095	1.067	1.026	1.013	.9999	.9620
1.05	1.120	1.084	1.033	1.016	.9999	.9531
1.06	1.144	1.101	1.039	1.019	.9998	.9444
1.07	1.169	1.118	1.046	1.023	.9996	.9360
1.08	1.194	1.135	1.052	1.026	.9994	.9277
1.09	1.219	1.152	1.059	1.029	.9992	.9196
1.10	1.245	1.169	1.065	1.032	.9989	.9118
1.11	1.271	1.186	1.071	1.035	.9986	.9041
1.12	1.297	1.203	1.078	1.038	.9982	.8966
1.13	1.323	1.221	1.084	1.041	.9978	.8892
1.14	1.350	1.238	1.090	1.044	.9973	.8820
1.15	1.376	1.255	1.097	1.047	.9967	.8750
1.16	1.403	1.272	1.103	1.050	.9961	.8682
1.17	1.430	1.290	1.109	1.053	.9953	.8615
1.18	1.458	1.307	1.115	1.056	.9946	.8549
1.19	1.485	1.324	1.122	1.059	.9937	.8485
1.20	1.513	1.342	1.128	1.062	.9928	.8422
1.21	1.541	1.359	1.134	1.065	.9918	.8360
1.22	1.570	1.376	1.141	1.068	.9907	.8300
1.23	1.598	1.394	1.147	1.071	.9896	.8241
1.24	1.627	1.411	1.153	1.074	.9884	.8183
1.25	1.656	1.429	1.159	1.077	.9871	.8126
1.26	1.686	1.446	1.166	1.080	.9857	.8071
1.27	1.715	1.463	1.172	1.083	.9842	.8016
1.28	1.745	1.481	1.178	1.085	.9827	.7963
1.29	1.775	1.498	1.185	1.088	.9811	.7911
1.30	1.805	1.516	1.191	1.091	.9794	.7860
1.31	1.835	1.533	1.197	1.094	.9776	.7809
1.32	1.866	1.551	1.204	1.097	.9758	.7760
1.33	1.897	1.568	1.210	1.100	.9738	.7712
1.34	1.928	1.585	1.216	1.103	.9718	.7664
1.35	1.960	1.603	1.223	1.106	.9697	.7618
1.36	1.991	1.620	1.229	1.109	.9676	.7572
1.37	2.023	1.638	1.235	1.111	.9653	.7527
1.38	2.055	1.655	1.242	1.114	.9630	.7483
1.39	2.087	1.672	1.248	1.117	.9606	.7440
1.40	2.120	1.690	1.255	1.120	.9582	.7397
1.41	2.153	1.707	1.261	1.123	.9557	.7355
1.42	2.186	1.724	1.268	1.126	.9531	.7314
1.43	2.219	1.742	1.274	1.129	.9504	.7274
1.44	2.253	1.759	1.281	1.132	.9476	.7235
1.45	2.286	1.776	1.287	1.135	.9448	.7196
1.46	2.320	1.793	1.294	1.137	.9420	.7157
1.47	2.354	1.811	1.300	1.140	.9390	.7120
1.48	2.389	1.828	1.307	1.143	.9360	.7083
1.49	2.423	1.845	1.314	1.146	.9329	.7047
1.50	2.458	1.862	1.320	1.149	.9298	.7011
1.51	2.493	1.879	1.327	1.152	.9266	.6976
1.52	2.529	1.896	1.334	1.155	.9233	.6941
1.53	2.564	1.913	1.340	1.158	.9200	.6907
1.54	2.600	1.930	1.347	1.161	.9166	.6874
1.55	2.636	1.947	1.354	1.164	.9132	.6841
1.56	2.673	1.964	1.361	1.166	.9097	.6809
1.57	2.709	1.981	1.367	1.169	.9061	.6777
1.58	2.746	1.998	1.374	1.172	.9026	.6746
1.59	2.783	2.015	1.381	1.175	.8989	.6715
1.60	2.820	2.032	1.388	1.178	.8952	.6684
1.61	2.857	2.049	1.395	1.181	.8914	.6655
1.62	2.895	2.065	1.402	1.184	.8877	.6625
1.63	2.933	2.082	1.409	1.187	.8838	.6596
1.64	2.971	2.099	1.416	1.190	.8799	.6568
1.65	3.010	2.115	1.423	1.193	.8760	.6540
1.66	3.048	2.132	1.430	1.196	.8720	.6512
1.67	3.087	2.148	1.437	1.199	.8680	.6485
1.68	3.126	2.165	1.444	1.202	.8640	.6458
1.69	3.165	2.181	1.451	1.205	.8599	.6431

Table 3 (Continued)

M_{1n}	p_2/p_1	ρ_2/ρ_1	T_2/T_1	a_2/a_1	p_2^0/p_1^0	M_2 for Normal Shocks Only
1.70	3.205	2.198	1.458	1.208	.8557	.6405
1.71	3.245	2.214	1.466	1.211	.8516	.6380
1.72	3.285	2.230	1.473	1.214	.8474	.6355
1.73	3.325	2.247	1.480	1.217	.8431	.6330
1.74	3.366	2.263	1.487	1.220	.8389	.6305
1.75	3.406	2.279	1.495	1.223	.8346	.6281
1.76	3.447	2.295	1.502	1.226	.8302	.6257
1.77	3.488	2.311	1.509	1.229	.8259	.6234
1.78	3.530	2.327	1.517	1.232	.8215	.6210
1.79	3.571	2.343	1.524	1.235	.8171	.6188
1.80	3.613	2.359	1.532	1.238	.8127	.6165
1.81	3.655	2.375	1.539	1.241	.8082	.6143
1.82	3.698	2.391	1.547	1.244	.8038	.6121
1.83	3.740	2.407	1.554	1.247	.7993	.6099
1.84	3.783	2.422	1.562	1.250	.7948	.6078
1.85	3.826	2.438	1.569	1.253	.7902	.6057
1.86	3.870	2.454	1.577	1.256	.7857	.6036
1.87	3.913	2.469	1.585	1.259	.7811	.6016
1.88	3.957	2.485	1.592	1.262	.7765	.5996
1.89	4.001	2.500	1.600	1.265	.7720	.5976
1.90	4.045	2.516	1.608	1.268	.7674	.5956
1.91	4.089	2.531	1.616	1.271	.7628	.5937
1.92	4.134	2.546	1.624	1.274	.7581	.5918
1.93	4.179	2.562	1.631	1.277	.7535	.5899
1.94	4.224	2.577	1.639	1.280	.7488	.5880
1.95	4.270	2.592	1.647	1.283	.7442	.5862
1.96	4.315	2.607	1.655	1.287	.7395	.5844
1.97	4.361	2.622	1.663	1.290	.7349	.5826
1.98	4.407	2.637	1.671	1.293	.7302	.5808
1.99	4.453	2.652	1.679	1.296	.7255	.5791
2.00	4.500	2.667	1.688	1.299	.7209	.5773
2.01	4.547	2.681	1.696	1.302	.7162	.5757
2.02	4.594	2.696	1.704	1.305	.7115	.5740
2.03	4.641	2.711	1.712	1.308	.7069	.5723
2.04	4.689	2.725	1.720	1.312	.7022	.5707

M_{1n}	p_2/p_1	ρ_2/ρ_1	T_2/T_1	a_2/a_1	p_2^0/p_1^0	M_2 for Normal Shocks Only
2.05	4.736	2.740	1.729	1.315	.6975	.5691
2.06	4.784	2.755	1.737	1.318	.6928	.5675
2.07	4.832	2.769	1.745	1.321	.6882	.5659
2.08	4.881	2.783	1.754	1.324	.6835	.5643
2.09	4.929	2.798	1.762	1.327	.6789	.5628
2.10	4.978	2.812	1.770	1.331	.6742	.5613
2.11	5.027	2.826	1.779	1.334	.6696	.5598
2.12	5.077	2.840	1.787	1.337	.6649	.5583
2.13	5.126	2.854	1.796	1.340	.6603	.5568
2.14	5.176	2.868	1.805	1.343	.6557	.5554
2.15	5.226	2.882	1.813	1.347	.6511	.5540
2.16	5.277	2.896	1.822	1.350	.6464	.5525
2.17	5.327	2.910	1.831	1.353	.6419	.5511
2.18	5.378	2.924	1.839	1.356	.6373	.5498
2.19	5.429	2.938	1.848	1.359	.6327	.5484
2.20	5.480	2.951	1.857	1.363	.6281	.5471
2.21	5.531	2.965	1.866	1.366	.6236	.5457
2.22	5.583	2.978	1.875	1.369	.6191	.5444
2.23	5.635	2.992	1.883	1.372	.6145	.5431
2.24	5.687	3.005	1.892	1.376	.6100	.5418
2.25	5.740	3.019	1.901	1.379	.6055	.5406
2.26	5.792	3.032	1.910	1.382	.6011	.5393
2.27	5.845	3.045	1.919	1.385	.5966	.5381
2.28	5.898	3.058	1.929	1.389	.5921	.5368
2.29	5.951	3.071	1.938	1.392	.5877	.5356
2.30	6.005	3.085	1.947	1.395	.5833	.5344
2.31	6.059	3.098	1.956	1.399	.5789	.5332
2.32	6.113	3.110	1.965	1.402	.5745	.5321
2.33	6.167	3.123	1.974	1.405	.5702	.5309
2.34	6.222	3.136	1.984	1.408	.5658	.5297
2.35	6.276	3.149	1.993	1.412	.5615	.5286
2.36	6.331	3.162	2.002	1.415	.5572	.5275
2.37	6.386	3.174	2.012	1.418	.5529	.5264
2.38	6.442	3.187	2.021	1.422	.5486	.5253
2.39	6.497	3.199	2.031	1.425	.5444	.5242

Table 3 (Continued)

M_{1n}	p_2/p_1	ρ_2/ρ_1	T_2/T_1	a_2/a_1	p_2^0/p_1^0	M_2 for Normal Shocks Only
2.40	6.553	3.212	2.040	1.428	.5401	.5231
2.41	6.609	3.224	2.050	1.432	.5359	.5221
2.42	6.666	3.237	2.059	1.435	.5317	.5210
2.43	6.722	3.249	2.069	1.438	.5276	.5200
2.44	6.779	3.261	2.079	1.442	.5234	.5189
2.45	6.836	3.273	2.088	1.445	.5193	.5179
2.46	6.894	3.285	2.098	1.449	.5152	.5169
2.47	6.951	3.298	2.108	1.452	.5111	.5159
2.48	7.009	3.310	2.118	1.455	.5071	.5149
2.49	7.067	3.321	2.128	1.459	.5030	.5140
2.50	7.125	3.333	2.138	1.462	.4990	.5130
2.51	7.183	3.345	2.147	1.465	.4950	.5120
2.52	7.242	3.357	2.157	1.469	.4911	.5111
2.53	7.301	3.369	2.167	1.472	.4871	.5102
2.54	7.360	3.380	2.177	1.476	.4832	.5092
2.55	7.420	3.392	2.187	1.479	.4793	.5083
2.56	7.479	3.403	2.198	1.482	.4754	.5074
2.57	7.539	3.415	2.208	1.486	.4715	.5065
2.58	7.599	3.426	2.218	1.489	.4677	.5056
2.59	7.659	3.438	2.228	1.493	.4639	.5047
2.60	7.720	3.449	2.238	1.496	.4601	.5039
2.61	7.781	3.460	2.249	1.500	.4564	.5030
2.62	7.842	3.471	2.259	1.503	.4526	.5022
2.63	7.903	3.483	2.269	1.506	.4489	.5013
2.64	7.965	3.494	2.280	1.510	.4452	.5005
2.65	8.026	3.505	2.290	1.513	.4416	.4996
2.66	8.088	3.516	2.301	1.517	.4379	.4988
2.67	8.150	3.527	2.311	1.520	.4343	.4980
2.68	8.213	3.537	2.322	1.524	.4307	.4972
2.69	8.275	3.548	2.332	1.527	.4271	.4964
2.70	8.338	3.559	2.343	1.531	.4236	.4956
2.71	8.401	3.570	2.354	1.534	.4201	.4949
2.72	8.465	3.580	2.364	1.538	.4166	.4941
2.73	8.528	3.591	2.375	1.541	.4131	.4933
2.74	8.592	3.601	2.386	1.545	.4097	.4926

M_{1n}	p_2/p_1	ρ_2/ρ_1	T_2/T_1	a_2/a_1	p_2^0/p_1^0	M_2 for Normal Shocks Only
2.75	8.656	3.612	2.397	1.548	.4062	.4918
2.76	8.721	3.622	2.407	1.552	.4028	.4911
2.77	8.785	3.633	2.418	1.555	.3994	.4903
2.78	8.850	3.643	2.429	1.559	.3961	.4896
2.79	8.915	3.653	2.440	1.562	.3928	.4889
2.80	8.980	3.664	2.451	1.566	.3895	.4882
2.81	9.045	3.674	2.462	1.569	.3862	.4875
2.82	9.111	3.684	2.473	1.573	.3829	.4868
2.83	9.177	3.694	2.484	1.576	.3797	.4861
2.84	9.243	3.704	2.496	1.580	.3765	.4854
2.85	9.310	3.714	2.507	1.583	.3733	.4847
2.86	9.376	3.724	2.518	1.587	.3701	.4840
2.87	9.443	3.734	2.529	1.590	.3670	.4833
2.88	9.510	3.743	2.540	1.594	.3639	.4827
2.89	9.577	3.753	2.552	1.597	.3608	.4820
2.90	9.645	3.763	2.563	1.601	.3577	.4814
2.91	9.713	3.773	2.575	1.605	.3547	.4807
2.92	9.781	3.782	2.586	1.608	.3517	.4801
2.93	9.849	3.792	2.598	1.612	.3487	.4795
2.94	9.918	3.801	2.609	1.615	.3457	.4788
2.95	9.986	3.811	2.621	1.619	.3428	.4782
2.96	10.06	3.820	2.632	1.622	.3398	.4776
2.97	10.12	3.829	2.644	1.626	.3369	.4770
2.98	10.19	3.839	2.656	1.630	.3340	.4764
2.99	10.26	3.848	2.667	1.633	.3312	.4758
3.00	10.33	3.857	2.679	1.637	.3283	.4752
3.10	11.05	3.947	2.799	1.673	.3012	.4695
3.20	11.78	4.031	2.922	1.709	.2762	.4643
3.30	12.54	4.112	3.049	1.746	.2533	.4596
3.40	13.32	4.188	3.180	1.783	.2322	.4552
3.50	14.13	4.261	3.315	1.821	.2129	.4512
3.60	14.95	4.330	3.454	1.858	.1953	.4474
3.70	15.80	4.395	3.596	1.896	.1792	.4439
3.80	16.68	4.457	3.743	1.935	.1645	.4407
3.90	17.58	4.516	3.893	1.973	.1510	.4377

Table 3 (Continued)

M_{1n}	p_2/p_1	ρ_2/ρ_1	T_2/T_1	a_2/a_1	p_2^0/p_1^0	M_2 for Normal Shocks Only
4.00	18.50	4.571	4.047	2.012	.1388	.4350
5.00	29.00	5.000	5.800	2.408	.06172	.4152
6.00	41.83	5.268	7.941	2.818	.02965	.4042
7.00	57.00	5.444	10.47	3.236	.01535	.3974
8.00	74.50	5.565	13.39	3.659	8.488×10^{-3}	.3929
9.00	94.33	5.651	16.69	4.086	4.964×10^{-3}	.3898
10.00	116.5	5.714	20.39	4.515	3.045×10^{-3}	.3876
100.00	11,666.5	5.997	1945.4	44.11	3.593×10^{-8}	.3781
∞	∞	6	∞	∞	0	.3780

Table 4

Fanno line—one-dimensional, adiabatic, constant-area flow of a perfect gas.
(Constant specific heat and molecular weight)
$\gamma = 1\cdot4$

M	$\dfrac{T}{T^*}$	$\dfrac{p}{p^*}$	$\dfrac{p_0}{p_0{}^*}$	$\dfrac{V}{V^*}$	$\dfrac{I}{I^*}$	$\dfrac{fL_{max}}{D}$
0	1.2000	∞	∞	0	∞	∞
0.01	1.2000	109.544	57.874	.01095	45.650	7134.40
.02	1.1999	54.770	28.942	.02191	22.834	1778.45
.03	1.1998	36.511	19.300	.03286	15.232	787.08
.04	1.1996	27.382	14.482	.04381	11.435	440.35
.05	1.1994	21.903	11.5914	.05476	9.1584	280.02
.06	1.1991	18.251	9.6659	.06570	7.6428	193.03
.07	1.1988	15.642	8.2915	.07664	6.5620	140.66
.08	1.1985	13.684	7.2616	.08758	5.7529	106.72
.09	1.1981	12.162	6.4614	.09851	5.1249	83.496
.10	1.1976	10.9435	5.8218	.10943	4.6236	66.922
.11	1.1971	9.9465	5.2992	.12035	4.2146	54.688
.12	1.1966	9.1156	4.8643	.13126	3.8747	45.408
.13	1.1960	8.4123	4.4968	.14216	3.5880	38.207
.14	1.1953	7.8093	4.1824	.15306	3.3432	32.511
.15	1.1946	7.2866	3.9103	.16395	3.1317	27.932
.16	1.1939	6.8291	3.6727	.17482	2.9474	24.198
.17	1.1931	6.4252	3.4635	.18568	2.7855	21.115
.18	1.1923	6.0662	3.2779	.19654	2.6422	18.543
.19	1.1914	5.7448	3.1123	.20739	2.5146	16.375
.20	1.1905	5.4555	2.9635	.21822	2.4004	14.533
.21	1.1895	5.1936	2.8293	.22904	2.2976	12.956
.22	1.1885	4.9554	2.7076	.23984	2.2046	11.596
.23	1.1874	4.7378	2.5968	.25063	2.1203	10.416
.24	1.1863	4.5383	2.4956	.26141	2.0434	9.3865
.25	1.1852	4.3546	2.4027	.27217	1.9732	8.4834
.26	1.1840	4.1850	2.3173	.28291	1.9088	7.6876
.27	1.1828	4.0280	2.2385	.29364	1.8496	6.9832
.28	1.1815	3.8820	2.1656	.30435	1.7950	6.3572
.29	1.1802	3.7460	2.0979	.31504	1.7446	5.7989
.30	1.1788	3.6190	2.0351	.32572	1.6979	5.2992
.31	1.1774	3.5002	1.9765	.33637	1.6546	4.8507
.32	1.1759	3.3888	1.9219	.34700	1.6144	4.4468
.33	1.1744	3.2840	1.8708	.35762	1.5769	4.0821
.34	1.1729	3.1853	1.8229	.36822	1.5420	3.7520
.35	1.1713	3.0922	1.7780	.37880	1.5094	3.4525
.36	1.1697	3.0042	1.7358	.38935	1.4789	3.1801
.37	1.1680	2.9209	1.6961	.39988	1.4503	2.9320
.38	1.1663	2.8420	1.6587	.41039	1.4236	2.7055
.39	1.1646	2.7671	1.6234	.42087	1.3985	2.4983

Table 4 (Continued)

$\gamma = 1\cdot4$

M	$\dfrac{T}{T^*}$	$\dfrac{p}{p^*}$	$\dfrac{p_0}{p_0{}^*}$	$\dfrac{V}{V^*}$	$\dfrac{I}{I^*}$	$\dfrac{fL_{\max}}{D}$
0.40	1.1628	2.6958	1.5901	.43133	1.3749	2.3085
.41	1.1610	2.6280	1.5587	.44177	1.3527	2.1344
.42	1.1591	2.5634	1.5289	.45218	1.3318	1.9744
.43	1.1572	2.5017	1.5007	.46257	1.3122	1.8272
.44	1.1553	2.4428	1.4739	.47293	1.2937	1.6915
.45	1.1533	2.3865	1.4486	.48326	1.2763	1.5664
.46	1.1513	2.3326	1.4246	.49357	1.2598	1.4509
.47	1.1492	2.2809	1.4018	.50385	1.2443	1.3442
.48	1.1471	2.2314	1.3801	.51410	1.2296	1.2453
.49	1.1450	2.1838	1.3595	.52433	1.2158	1.1539
.50	1.1429	2.1381	1.3399	.53453	1.2027	1.06908
.51	1.1407	2.0942	1.3212	.54469	1.1903	.99042
.52	1.1384	2.0519	1.3034	.55482	1.1786	.91741
.53	1.1362	2.0112	1.2864	.56493	1.1675	.84963
.54	1.1339	1.9719	1.2702	.57501	1.1571	.78662
.55	1.1315	1.9341	1.2549	.58506	1.1472	.72805
.56	1.1292	1.8976	1.2403	.59507	1.1378	.67357
.57	1.1268	1.8623	1.2263	.60505	1.1289	.62286
.58	1.1244	1.8282	1.2130	.61500	1.1205	.57568
.59	1.1219	1.7952	1.2003	.62492	1.1126	.53174
.60	1.1194	1.7634	1.1882	.63481	1.10504	.49081
.61	1.1169	1.7325	1.1766	.64467	1.09793	.45270
.62	1.1144	1.7026	1.1656	.65449	1.09120	.41720
.63	1.1118	1.6737	1.1551	.66427	1.08485	.38411
.64	1.1091	1.6456	1.1451	.67402	1.07883	.35330
.65	1.10650	1.6183	1.1356	.68374	1.07314	.32460
.66	1.10383	1.5919	1.1265	.69342	1.06777	.29785
.67	1.10114	1.5662	1.1179	.70306	1.06271	.27295
.68	1.09842	1.5413	1.1097	.71267	1.05792	.24978
.69	1.09567	1.5170	1.1018	.72225	1.05340	.22821
.70	1.09290	1.4934	1.09436	.73179	1.04915	.20814
.71	1.09010	1.4705	1.08729	.74129	1.04514	.18949
.72	1.08727	1.4482	1.08057	.75076	1.04137	.17215
.73	1.08442	1.4265	1.07419	.76019	1.03783	.15606
.74	1.08155	1.4054	1.06815	.76958	1.03450	.14113
.75	1.07865	1.3848	1.06242	.77893	1.03137	.12728
.76	1.07573	1.3647	1.05700	.78825	1.02844	.11446
.77	1.07279	1.3451	1.05188	.79753	1.02570	.10262
.78	1.06982	1.3260	1.04705	.80677	1.02314	.09167
.79	1.06684	1.3074	1.04250	.81598	1.02075	.08159

$\gamma = 1\cdot4$

M	$\dfrac{T}{T^*}$	$\dfrac{p}{p^*}$	$\dfrac{p_0}{p_0{}^*}$	$\dfrac{V}{V^*}$	$\dfrac{I}{I^*}$	$\dfrac{fL_{\max}}{D}$
0.80	1.06383	1.2892	1.03823	.82514	1.01853	.07229
.81	1.06080	1.2715	1.03422	.83426	1.01646	.06375
.82	1.05775	1.2542	1.03047	.84334	1.01455	.05593
.83	1.05468	1.2373	1.02696	.85239	1.01278	.04878
.84	1.05160	1.2208	1.02370	.86140	1.01115	.04226
.85	1.04849	1.2047	1.02067	.87037	1.00966	.03632
.86	1.04537	1.1889	1.01787	.87929	1.00829	.03097
.87	1.04223	1.1735	1.01529	.88818	1.00704	.02613
.88	1.03907	1.1584	1.01294	.89703	1.00591	.02180
.89	1.03589	1.1436	1.01080	.90583	1.00490	.01793
.90	1.03270	1.12913	1.00887	.91459	1.00399	.014513
.91	1.02950	1.11500	1.00714	.92332	1.00318	.011519
.92	1.02627	1.10114	1.00560	.93201	1.00248	.008916
.93	1.02304	1.08758	1.00426	.94065	1.00188	.006694
.94	1.01978	1.07430	1.00311	.94925	1.00136	.004815
.95	1.01652	1.06129	1.00215	.95782	1.00093	.003280
.96	1.01324	1.04854	1.00137	.96634	1.00059	.002056
.97	1.00995	1.03605	1.00076	.97481	1.00033	.001135
.98	1.00664	1.02379	1.00033	.98324	1.00014	.000493
.99	1.00333	1.01178	1.00008	.99164	1.00003	.000120
1.00	1.00000	1.00000	1.00000	1.00000	1.00000	0
1.01	.99666	.98844	1.00008	1.00831	1.00003	.000114
1.02	.99331	.97711	1.00033	1.01658	1.00013	.000458
1.03	.98995	.96598	1.00073	1.02481	1.00030	.001013
1.04	.98658	.95506	1.00130	1.03300	1.00053	.001771
1.05	.98320	.94435	1.00203	1.04115	1.00082	.002712
1.06	.97982	.93383	1.00291	1.04925	1.00116	.003837
1.07	.97642	.92350	1.00394	1.05731	1.00155	.005129
1.08	.97302	.91335	1.00512	1.06533	1.00200	.006582
1.09	.96960	.90338	1.00645	1.07331	1.00250	.008185
1.10	.96618	.89359	1.00793	1.08124	1.00305	.009933
1.11	.96276	.88397	1.00955	1.08913	1.00365	.011813
1.12	.95933	.87451	1.01131	1.09698	1.00429	.013824
1.13	.95589	.86522	1.01322	1.10479	1.00497	.015949
1.14	.95244	.85608	1.01527	1.11256	1.00569	.018187
1.15	.94899	.84710	1.01746	1.1203	1.00646	.02053
1.16	.94554	.83827	1.01978	1.1280	1.00726	.02298
1.17	.94208	.82958	1.02224	1.1356	1.00810	.02552
1.18	.93862	.82104	1.02484	1.1432	1.00897	.02814
1.19	.93515	.81263	1.02757	1.1508	1.00988	.03085

Table 4 (Continued)

$\gamma = 1\cdot4$

M	$\dfrac{T}{T^*}$	$\dfrac{p}{p^*}$	$\dfrac{p_0}{p_0{}^*}$	$\dfrac{V}{V^*}$	$\dfrac{I}{I^*}$	$\dfrac{fL_{\max}}{D}$
1.20	.93168	.80436	1.03044	1.1583	1.01082	.03364
1.21	.92820	.79623	1.03344	1.1658	1.01178	.03650
1.22	.92473	.78822	1.03657	1.1732	1.01278	.03942
1.23	.92125	.78034	1.03983	1.1806	1.01381	.04241
1.24	.91777	.77258	1.04323	1.1879	1.01486	.04547
1.25	.91429	.76495	1.04676	1.1952	1.01594	.04858
1.26	.91080	.75743	1.05041	1.2025	1.01705	.05174
1.27	.90732	.75003	1.05419	1.2097	1.01818	.05494
1.28	.90383	.74274	1.05809	1.2169	1.01933	.05820
1.29	.90035	.73556	1.06213	1.2240	1.02050	.06150
1.30	.89686	.72848	1.06630	1.2311	1.02169	.06483
1.31	.89338	.72152	1.07060	1.2382	1.02291	.06820
1.32	.88989	.71465	1.07502	1.2452	1.02415	.07161
1.33	.88641	.70789	1.07957	1.2522	1.02540	.07504
1.34	.88292	.70123	1.08424	1.2591	1.02666	.07850
1.35	.87944	.69466	1.08904	1.2660	1.02794	.08199
1.36	.87596	.68818	1.09397	1.2729	1.02924	.08550
1.37	.87249	.68180	1.09902	1.2797	1.03056	.08904
1.38	.86901	.67551	1.10419	1.2864	1.03189	.09259
1.39	.86554	.66931	1.10948	1.2932	1.03323	.09616
1.40	.86207	.66320	1.1149	1.2999	1.03458	.09974
1.41	.85860	.65717	1.1205	1.3065	1.03595	.10333
1.42	.85514	.65122	1.1262	1.3131	1.03733	.10694
1.43	.85168	.64536	1.1320	1.3197	1.03872	.11056
1.44	.84822	.63958	1.1379	1.3262	1.04012	.11419
1.45	.84477	.63387	1.1440	1.3327	1.04153	.11782
1.46	.84133	.62824	1.1502	1.3392	1.04295	.12146
1.47	.83788	.62269	1.1565	1.3456	1.04438	.12510
1.48	.83445	.61722	1.1629	1.3520	1.04581	.12875
1.49	.83101	.61181	1.1695	1.3583	1.04725	.13240
1.50	.82759	.60648	1.1762	1.3646	1.04870	.13605
1.51	.82416	.60122	1.1830	1.3708	1.05016	.13970
1.52	.82075	.59602	1.1899	1.3770	1.05162	.14335
1.53	.81734	.59089	1.1970	1.3832	1.05309	.14699
1.54	.81394	.58583	1.2043	1.3894	1.05456	.15063
1.55	.81054	.58084	1.2116	1.3955	1.05604	.15427
1.56	.80715	.57591	1.2190	1.4015	1.05752	.15790
1.57	.80376	.57104	1.2266	1.4075	1.05900	.16152
1.58	.80038	.56623	1.2343	1.4135	1.06049	.16514
1.59	.79701	.56148	1.2422	1.4195	1.06198	.16876

$\gamma = 1\cdot4$

M	$\dfrac{T}{T^*}$	$\dfrac{p}{p^*}$	$\dfrac{p_0}{p_0{}^*}$	$\dfrac{V}{V^*}$	$\dfrac{I}{I^*}$	$\dfrac{fL_{\max}}{D}$
1.60	.79365	.55679	1.2502	1.4254	1.06348	.17236
1.61	.79030	.55216	1.2583	1.4313	1.06498	.17595
1.62	.78695	.54759	1.2666	1.4371	1.06648	.17953
1.63	.78361	.54308	1.2750	1.4429	1.06798	.18311
1.64	.78028	.53862	1.2835	1.4487	1.06948	.18667
1.65	.77695	.53421	1.2922	1.4544	1.07098	.19022
1.66	.77363	.52986	1.3010	1.4601	1.07249	.19376
1.67	.77033	.52556	1.3099	1.4657	1.07399	.19729
1.68	.76703	.52131	1.3190	1.4713	1.07550	.20081
1.69	.76374	.51711	1.3282	1.4769	1.07701	.20431
1.70	.76046	.51297	1.3376	1.4825	1.07851	.20780
1.71	.75718	.50887	1.3471	1.4880	1.08002	.21128
1.72	.75392	.50482	1.3567	1.4935	1.08152	.21474
1.73	.75067	.50082	1.3665	1.4989	1.08302	.21819
1.74	.74742	.49686	1.3764	1.5043	1.08453	.22162
1.75	.74419	.49295	1.3865	1.5097	1.08603	.22504
1.76	.74096	.48909	1.3967	1.5150	1.08753	.22844
1.77	.73774	.48527	1.4070	1.5203	1.08903	.23183
1.78	.73453	.48149	1.4175	1.5256	1.09053	.23520
1.79	.73134	.47776	1.4282	1.5308	1.09202	.23855
1.80	.72816	.47407	1.4390	1.5360	1.09352	.24189
1.81	.72498	.47042	1.4499	1.5412	1.09500	.24521
1.82	.72181	.46681	1.4610	1.5463	1.09649	.24851
1.83	.71865	.46324	1.4723	1.5514	1.09798	.25180
1.84	.71551	.45972	1.4837	1.5564	1.09946	.25507
1.85	.71238	.45623	1.4952	1.5614	1.1009	.25832
1.86	.70925	.45278	1.5069	1.5664	1.1024	.26156
1.87	.70614	.44937	1.5188	1.5714	1.1039	.26478
1.88	.70304	.44600	1.5308	1.5763	1.1054	.26798
1.89	.69995	.44266	1.5429	1.5812	1.1068	.27116
1.90	.69686	.43936	1.5552	1.5861	1.1083	.27433
1.91	.69379	.43610	1.5677	1.5909	1.1097	.27748
1.92	.69074	.43287	1.5804	1.5957	1.1112	.28061
1.93	.68769	.42967	1.5932	1.6005	1.1126	.28372
1.94	.68465	.42651	1.6062	1.6052	1.1141	.28681
1.95	.68162	.42339	1.6193	1.6099	1.1155	.28989
1.96	.67861	.42030	1.6326	1.6146	1.1170	.29295
1.97	.67561	.41724	1.6461	1.6193	1.1184	.29599
1.98	.67262	.41421	1.6597	1.6239	1.1198	.29901
1.99	.66964	.41121	1.6735	1.6284	1.1213	.30201

Table 4 (Continued)
$\gamma = 1\cdot4$

M	$\dfrac{T}{T^*}$	$\dfrac{p}{p^*}$	$\dfrac{p_0}{p_0{}^*}$	$\dfrac{V}{V^*}$	$\dfrac{I}{I^*}$	$\dfrac{fL_{max}}{D}$
2.00	.66667	.40825	1.6875	1.6330	1.1227	.30499
2.01	.66371	.40532	1.7017	1.6375	1.1241	.30796
2.02	.66076	.40241	1.7160	1.6420	1.1255	.31091
2.03	.65783	.39954	1.7305	1.6465	1.1269	.31384
2.04	.65491	.39670	1.7452	1.6509	1.1283	.31675
2.05	.65200	.39389	1.7600	1.6553	1.1297	.31965
2.06	.64910	.39110	1.7750	1.6597	1.1311	.32253
2.07	.64621	.38834	1.7902	1.6640	1.1325	.32538
2.08	.64333	.38562	1.8056	1.6683	1.1339	.32822
2.09	.64047	.38292	1.8212	1.6726	1.1352	.33104
2.10	.63762	.38024	1.8369	1.6769	1.1366	.33385
2.11	.63478	.37760	1.8528	1.6811	1.1380	.33664
2.12	.63195	.37498	1.8690	1.6853	1.1393	.33940
2.13	.62914	.37239	1.8853	1.6895	1.1407	.34215
2.14	.62633	.36982	1.9018	1.6936	1.1420	.34488
2.15	.62354	.36728	1.9185	1.6977	1.1434	.34760
2.16	.62076	.36476	1.9354	1.7018	1.1447	.35030
2.17	.61799	.36227	1.9525	1.7059	1.1460	.35298
2.18	.61523	.35980	1.9698	1.7099	1.1474	.35564
2.19	.61249	.35736	1.9873	1.7139	1.1487	.35828
2.20	.60976	.35494	2.0050	1.7179	1.1500	.36091
2.21	.60704	.35254	2.0228	1.7219	1.1513	.36352
2.22	.60433	.35017	2.0409	1.7258	1.1526	.36611
2.23	.60163	.34782	2.0592	1.7297	1.1539	.36868
2.24	.59895	.34550	2.0777	1.7336	1.1552	.37124
2.25	.59627	.34319	2.0964	1.7374	1.1565	.37378
2.26	.59361	.34091	2.1154	1.7412	1.1578	.37630
2.27	.59096	.33865	2.1345	1.7450	1.1590	.37881
2.28	.58833	.33641	2.1538	1.7488	1.1603	.38130
2.29	.58570	.33420	2.1733	1.7526	1.1616	.38377
2.30	.58309	.33200	2.1931	1.7563	1.1629	.38623
2.31	.58049	.32983	2.2131	1.7600	1.1641	.38867
2.32	.57790	.32767	2.2333	1.7637	1.1653	.39109
2.33	.57532	.32554	2.2537	1.7673	1.1666	.39350
2.34	.57276	.32342	2.2744	1.7709	1.1678	.39589
2.35	.57021	.32133	2.2953	1.7745	1.1690	.39826
2.36	.56767	.31925	2.3164	1.7781	1.1703	.40062
2.37	.56514	.31720	2.3377	1.7817	1.1715	.40296
2.38	.56262	.31516	2.3593	1.7852	1.1727	.40528
2.39	.56011	.31314	2.3811	1.7887	1.1739	.40760

$\gamma = 1\cdot4$

M	$\dfrac{T}{T^*}$	$\dfrac{p}{p^*}$	$\dfrac{p_0}{p_0{}^*}$	$\dfrac{V}{V^*}$	$\dfrac{I}{I^*}$	$\dfrac{fL_{max}}{D}$
2.40	.55762	.31114	2.4031	1.7922	1.1751	.40989
2.41	.55514	.30916	2.4254	1.7956	1.1763	.41216
2.42	.55267	.30720	2.4479	1.7991	1.1775	.41442
2.43	.55021	.30525	2.4706	1.8025	1.1786	.41667
2.44	.54776	.30332	2.4936	1.8059	1.1798	.41891
2.45	.54533	.30141	2.5168	1.8092	1.1810	.42113
2.46	.54291	.29952	2.5403	1.8126	1.1821	.42333
2.47	.54050	.29765	2.5640	1.8159	1.1833	.42551
2.48	.53810	.29579	2.5880	1.8192	1.1844	.42768
2.49	.53571	.29395	2.6122	1.8225	1.1856	.42983
2.50	.53333	.29212	2.6367	1.8257	1.1867	.43197
2.51	.53097	.29031	2.6615	1.8290	1.1879	.43410
2.52	.52862	.28852	2.6865	1.8322	1.1890	.43621
2.53	.52627	.28674	2.7117	1.8354	1.1901	.43831
2.54	.52394	.28498	2.7372	1.8386	1.1912	.44040
2.55	.52163	.28323	2.7630	1.8417	1.1923	.44247
2.56	.51932	.28150	2.7891	1.8448	1.1934	.44452
2.57	.51702	.27978	2.8154	1.8479	1.1945	.44655
2.58	.51474	.27808	2.8420	1.8510	1.1956	.44857
2.59	.51247	.27640	2.8689	1.8541	1.1967	.45059
2.60	.51020	.27473	2.8960	1.8571	1.1978	.45259
2.61	.50795	.27307	2.9234	1.8602	1.1989	.45457
2.62	.50571	.27143	2.9511	1.8632	1.2000	.45654
2.63	.50349	.26980	2.9791	1.8662	1.2011	.45850
2.64	.50127	.26818	3.0074	1.8691	1.2021	.46044
2.65	.49906	.26658	3.0359	1.8721	1.2031	.46237
2.66	.49687	.26499	3.0647	1.8750	1.2042	.46429
2.67	.49469	.26342	3.0938	1.8779	1.2052	.46619
2.68	.49251	.26186	3.1234	1.8808	1.2062	.46807
2.69	.49035	.26032	3.1530	1.8837	1.2073	.46996
2.70	.48820	.25878	3.1830	1.8865	1.2083	.47182
2.71	.48606	.25726	3.2133	1.8894	1.2093	.47367
2.72	.48393	.25575	3.2440	1.8922	1.2103	.47551
2.73	.48182	.25426	3.2749	1.8950	1.2113	.47734
2.74	.47971	.25278	3.3061	1.8978	1.2123	.47915
2.75	.47761	.25131	3.3376	1.9005	1.2133	.48095
2.76	.47553	.24985	3.3695	1.9032	1.2143	.48274
2.77	.47346	.24840	3.4017	1.9060	1.2153	.48452
2.78	.47139	.24697	3.4342	1.9087	1.2163	.48628
2.79	.46933	.24555	3.4670	1.9114	1.2173	.48803

Table 4 (Continued)
$\gamma = 1 \cdot 4$

M	$\dfrac{T}{T^*}$	$\dfrac{p}{p^*}$	$\dfrac{p_0}{p_0{}^*}$	$\dfrac{V}{V^*}$	$\dfrac{I}{I^*}$	$\dfrac{fL_{max}}{D}$
2.80	.46729	.24414	3.5001	1.9140	1.2182	.48976
2.81	.46526	.24274	3.5336	1.9167	1.2192	.49148
2.82	.46324	.24135	3.5674	1.9193	1.2202	.49321
2.83	.46122	.23997	3.6015	1.9220	1.2211	.49491
2.84	.45922	.23861	3.6359	1.9246	1.2221	.49660
2.85	.45723	.23726	3.6707	1.9271	1.2230	.49828
2.86	.45525	.23592	3.7058	1.9297	1.2240	.49995
2.87	.45328	.23458	3.7413	1.9322	1.2249	.50161
2.88	.45132	.23326	3.7771	1.9348	1.2258	.50326
2.89	.44937	.23196	3.8133	1.9373	1.2268	.50489
2.90	.44743	.23066	3.8498	1.9398	1.2277	.50651
2.91	.44550	.22937	3.8866	1.9423	1.2286	.50812
2.92	.44358	.22809	3.9238	1.9448	1.2295	.50973
2.93	.44167	.22682	3.9614	1.9472	1.2304	.51133
2.94	.43977	.22556	3.9993	1.9497	1.2313	.51291
2.95	.43788	.22431	4.0376	1.9521	1.2322	.51447
2.96	.43600	.22307	4.0763	1.9545	1.2331	.51603
2.97	.43413	.22185	4.1153	1.9569	1.2340	.51758
2.98	.43226	.22063	4.1547	1.9592	1.2348	.51912
2.99	.43041	.21942	4.1944	1.9616	1.2357	.52064
3.0	.42857	.21822	4.2346	1.9640	1.2366	.52216
3.5	.34783	.16850	6.7896	2.0642	1.2743	.53643
4.0	.28571	.13363	10.719	2.1381	1.3029	.63306
4.5	.23762	.10833	16.562	2.1936	1.3247	.66764
5.0	.20000	.08944	25.000	2.2361	1.3416	.69381
6.0	.14634	.06376	53.180	2.2953	1.3655	.72987
7.0	.11111	.04762	104.14	2.3333	1.3810	.75281
8.0	.08696	.03686	190.11	2.3591	1.3915	.76820
9.0	.06977	.02935	327.19	2.3772	1.3989	.77898
10.0	.05714	.02390	535.94	2.3905	1.4044	.78683
∞	0	0	∞	2.4495	1.4289	.82153

$\gamma = 1 \cdot 0$

M	$\dfrac{T}{T^*}$	$\dfrac{p}{p^*}$	$\dfrac{p_0}{p_0{}^*}$	$\dfrac{V}{V^*}$	$\dfrac{I}{I^*}$	$\dfrac{fL_{max}}{D}$
0	1.000	∞	∞	0	∞	∞
0.05		20.000	12.146	.0500	10.025	393.01
.10		10.000	6.096	.1000	5.050	94.39
.15		6.667	4.089	.1500	3.408	39.65
.20		5.000	3.094	.2000	2.600	20.78
.25		4.000	2.503	.2500	2.125	12.227
.30		3.333	2.115	.3000	1.817	7.703
.35		2.857	1.842	.3500	1.604	5.064
.40		2.500	1.643	.4000	1.450	3.417
.45		2.222	1.492	.4500	1.336	2.341
.50		2.000	1.375	.5000	1.250	1.614
.55		1.818	1.283	.5500	1.184	1.110
.60		1.667	1.210	.6000	1.133	.7561
.65		1.539	1.153	.6500	1.0942	.5053
.70		1.429	1.107	.7000	1.0643	.3275
.75		1.333	1.0714	.7500	1.0417	.2024
.80		1.250	1.0441	.8000	1.0250	.1162
.85		1.176	1.0240	.8500	1.0132	.05904
.90		1.111	1.0104	.9000	1.0056	.02385
.95		1.0526	1.0026	.9500	1.0013	.00545
1.00		1.0000	1.0000	1.0000	1.0000	0
1.05		.9524	1.0025	1.0500	1.0012	.00461
1.10		.9091	1.0097	1.100	1.0045	.01707
1.15		.8695	1.0217	1.150	1.0098	.03567
1.20		.8333	1.0384	1.200	1.0167	.05909
1.25		.8000	1.0598	1.250	1.0250	.08629
1.30		.7692	1.0862	1.300	1.0346	.1164
1.35		.7407	1.118	1.350	1.0453	.1489
1.40		.7143	1.154	1.400	1.0571	.1831
1.45		.6897	1.196	1.450	1.0698	.2188
1.50		.6667	1.245	1.500	1.0833	.2554
1.55		.6452	1.300	1.550	1.0976	.2927
1.60		.6250	1.363	1.600	1.112	.3306
1.65		.6061	1.434	1.650	1.128	.3689
1.70		.5882	1.514	1.700	1.144	.4073

Table 4 (Continued)

$\gamma = 1 \cdot 0$

M	$\dfrac{T}{T^*}$	$\dfrac{p}{p^*}$	$\dfrac{p_0}{p_0{}^*}$	$\dfrac{V}{V^*}$	$\dfrac{I}{I^*}$	$\dfrac{fL_{max}}{D}$
1.75	1.000	.5714	1.603	1.750	1.161	.4458
1.80		.5556	1.703	1.800	1.178	.4842
1.85		.5406	1.815	1.850	1.195	.5225
1.90		.5263	1.941	1.900	1.213	.5607
1.95		.5128	2.082	1.950	1.231	.5986
2.00		.5000	2.241	2.000	1.250	.6363
2.05		.4878	2.419	2.050	1.269	.6736
2.10		.4762	2.620	2.100	1.288	.7106
2.15		.4651	2.846	2.150	1.308	.7472
2.20		.4545	3.100	2.200	1.327	7835
2.25		.4444	3.388	2.250	1.347	.8194
2.30		.4348	3.714	2.300	1.367	.8549
2.35		.4256	4.083	2.350	1.388	.8900
2.40		.4167	4.502	2.400	1.408	.9246
2.45		.4082	4.979	2.450	1.429	.9588
2.50		.4000	5.522	2.500	1.450	.9926
2.55		.3922	6.142	2.550	1.471	1.0260
2.60		.3847	6.852	2.600	1.492	1.0590
2.65		.3774	7.665	2.650	1.514	1.0916
2.70		.3704	8.600	2.700	1.535	1.1237
2.75		.3636	9.676	2.750	1.557	1.155
2.80		.3571	10.92	2.800	1.579	1.187
2.85		.3509	12.35	2.850	1.600	1.218
2.90		.3449	14.02	2.900	1.622	1.248
2.95		.3390	15.95	2.950	1.644	1.279
3.00		.3333	18.20	3.000	1.667	1.308
3.50		.2857	79.22	3.500	1.893	1.587
4.00		.2500	452.01	4.000	2.125	1.835
4.50		.2222	3364	4.500	2.361	2.058
5.00		.2000	32550	5.000	2.600	2.259
6.00		.1667	$664(10)^4$	6.000	3.083	2.611
7.00		.1429	$378(10)^7$	7.000	3.571	2.912
8.00		.1250	$599(10)^{10}$	8.000	4.062	3.174
9.00		.1111	$262(10)^{14}$	9.000	4.556	3.407
10.00		.1000	$314(10)^{18}$	10.000	5.050	3.615
∞	1.000	0	∞	∞	∞	∞

$\gamma = 1 \cdot 1$

M	$\dfrac{T}{T^*}$	$\dfrac{p}{p^*}$	$\dfrac{p_0}{p_0{}^*}$	$\dfrac{V}{V^*}$	$\dfrac{I}{I^*}$	$\dfrac{fL_{max}}{D}$
0	1.0500	∞	∞	0	∞	∞
0.05	1.0499	20.493	11.999	.05123	9.785	357.05
.10	1.0495	10.244	6.023	.1024	4.932	85.65
.15	1.0488	6.828	4.042	.1536	3.332	35.92
.20	1.0479	5.118	3.059	.2047	2.545	18.79
.25	1.0467	4.092	2.476	.2558	2.083	11.03
.30	1.0453	3.408	2.094	.3067	1.784	6.936
.35	1.0436	2.919	1.825	.3575	1.577	4.549
.40	1.0417	2.552	1.628	.4082	1.429	3.062
.45	1.0395	2.266	1.480	.4588	1.319	2.093
.50	1.0370	2.037	1.365	.5092	1.237	1.439
.55	1.0343	1.849	1.275	.5594	1.174	.9871
.60	1.0314	1.693	1.204	.6094	1.125	.6705
.65	1.0283	1.560	1.148	.6591	1.0882	.4468
.70	1.0249	1.446	1.104	.7086	1.0599	.2887
.75	1.0213	1.347	1.0689	.7579	1.0386	.1780
.80	1.0174	1.261	1.0425	.8069	1.0231	.1019
.85	1.0133	1.184	1.0231	.8557	1.0122	.05160
.90	1.0091	1.116	1.0100	.9041	1.0051	.02078
.95	1.0047	1.0551	1.0024	.9522	1.0012	.00472
1.00	1.0000	1.0000	1.0000	1.0000	1.0000	0
1.05	.9951	.9501	1.0023	1.0474	1.0011	.00398
1.10	.9901	.9046	1.0092	1.0945	1.0041	.01468
1.15	.9849	.8630	1.0204	1.1412	1.0087	.03058
1.20	.9795	.8247	1.0360	1.1876	1.0148	.05050
1.25	.9739	.7895	1.0559	1.234	1.0221	.07350
1.30	.9682	.7569	1.0801	1.279	1.0304	.09885
1.35	.9623	.7266	1.109	1.324	1.0397	.1260
1.40	.9563	.6985	1.142	1.369	1.0498	.1544
1.45	.9501	.6722	1.180	1.413	1.0605	.1838
1.50	.9438	.6476	1.223	1.457	1.0717	.2138
1.55	.9374	.6246	1.272	1.501	1.0835	.2443
1.60	.9309	.6030	1.326	1.544	1.0958	.2749
1.65	.9242	.5826	1.387	1.586	1.108	.3056
1.70	.9174	.5634	1.454	1.628	1.121	.3362

Table 4 (Continued)

$\gamma = 1\cdot1$

M	$\dfrac{T}{T^*}$	$\dfrac{p}{p^*}$	$\dfrac{p_0}{p_0{}^*}$	$\dfrac{V}{V^*}$	$\dfrac{I}{I^*}$	$\dfrac{fL_{max}}{D}$
1.75	.9105	.5453	1.528	1.670	1.134	.3667
1.80	.9036	.5281	1.610	1.711	1.148	.3969
1.85	.8966	.5118	1.701	1.752	1.161	.4268
1.90	.8895	.4964	1.801	1.792	1.175	.4563
1.95	.8823	.4817	1.911	1.832	1.189	.4854
2.00	.8750	.4677	2.032	1.871	1.203	.5140
2.05	.8677	.4544	2.165	1.910	1.217	.5422
2.10	.8603	.4417	2.312	1.948	1.231	.5698
2.15	.8529	.4295	2.473	1.986	1.245	.5970
2.20	.8454	.4179	2.651	2.023	1.259	.6237
2.25	.8379	.4068	2.846	2.060	1.273	.6498
2.30	.8304	.3962	3.061	2.096	1.286	.6754
2.35	.8228	.3860	3.299	2.132	1.300	.7005
2.40	.8152	.3762	3.560	2.167	1.314	.7251
2.45	.8076	.3668	3.848	2.202	1.328	.7491
2.50	.8000	.3578	4.165	2.236	1.342	.7726
2.55	.7924	.3491	4.515	2.270	1.355	.7957
2.60	.7848	.3407	4.902	2.303	1.369	.8182
2.65	.7771	.3327	5.328	2.336	1.382	.8402
2.70	.7695	.3249	5.799	2.368	1.395	.8617
2.75	.7619	.3174	6.320	2.400	1.409	.8828
2.80	.7543	.3102	6.895	2.432	1.422	.9034
2.85	.7467	.3032	7.532	2.463	1.434	.9235
2.90	.7392	.2965	8.237	2.493	1.447	.9432
2.95	.7316	.2900	9.016	2.523	1.460	.9624
3.00	.7241	.2837	9.880	2.553	1.472	.9812
3.50	.6512	.2305	25.83	2.824	1.589	1.147
4.00	.5833	.1909	71.74	3.055	1.691	1.280
4.50	.5217	.1605	205.7	3.250	1.779	1.386
5.00	.4667	.1366	597.3	3.416	1.854	1.472
6.00	.3750	.1021	4911	3.674	1.973	1.601
7.00	.3043	.07881	37919	3.862	2.060	1.689
8.00	.2500	.06250	$263(10)^3$	4.000	2.125	1.752
9.00	.2079	.05067	$161(10)^4$	4.104	2.174	1.798
10.00	.1750	.04183	$889(10)^4$	4.183	2.211	1.832
∞	0	0	∞	4.583	2.400	1.997

$\gamma = 1\cdot2$

M	$\dfrac{T}{T^*}$	$\dfrac{p}{p^*}$	$\dfrac{p_0}{p_0{}^*}$	$\dfrac{V}{V^*}$	$\dfrac{I}{I^*}$	$\dfrac{fL_{max}}{D}$
0	1.1000	∞	∞	0	∞	∞
0.05	1.0997	20.974	11.857	.05243	9.562	327.09
.10	1.0989	10.483	5.953	.1048	4.822	78.36
.15	1.0975	6.984	3.996	.1571	3.260	32.81
.20	1.0956	5.234	3.026	.2093	2.493	17.13
.25	1.0932	4.182	2.451	.2614	2.044	10.04
.30	1.0902	3.480	2.073	.3133	1.753	6.298
.35	1.0867	2.978	1.809	.3649	1.553	4.121
.40	1.0827	2.601	1.615	.4162	1.409	2.768
.45	1.0782	2.307	1.469	.4672	1.304	1.887
.50	1.0732	2.072	1.356	.5179	1.224	1.294
.55	1.0677	1.879	1.268	.5683	1.164	.8855
.60	1.0618	1.717	1.199	.6183	1.118	.5999
.65	1.0554	1.581	1.144	.6678	1.0826	.3987
.70	1.0486	1.463	1.100	.7168	1.0561	.2570
.75	1.0414	1.361	1.0666	.7654	1.0360	.1579
.80	1.0338	1.271	1.0410	.8134	1.0214	.09016
.85	1.0259	1.192	1.0222	.8609	1.0112	.04554
.90	1.0176	1.121	1.0096	.9078	1.0047	.01829
.95	1.0089	1.0573	1.0023	.9542	1.0011	.00414
1.00	1.0000	1.0000	1.0000	1.0000	1.0000	0
1.05	.9908	.9480	1.0022	1.0451	1.0010	.00347
1.10	.9813	.9005	1.0087	1.0896	1.0037	.01277
1.15	.9715	.8571	1.0194	1.134	1.0079	.02657
1.20	.9615	.8172	1.0340	1.177	1.0134	.04368
1.25	.9514	.7803	1.0525	1.219	1.0197	.06338
1.30	.9410	.7462	1.0749	1.261	1.0270	.08500
1.35	.9304	.7145	1.101	1.302	1.0351	.1080
1.40	.9197	.6850	1.132	1.342	1.0437	.1320
1.45	.9089	.6575	1.166	1.382	1.0529	.1567
1.50	.8980	.6317	1.205	1.421	1.0625	.1817
1.55	.8869	.6076	1.248	1.459	1.0724	.2069
1.60	.8758	.5849	1.296	1.497	1.0826	.2323
1.65	.8646	.5635	1.349	1.534	1.0930	.2575
1.70	.8534	.5434	1.407	1.570	1.1036	.2825

Table 4 (Continued)

$\gamma = 1 \cdot 2$

M	$\dfrac{T}{T^*}$	$\dfrac{p}{p^*}$	$\dfrac{p_0}{p_0^*}$	$\dfrac{V}{V^*}$	$\dfrac{I}{I^*}$	$\dfrac{fL_{max}}{D}$
1.75	.8421	.5244	1.471	1.606	1.114	.3072
1.80	.8308	.5064	1.540	1.641	1.125	.3316
1.85	.8195	.4894	1.615	1.675	1.136	.3556
1.90	.8082	.4732	1.697	1.708	1.147	.3791
1.95	.7970	.4578	1.787	1.741	1.158	.4021
2.00	.7857	.4432	1.884	1.773	1.168	.4247
2.05	.7745	.4293	1.989	1.804	1.179	.4468
2.10	.7634	.4160	2.103	1.835	1.190	.4684
2.15	.7523	.4034	2.226	1.865	1.201	.4894
2.20	.7413	.3913	2.359	1.894	1.211	.5099
2.25	.7303	.3798	2.504	1.923	1.221	.5299
2.30	.7194	.3688	2.660	1.951	1.232	.5493
2.35	.7086	.3582	2.829	1.978	1.242	.5683
2.40	.6980	.3481	3.011	2.005	1.252	.5868
2.45	.6874	.3384	3.208	2.031	1.262	.6047
2.50	.6769	.3291	3.420	2.057	1.272	.6222
2.55	.6665	.3202	3.650	2.082	1.281	.6392
2.60	.6563	.3116	3.898	2.106	1.291	.6557
2.65	.6462	.3033	4.166	2.130	1.300	.6718
2.70	.6362	.2954	4.455	2.154	1.309	.6874
2.75	.6263	.2878	4.767	2.176	1.318	.7026
2.80	.6166	.2804	5.103	2.199	1.327	.7173
2.85	.6070	.2733	5.466	2.220	1.335	.7316
2.90	.5975	.2665	5.858	2.242	1.344	.7456
2.95	.5882	.2600	6.280	2.263	1.352	.7592
3.00	.5789	.2536	6.735	2.283	1.360	.7724
3.50	.4944	.2009	13.76	2.461	1.434	.8857
4.00	.4231	.1626	28.35	2.602	1.493	.9718
4.50	.3636	.1340	57.96	2.714	1.541	1.0380
5.00	.3143	.1121	116.31	2.803	1.580	1.0896
6.00	.2391	.08150	434.7	2.934	1.637	1.163
7.00	.1864	.06168	1458	3.023	1.677	1.212
8.00	.1486	.04819	4353	3.084	1.704	1.245
9.00	.1209	.03863	13156	3.129	1.724	1.268
10.00	.1000	.03162	29601	3.162	1.739	1.286
∞	0	0	∞	3.317	1.809	1.365

$\gamma = 1 \cdot 3$

M	$\dfrac{T}{T^*}$	$\dfrac{p}{p^*}$	$\dfrac{p_0}{p_0^*}$	$\dfrac{V}{V^*}$	$\dfrac{I}{I^*}$	$\dfrac{fL_{max}}{D}$
0	1.150	∞	∞	0	∞	∞
0.05	1.149	21.444	11.721	.05361	9.354	301.74
.10	1.148	10.716	5.885	.1072	4.720	72.20
.15	1.146	7.137	3.952	.1606	3.194	30.18
.20	1.143	5.346	2.994	.2138	2.445	15.73
.25	1.139	4.270	2.426	.2668	2.007	9.201
.30	1.134	3.551	2.054	.3195	1.724	5.759
.35	1.129	3.036	1.793	.3719	1.530	3.760
.40	1.123	2.649	1.602	.4239	1.391	2.520
.45	1.116	2.348	1.459	.4754	1.289	1.714
.50	1.1084	2.106	1.348	.5264	1.213	1.172
.55	1.1001	1.907	1.261	.5769	1.155	.8004
.60	1.0911	1.741	1.193	.6267	1.111	.5409
.65	1.0815	1.600	1.140	.6759	1.0777	.3586
.70	1.0713	1.479	1.0972	.7245	1.0524	.2305
.75	1.0605	1.373	1.0644	.7724	1.0336	.14131
.80	1.0493	1.280	1.0395	.8195	1.0199	.08044
.85	1.0376	1.198	1.0214	.8658	1.0104	.04053
.90	1.0254	1.125	1.0092	.9113	1.0043	.01623
.95	1.0129	1.0594	1.0022	.9561	1.0010	.00367
1.00	1.0000	1.0000	1.0000	1.0000	1.0000	0
1.05	.9868	.9461	1.0021	1.0430	1.0009	.00305
1.10	.9733	.8969	1.0083	1.0852	1.0033	.01122
1.15	.9596	.8518	1.0183	1.1266	1.0071	.02324
1.20	.9457	.8104	1.0321	1.1670	1.0120	.03820
1.25	.9316	.7722	1.0495	1.206	1.0177	.05524
1.30	.9174	.7368	1.0704	1.245	1.0241	.07388
1.35	.9031	.7039	1.0948	1.283	1.0312	.09365
1.40	.8887	.6734	1.1227	1.320	1.0388	.11417
1.45	.8743	.6448	1.1543	1.356	1.0467	.13513
1.50	.8598	.6182	1.189	1.391	1.0549	.1564
1.55	.8454	.5932	1.228	1.425	1.0634	.1777
1.60	.8309	.5697	1.271	1.458	1.0721	.1989
1.65	.8165	.5477	1.318	1.491	1.0808	.2200
1.70	.8022	.5269	1.369	1.523	1.0897	.2408

Table 4 (Continued)
$\gamma = 1 \cdot 3$

M	$\dfrac{T}{T^*}$	$\dfrac{p}{p^*}$	$\dfrac{p_0}{p_0{}^*}$	$\dfrac{V}{V^*}$	$\dfrac{I}{I^*}$	$\dfrac{fL_{max}}{D}$
1.75	.7880	.5073	1.424	1.554	1.0986	.2613
1.80	.7739	.4887	1.484	1.584	1.108	.2814
1.85	.7599	.4712	1.549	1.613	1.116	.3010
1.90	.7460	.4546	1.618	1.641	1.125	.3202
1.95	.7323	.4388	1.693	1.669	1.134	.3390
2.00	.7188	.4239	1.773	1.696	1.143	.3573
2.05	.7054	.4097	1.859	1.722	1.151	.3751
2.10	.6922	.3962	1.951	1.747	1.160	.3924
2.15	.6791	.3833	2.050	1.772	1.168	.4092
2.20	.6662	.3710	2.156	1.796	1.176	.4255
2.25	.6536	.3593	2.268	1.819	1.184	.4413
2.30	.6412	.3482	2.388	1.842	1.192	.4566
2.35	.6290	.3375	2.517	1.864	1.200	.4715
2.40	.6170	.3273	2.654	1.885	1.208	.4860
2.45	.6051	.3175	2.800	1.906	1.215	.5000
2.50	.5935	.3082	2.954	1.926	1.223	.5136
2.55	.5822	.2992	3.119	1.946	1.230	.5267
2.60	.5711	.2906	3.295	1.965	1.237	.5394
2.65	.5601	.2824	3.482	1.983	1.244	.5517
2.70	.5493	.2745	3.681	2.001	1.250	.5636
2.75	.5388	.2669	3.892	2.019	1.257	.5752
2.80	.5285	.2596	4.116	2.036	1.263	.5864
2.85	.5184	.2526	4.354	2.052	1.270	.5972
2.90	.5085	.2459	4.607	2.068	1.276	.6077
2.95	.4938	.2394	4.875	2.084	1.282	.6179
3.00	.4894	.2332	5.160	2.099	1.288	.6277
3.50	.4053	.1819	9.110	2.228	1.338	.7110
4.00	.3332	.1454	15.94	2.326	1.378	.7726
4.50	.2848	.1186	27.39	2.402	1.409	.8189
5.00	.2421	.09841	45.95	2.460	1.433	.8543
6.00	.1797	.07065	120.1	2.543	1.468	.9037
7.00	.1377	.05302	285.3	2.598	1.491	.9355
8.00	.1085	.04117	625.2	2.635	1.507	.9570
9.00	.08745	.03286	1275	2.662	1.519	.9722
10.00	.07188	.02769	2438	2.681	1.527	.9832
∞	0	0	∞	2.769	1.565	1.0326

$\gamma = 1 \cdot 67$

M	$\dfrac{T}{T^*}$	$\dfrac{p}{p^*}$	$\dfrac{p_0}{p_0{}^*}$	$\dfrac{V}{V^*}$	$\dfrac{I}{I^*}$	$\dfrac{fL_{max}}{D}$
0	1.335	∞	∞	0	∞	∞
0.05	1.334	23.099	11.265	.05775	8.687	234.36
.10	1.331	11.535	5.661	.1154	4.392	55.83
.15	1.325	7.674	3.805	.1727	2.982	23.21
.20	1.317	5.739	2.887	.2296	2.293	12.11
.25	1.308	4.574	2.344	.2859	1.892	6.980
.30	1.296	3.795	1.989	.3415	1.635	4.337
.35	1.282	3.235	1.741	.3963	1.460	2.810
.40	1.267	2.814	1.560	.4502	1.336	1.868
.45	1.250	2.485	1.424	.5031	1.245	1.260
.50	1.232	2.220	1.320	.5549	1.178	.8549
.55	1.212	2.002	1.239	.6056	1.128	.5787
.60	1.191	1.819	1.176	.6548	1.0909	.3877
.65	1.169	1.664	1.126	.7029	1.0628	.2548
.70	1.146	1.530	1.0874	.7496	1.0418	.1625
.75	1.1233	1.413	1.0576	.7949	1.0265	.09870
.80	1.0993	1.311	1.0351	.8388	1.0155	.05576
.85	1.0748	1.220	1.0189	.8812	1.0080	.02780
.90	1.0501	1.139	1.0081	.9222	1.0033	.01106
.95	1.0251	1.0657	1.0019	.9618	1.0008	.00248
1.00	1.0000	1.0000	1.0000	1.0000	1.0000	0
1.05	.9749	.9404	1.0018	1.0368	1.0006	.00203
1.10	.9499	.8860	1.0070	1.0721	1.0024	.00740
1.15	.9251	.8364	1.0154	1.1061	1.0051	.01522
1.20	.9006	.7908	1.0266	1.1388	1.0084	.02481
1.25	.8763	.7489	1.0406	1.170	1.0124	.03564
1.30	.8524	.7102	1.0573	1.200	1.0167	.04733
1.35	.8289	.6744	1.0765	1.229	1.0213	.05957
1.40	.8059	.6412	1.0981	1.257	1.0262	.07212
1.45	.7833	.6104	1.1220	1.284	1.0313	.08481
1.50	.7612	.5817	1.148	1.309	1.0364	.09749
1.55	.7397	.5549	1.176	1.333	1.0416	.1101
1.60	.7187	.5298	1.207	1.356	1.0468	.1225
1.65	.6982	.5064	1.240	1.378	1.0520	.1346
1.70	.6783	.4845	1.275	1.400	1.0572	.1465

Table 4 (Continued)

$\gamma = 1\cdot67$

M	$\dfrac{T}{T^*}$	$\dfrac{p}{p^*}$	$\dfrac{p_0}{p_0{}^*}$	$\dfrac{V}{V^*}$	$\dfrac{I}{I^*}$	$\dfrac{fL_{max}}{D}$
1.75	.6590	.4639	1.312	1.421	1.0623	.1580
1.80	.6402	.4445	1.351	1.440	1.0673	.1692
1.85	.6219	.4263	1.392	1.459	1.0722	.1800
1.90	.6042	.4091	1.436	1.477	1.0770	.1905
1.95	.5871	.3929	1.482	1.494	1.0817	.2007
2.00	.5705	.3776	1.530	1.510	1.0863	.2105
2.05	.5544	.3632	1.580	1.526	1.0908	.2199
2.10	.5388	.3496	1.632	1.541	1.0952	.2290
2.15	.5238	.3367	1.687	1.556	1.0994	.2377
2.20	.5093	.3244	1.744	1.570	1.1035	.2461
2.25	.4952	.3128	1.803	1.583	1.107	.2542
2.30	.4816	.3017	1.865	1.596	1.111	.2620
2.35	.4684	.2912	1.929	1.608	1.115	.2694
2.40	.4557	.2813	1.995	1.620	1.119	.2766
2.45	.4434	.2718	2.064	1.631	1.122	.2835
2.50	.4315	.2628	2.135	1.642	1.126	.2901
2.55	.4200	.2542	2.209	1.653	1.129	.2965
2.60	.4089	.2460	2.285	1.663	1.132	.3026
2.65	.3982	.2381	2.364	1.672	1.135	.3085
2.70	.3878	.2306	2.445	1.682	1.138	.3141
2.75	.3778	.2235	2.529	1.691	1.141	.3196
2.80	.3681	.2167	2.616	1.699	1.144	.3248
2.85	.3587	.2102	2.705	1.707	1.146	.3299
2.90	.3497	.2039	2.797	1.715	1.149	.3348
2.95	.3410	.1979	2.892	1.723	1.152	.3395
3.00	.3325	.1922	2.990	1.730	1.154	.3440
3.50	.2616	.1461	4.134	1.790	1.174	.3810
4.00	.2099	.1145	5.608	1.833	1.189	.4071
4.50	.1715	.09203	7.456	1.864	1.200	.4261
5.00	.1424	.07547	9.721	1.887	1.208	.4402
6.00	.10222	.05329	15.68	1.918	1.220	.4594
7.00	.07666	.03955	23.85	1.938	1.227	.4714
8.00	.05949	.03049	34.58	1.951	1.232	.4793
9.00	.04745	.02420	48.24	1.960	1.235	.4849
10.00	.03870	.01996	65.18	1.967	1.238	.4889
∞	0	0	∞	1.996	1.249	.5064

Table 5

Rayleigh line—one-dimensional, frictionless, constant-area flow with stagnation temperature change for a perfect gas

$\gamma = 1\cdot4$

M	$\dfrac{T_0}{T_0{}^*}$	$\dfrac{T}{T^*}$	$\dfrac{p}{p^*}$	$\dfrac{p_0}{p_0{}^*}$	$\dfrac{V}{V^*}$	$\left(\dfrac{T_0}{T_0{}^*}\right)_{isoth}$
0	0	0	2.4000	1.2679	0	.87500
0.01	.000480	.000576	2.3997	1.2678	.000240	.87502
0.02	.00192	.00230	2.3987	1.2675	.000959	.87507
0.03	.00431	.00516	2.3970	1.2671	.00216	.87516
0.04	.00765	.00917	2.3946	1.2665	.00383	.87528
0.05	.01192	.01430	2.3916	1.2657	.00598	.87544
0.06	.01712	.02053	2.3880	1.2647	.00860	.87563
0.07	.02322	.02784	2.3837	1.2636	.01168	.87586
0.08	.03021	.03621	2.3787	1.2623	.01522	.87612
0.09	.03807	.04562	2.3731	1.2608	.01922	.87642
0.10	.04678	.05602	2.3669	1.2591	.02367	.87675
0.11	.05630	.06739	2.3600	1.2573	.02856	.87712
0.12	.06661	.07970	2.3526	1.2554	.03388	.87752
0.13	.07768	.09290	2.3445	1.2533	.03962	.87796
0.14	.08947	.10695	2.3359	1.2510	.04578	.87843
0.15	.10196	.12181	2.3267	1.2486	.05235	.87894
0.16	.11511	.13743	2.3170	1.2461	.05931	.87948
0.17	.12888	.15377	2.3067	1.2434	.06666	.88006
0.18	.14324	.17078	2.2959	1.2406	.07438	.88067
0.19	.15814	.18841	2.2845	1.2377	.08247	.88132
0.20	.17355	.20661	2.2727	1.2346	.09091	.88200
0.21	.18943	.22533	2.2604	1.2314	.09969	.88272
0.22	.20574	.24452	2.2477	1.2281	.10879	.88347
0.23	.22244	.26413	2.2345	1.2248	.11820	.88426
0.24	.23948	.28411	2.2209	1.2213	.12792	.88508
0.25	.25684	.30440	2.2069	1.2177	.13793	.88594
0.26	.27446	.32496	2.1925	1.2140	.14821	.88683
0.27	.29231	.34573	2.1777	1.2102	.15876	.88776
0.28	.31035	.36667	2.1626	1.2064	.16955	.88872
0.29	.32855	.38773	2.1472	1.2025	.18058	.88972

Table 5 (Continued)

$\gamma = 1\cdot4$

M	$\dfrac{T_0}{T_0{}^*}$	$\dfrac{T}{T^*}$	$\dfrac{p}{p^*}$	$\dfrac{p_0}{p_0{}^*}$	$\dfrac{V}{V^*}$	$\left(\dfrac{T_0}{T_0{}^*}\right)_{\text{isoth}}$
0.30	.34686	.40887	2.1314	1.1985	0.19183	.89075
0.31	.36525	.43004	2.1154	1.1945	0.20329	.89182
0.32	.38369	.45119	2.0991	1.1904	0.21494	.89292
0.33	.40214	.47228	2.0825	1.1863	0.22678	.89406
0.34	.42057	.49327	2.0657	1.1821	0.23879	.89523
0.35	.43894	.51413	2.0487	1.1779	0.25096	.89644
0.36	.45723	.53482	2.0314	1.1737	0.26327	.89768
0.37	.47541	.55530	2.0140	1.1695	0.27572	.89896
0.38	.49346	.57553	1.9964	1.1652	0.28828	.90027
0.39	.51134	.59549	1.9787	1.1609	0.30095	.90162
0.40	.52903	.61515	1.9608	1.1566	0.31372	.90300
0.41	.54651	.63448	1.9428	1.1523	0.32658	.90442
0.42	.56376	.65345	1.9247	1.1480	0.33951	.90587
0.43	.58075	.67205	1.9065	1.1437	0.35251	.90736
0.44	.59748	.69025	1.8882	1.1394	0.36556	.90888
0.45	.61393	.70803	1.8699	1.1351	0.37865	.91044
0.46	.63007	.72538	1.8515	1.1308	0.39178	.91203
0.47	.64589	.74228	1.8331	1.1266	0.40493	.91366
0.48	.66139	.75871	1.8147	1.1224	0.41810	.91532
0.49	.67655	.77466	1.7962	1.1182	0.43127	.91702
0.50	.69136	.79012	1.7778	1.1140	0.4445	.91875
0.51	.70581	.80509	1.7594	1.1099	0.45761	.92052
0.52	.71990	.81955	1.7410	1.1059	0.47075	.92232
0.53	.73361	.83351	1.7226	1.1019	0.48387	.92416
0.54	.74695	.84695	1.7043	1.0979	0.49696	.92603
0.55	.75991	.85987	1.6860	1.09397	0.51001	.92794
0.56	.77248	.87227	1.6678	1.09010	0.52302	.92988
0.57	.78467	.88415	1.6496	1.08630	0.53597	.93186
0.58	.79647	.89552	1.6316	1.08255	0.54887	.93387
0.59	.80789	.90637	1.6136	1.07887	0.56170	.93592
0.60	.81892	.91670	1.5957	1.07525	0.57447	.93800
0.61	.82956	.92653	1.5780	1.07170	0.58716	.94012
0.62	.83982	.93585	1.5603	1.06821	0.59978	.94227
0.63	.84970	.94466	1.5427	1.06480	0.61232	.94446
0.64	.85920	.95298	1.5253	1.06146	0.62477	.94668

$\gamma = 1\cdot4$

M	$\dfrac{T_0}{T_0{}^*}$	$\dfrac{T}{T^*}$	$\dfrac{p}{p^*}$	$\dfrac{p_0}{p_0{}^*}$	$\dfrac{V}{V^*}$	$\left(\dfrac{T_0}{T_0{}^*}\right)_{\text{isoth}}$
0.65	.86833	.96081	1.5080	1.05820	.63713	.94894
0.66	.87709	.96816	1.4908	1.05502	.64941	.95123
0.67	.88548	.97503	1.4738	1.05192	.66159	.95356
0.68	.89350	.98144	1.4569	1.04890	.67367	.95592
0.69	.90117	.98739	1.4401	1.04596	.68564	.95832
0.70	.90850	.99289	1.4235	1.04310	.69751	.96075
0.71	.91548	.99796	1.4070	1.04033	.70927	.96322
0.72	.92212	1.00260	1.3907	1.03764	.72093	.96572
0.73	.92843	1.00682	1.3745	1.03504	.73248	.96826
0.74	.93442	1.01062	1.3585	1.03253	.74392	.97083
0.75	.94009	1.01403	1.3427	1.03010	.75525	.97344
0.76	.94546	1.01706	1.3270	1.02776	.76646	.97608
0.77	.95052	1.01971	1.3115	1.02552	.77755	.97876
0.78	.95528	1.02198	1.2961	1.02337	.78852	.98147
0.79	.95975	1.02390	1.2809	1.02131	.79938	.98422
0.80	.96394	1.02548	1.2658	1.01934	.81012	.98700
0.81	.96786	1.02672	1.2509	1.01746	.82075	.98982
0.82	.97152	1.02763	1.2362	1.01569	.83126	.99267
0.83	.97492	1.02823	1.2217	1.01399	.84164	.99556
0.84	.97807	1.02853	1.2073	1.01240	.85190	.99848
0.85	.98097	1.02854	1.1931	1.01091	.86204	1.00144
0.86	.98363	1.02826	1.1791	1.00951	.87206	1.00443
0.87	.98607	1.02771	1.1652	1.00819	.88196	1.00746
0.88	.98828	1.02690	1.1515	1.00698	.89175	1.01052
0.89	.99028	1.02583	1.1380	1.00587	.90142	1.01362
0.90	.99207	1.02451	1.1246	1.00485	.91097	1.01675
0.91	.99366	1.02297	1.1114	1.00393	.92039	1.01992
0.92	.99506	1.02120	1.09842	1.00310	.92970	1.02312
0.93	.99627	1.01921	1.08555	1.00237	.93889	1.02636
0.94	.99729	1.01702	1.07285	1.00174	.94796	1.02963
0.95	.99814	1.01463	1.06030	1.00121	.95692	1.03294
0.96	.99883	1.01205	1.04792	1.00077	.96576	1.03628
0.97	.99935	1.00929	1.03570	1.00043	.97449	1.03966
0.98	.99972	1.00636	1.02364	1.00019	.98311	1.04307
0.99	.99993	1.00326	1.01174	1.00004	.99161	1.04652

Table 5 (Continued)
$\gamma = 1\cdot4$

M	$\dfrac{T_0}{T_0{}^*}$	$\dfrac{T}{T^*}$	$\dfrac{p}{p^*}$	$\dfrac{p_0}{p_0{}^*}$	$\dfrac{V}{V^*}$	$\left(\dfrac{T_0}{T_0{}^*}\right)_{\text{isoth}}$	M	$\dfrac{T_0}{T_0{}^*}$	$\dfrac{T}{T^*}$	$\dfrac{p}{p^*}$	$\dfrac{p_0}{p_0{}^*}$	$\dfrac{V}{V^*}$	$\left(\dfrac{T_0}{T_0{}^*}\right)_{\text{isoth}}$
1.00	1.00000	1.00000	1.00000	1.00000	1.00000	1.05000	1.35	.94636	.83227	.67577	1.05943	1.2316	1.19394
1.01	.99993	.99659	.98841	1.00004	1.00828	1.05352	1.36	.94397	.82698	.66863	1.06288	1.2367	1.19868
1.02	.99973	.99304	.97697	1.00019	1.01644	1.05707	1.37	.94157	.82151	.66159	1.06642	1.2417	1.20346
1.03	.99940	.98936	.96569	1.00043	1.02450	1.06066	1.38	.93915	.81613	.65464	1.07006	1.2467	1.20827
1.04	.99895	.98553	.95456	1.00077	1.03246	1.06428	1.39	.93671	.81076	.64778	1.07380	1.2516	1.21312
1.05	.99838	.98161	.94358	1.00121	1.04030	1.06794	1.40	.93425	.80540	.64102	1.07765	1.2564	1.21800
1.06	.99769	.97755	.93275	1.00175	1.04804	1.07163	1.41	.93178	.80004	.63436	1.08159	1.2612	1.22292
1.07	.99690	.97339	.92206	1.00238	1.05567	1.07536	1.42	.92931	.79469	.62779	1.08563	1.2659	1.22787
1.08	.99600	.96913	.91152	1.00311	1.06320	1.07912	1.43	.92683	.78936	.62131	1.08977	1.2705	1.23286
1.09	.99501	.96477	.90112	1.00394	1.07062	1.08292	1.44	.92434	.78405	.61491	1.09400	1.2751	1.23788
1.10	.99392	.96031	.89086	1.00486	1.07795	1.08675	1.45	.92184	.77875	.60860	1.0983	1.2796	1.24294
1.11	.99274	.95577	.88075	1.00588	1.08518	1.09062	1.46	.91933	.77346	.60237	1.1028	1.2840	1.24803
1.12	.99148	.95115	.87078	1.00699	1.09230	1.09452	1.47	.91682	.76819	.59623	1.1073	1.2884	1.25316
1.13	.99013	.94646	.86094	1.00820	1.09933	1.09846	1.48	.91431	.76294	.59018	1.1120	1.2927	1.25832
1.14	.98871	.94169	.85123	1.00951	1.10626	1.10243	1.49	.91179	.75771	.58421	1.1167	1.2970	1.26352
1.15	.98721	.93685	.84166	1.01092	1.1131	1.10644	1.50	.90928	.75250	.57831	1.1215	1.3012	1.26875
1.16	.98564	.93195	.83222	1.01243	1.1198	1.11048	1.51	.90676	.74731	.57250	1.1264	1.3054	1.27402
1.17	.98400	.92700	.82292	1.01403	1.1264	1.11456	1.52	.90424	.74215	.56677	1.1315	1.3095	1.27932
1.18	.98230	.92200	.81374	1.01572	1.1330	1.11867	1.53	.90172	.73701	.56111	1.1367	1.3135	1.28466
1.19	.98054	.91695	.80468	1.01752	1.1395	1.12282	1.54	.89920	.73189	.55553	1.1420	1.3175	1.29003
1.20	.97872	.91185	.79576	1.01941	1.1459	1.12700	1.55	.89669	.72680	.55002	1.1473	1.3214	1.29544
1.21	.97685	.90671	.78695	1.02140	1.1522	1.13122	1.56	.89418	.72173	.54458	1.1527	1.3253	1.30088
1.22	.97492	.90153	.77827	1.02348	1.1584	1.13547	1.57	.89167	.71669	.53922	1.1582	1.3291	1.30636
1.23	.97294	.89632	.76971	1.02566	1.1645	1.13976	1.58	.88917	.71168	.53393	1.1639	1.3329	1.31187
1.24	.97092	.89108	.76127	1.02794	1.1705	1.14408	1.59	.88668	.70669	.52871	1.1697	1.3366	1.31742
1.25	.96886	.88581	.75294	1.03032	1.1764	1.14844	1.60	.88419	.70173	.52356	1.1756	1.3403	1.32300
1.26	.96675	.88052	.74473	1.03280	1.1823	1.15283	1.61	.88170	.69680	.51848	1.1816	1.3439	1.32862
1.27	.96461	.87521	.73663	1.03536	1.1881	1.15726	1.62	.87922	.69190	.51346	1.1877	1.3475	1.33427
1.28	.96243	.86988	.72865	1.03803	1.1938	1.16172	1.63	.87675	.68703	.50851	1.1939	1.3511	1.33996
1.29	.96022	.86453	.72078	1.04080	1.1994	1.16622	1.64	.87429	.68219	.50363	1.2002	1.3546	1.34568
1.30	.95798	.85917	.71301	1.04365	1.2050	1.17075	1.65	.87184	.67738	.49881	1.2066	1.3580	1.35144
1.31	.95571	.85380	.70535	1.04661	1.2105	1.17532	1.66	.86940	.67259	.49405	1.2131	1.3614	1.35723
1.32	.95341	.84843	.69780	1.04967	1.2159	1.17992	1.67	.86696	.66784	.48935	1.2197	1.3648	1.36306
1.33	.95108	.84305	.69035	1.05283	1.2212	1.18456	1.68	.86453	.66312	.48471	1.2264	1.3681	1.36892
1.34	.94873	.83766	.68301	1.05608	1.2264	1.18923	1.69	.86211	.65843	.48014	1.2332	1.3713	1.37482

Table 5 (Continued)

$\gamma = 1 \cdot 4$

M	$\dfrac{T_0}{T_0^*}$	$\dfrac{T}{T^*}$	$\dfrac{p}{p^*}$	$\dfrac{p_0}{p_0^*}$	$\dfrac{V}{V^*}$	$\left(\dfrac{T_0}{T_0^*}\right)_{\text{isoth}}$
1.70	.85970	.65377	.47563	1.2402	1.3745	1.38075
1.71	.85731	.64914	.47117	1.2473	1.3777	1.38672
1.72	.85493	.64455	.46677	1.2545	1.3809	1.39272
1.73	.85256	.63999	.46242	1.2618	1.3840	1.39876
1.74	.85020	.63546	.45813	1.2692	1.3871	1.40483
1.75	.84785	.63096	.45390	1.2767	1.3901	1.41094
1.76	.84551	.62649	.44972	1.2843	1.3931	1.41708
1.77	.84318	.62205	.44559	1.2920	1.3960	1.42326
1.78	.84087	.61765	.44152	1.2998	1.3989	1.42947
1.79	.83857	.61328	.43750	1.3078	1.4018	1.43572
1.80	.83628	.60894	.43353	1.3159	1.4046	1.44200
1.81	.83400	.60463	.42960	1.3241	1.4074	1.44832
1.82	.83174	.60036	.42573	1.3324	1.4102	1.45467
1.83	.82949	.59612	.42191	1.3408	1.4129	1.46106
1.84	.82726	.59191	.41813	1.3494	1.4156	1.46748
1.85	.82504	.58773	.41440	1.3581	1.4183	1.47394
1.86	.82283	.58359	.41072	1.3669	1.4209	1.48043
1.87	.82064	.57948	.40708	1.3758	1.4235	1.48696
1.88	.81846	.57540	.40349	1.3848	1.4261	1.49352
1.89	.81629	.57135	.39994	1.3940	1.4286	1.50012
1.90	.81414	.56734	.39643	1.4033	1.4311	1.50675
1.91	.81200	.56336	.39297	1.4127	1.4336	1.51342
1.92	.80987	.55941	.38955	1.4222	1.4360	1.52012
1.93	.80776	.55549	.38617	1.4319	1.4384	1.52686
1.94	.80567	.55160	.38283	1.4417	1.4408	1.53363
1.95	.80359	.54774	.37954	1.4516	1.4432	1.54044
1.96	.80152	.54391	.37628	1.4616	1.4455	1.54728
1.97	.79946	.54012	.37306	1.4718	1.4478	1.55416
1.98	.79742	.53636	.36988	1.4821	1.4501	1.56107
1.99	.79540	.53263	.36674	1.4925	1.4523	1.56802
2.00	.79339	.52893	.36364	1.5031	1.4545	1.57500
2.01	.79139	.52526	.36057	1.5138	1.4567	1.58202
2.02	.78941	.52161	.35754	1.5246	1.4589	1.58907
2.03	.78744	.51800	.35454	1.5356	1.4610	1.59616
2.04	.78549	.51442	.35158	1.5467	1.4631	1.60328

$\gamma = 1 \cdot 4$

M	$\dfrac{T_0}{T_0^*}$	$\dfrac{T}{T^*}$	$\dfrac{p}{p^*}$	$\dfrac{p_0}{p_0^*}$	$\dfrac{V}{V^*}$	$\left(\dfrac{T_0}{T_0^*}\right)_{\text{isoth}}$
2.05	.78355	.51087	.34866	1.5579	1.4652	1.61044
2.06	.78162	.50735	.34577	1.5693	1.4673	1.61763
2.07	.77971	.50386	.34291	1.5808	1.4694	1.62486
2.08	.77781	.50040	.34009	1.5924	1.4714	1.63212
2.09	.77593	.49697	.33730	1.6042	1.4734	1.63942
2.10	.77406	.49356	.33454	1.6161	1.4753	1.64675
2.11	.77221	.49018	.33181	1.6282	1.4773	1.65412
2.12	.77037	.48683	.32912	1.6404	1.4792	1.66152
2.13	.76854	.48351	.32646	1.6528	1.4811	1.66896
2.14	.76673	.48022	.32383	1.6653	1.4830	1.67643
2.15	.76493	.47696	.32122	1.6780	1.4849	1.68394
2.16	.76314	.47373	.31864	1.6908	1.4867	1.69148
2.17	.76137	.47052	.31610	1.7037	1.4885	1.69906
2.18	.75961	.46734	.31359	1.7168	1.4903	1.70667
2.19	.75787	.46419	.31110	1.7300	1.4921	1.71432
2.20	.75614	.46106	.30864	1.7434	1.4939	1.72200
2.21	.75442	.45796	.30621	1.7570	1.4956	1.72972
2.22	.75271	.45489	.30381	1.7707	1.4973	1.73747
2.23	.75102	.45184	.30143	1.7846	1.4990	1.74526
2.24	.74934	.44882	.29908	1.7986	1.5007	1.75308
2.25	.74767	.44582	.29675	1.8128	1.5024	1.76094
2.26	.74602	.44285	.29445	1.8271	1.5040	1.76883
2.27	.74438	.43990	.29218	1.8416	1.5056	1.77676
2.28	.74275	.43698	.28993	1.8562	1.5072	1.78472
2.29	.74114	.43409	.28771	1.8710	1.5088	1.79272
2.30	.73954	.43122	.28551	1.8860	1.5104	1.80075
2.31	.73795	.42837	.28333	1.9012	1.5119	1.80882
2.32	.73638	.42555	.28118	1.9165	1.5134	1.81692
2.33	.73482	.42276	.27905	1.9320	1.5150	1.82506
2.34	.73327	.41999	.27695	1.9476	1.5165	1.83323
2.35	.73173	.41724	.27487	1.9634	1.5180	1.84144
2.36	.73020	.41451	.27281	1.9794	1.5195	1.84968
2.37	.72868	.41181	.27077	1.9955	1.5209	1.85796
2.38	.72718	.40913	.26875	2.0118	1.5223	1.86627
2.39	.72569	.40647	.26675	2.0283	1.5237	1.87462

Table 5 (Continued)

$\gamma = 1.4$

M	$\dfrac{T_0}{T_0{}^*}$	$\dfrac{T}{T^*}$	$\dfrac{p}{p^*}$	$\dfrac{p_0}{p_0{}^*}$	$\dfrac{V}{V^*}$	$\left(\dfrac{T_0}{T_0{}^*}\right)_{\text{isoth}}$
2.40	.72421	.40383	.26478	2.0450	1.5252	1.88300
2.41	.72274	.40122	.26283	2.0619	1.5266	1.89142
2.42	.72129	.39863	.26090	2.0789	1.5279	1.89987
2.43	.71985	.39606	.25899	2.0961	1.5293	1.90836
2.44	.71842	.39352	.25710	2.1135	1.5306	1.91688
2.45	.71700	.39100	.25523	2.1311	1.5320	1.92544
2.46	.71559	.38850	.25337	2.1489	1.5333	1.93403
2.47	.71419	.38602	.25153	2.1669	1.5346	1.94266
2.48	.71280	.38356	.24972	2.1850	1.5359	1.95132
2.49	.71142	.38112	.24793	2.2033	1.5372	1.96002
2.50	.71005	.37870	.24616	2.2218	1.5385	1.96875
2.51	.70870	.37630	.24440	2.2405	1.5398	1.97752
2.52	.70736	.37392	.24266	2.2594	1.5410	1.98632
2.53	.70603	.37157	.24094	2.2785	1.5422	1.99515
2.54	.70471	.36923	.23923	2.2978	1.5434	2.00403
2.55	.70340	.36691	.23754	2.3173	1.5446	2.01294
2.56	.70210	.36461	.23587	2.3370	1.5458	2.02188
2.57	.70081	.36233	.23422	2.3569	1.5470	2.03086
2.58	.69953	.36007	.23258	2.3770	1.5482	2.03987
2.59	.69825	.35783	.23096	2.3972	1.5494	2.04892
2.60	.69699	.35561	.22936	2.4177	1.5505	2.05800
2.61	.69574	.35341	.22777	2.4384	1.5516	2.06711
2.62	.69450	.35123	.22620	2.4593	1.5527	2.07627
2.63	.69327	.34906	.22464	2.4804	1.5538	2.08546
2.64	.69205	.34691	.22310	2.5017	1.5549	2.09468
2.65	.69084	.34478	.22158	2.5233	1.5560	2.10394
2.66	.68964	.34267	.22007	2.5451	1.5571	2.11323
2.67	.68845	.34057	.21857	2.5671	1.5582	2.12256
2.68	.68727	.33849	.21709	2.5892	1.5593	2.13192
2.69	.68610	.33643	.21562	2.6116	1.5603	2.14132
2.70	.68494	.33439	.21417	2.6342	1.5613	2.15075
2.71	.68378	.33236	.21273	2.6571	1.5623	2.16022
2.72	.68263	.33035	.21131	2.6802	1.5633	2.16972
2.73	.68150	.32836	.20990	2.7035	1.5644	2.17925
2.74	.68038	.32638	.20850	2.7270	1.5654	2.18883

$\gamma = 1.4$

M	$\dfrac{T_0}{T_0{}^*}$	$\dfrac{T}{T^*}$	$\dfrac{p}{p^*}$	$\dfrac{p_0}{p_0{}^*}$	$\dfrac{V}{V^*}$	$\left(\dfrac{T_0}{T_0{}^*}\right)_{\text{isoth}}$
2.75	.67926	.32442	.20712	2.7508	1.5663	2.19844
2.76	.67815	.32248	.20575	2.7748	1.5673	2.20808
2.77	.67704	.32055	.20439	2.7990	1.5683	2.21776
2.78	.67595	.31864	.20305	2.8235	1.5692	2.22747
2.79	.67487	.31674	.20172	2.8482	1.5702	2.23722
2.80	.67380	.31486	.20040	2.8731	1.5711	2.24700
2.81	.67273	.31299	.19909	2.8982	1.5721	2.25682
2.82	.67167	.31114	.19780	2.9236	1.5730	2.26667
2.83	.67062	.30931	.19652	2.9493	1.5739	2.27655
2.84	.66958	.30749	.19525	2.9752	1.5748	2.28648
2.85	.66855	.30568	.19399	3.0013	1.5757	2.29644
2.86	.66752	.30389	.19274	3.0277	1.5766	2.30643
2.87	.66650	.30211	.19151	3.0544	1.5775	2.31646
2.88	.66549	.30035	.19029	3.0813	1.5784	2.32652
2.89	.66449	.29860	.18908	3.1084	1.5792	2.33662
2.90	.66350	.29687	.18788	3.1358	1.5801	2.34675
2.91	.66252	.29515	.18669	3.1635	1.5809	2.35692
2.92	.66154	.29344	.18551	3.1914	1.5818	2.36712
2.93	.66057	.29175	.18435	3.2196	1.5826	2.37735
2.94	.65961	.29007	.18320	3.2481	1.5834	2.38763
2.95	.65865	.28841	.18205	3.2768	1.5843	2.39794
2.96	.65770	.28676	.18091	3.3058	1.5851	2.40828
2.97	.65676	.28512	.17978	3.3351	1.5859	2.41865
2.98	.65583	.28349	.17867	3.3646	1.5867	2.42907
2.99	.65490	.28188	.17757	3.3944	1.5875	2.43952
3.00	.65398	.28028	.17647	3.4244	1.5882	2.45000
3.50	.61580	.21419	.13223	5.3280	1.6198	3.01875
4.00	.58909	.16831	.10256	8.2268	1.6410	3.67500
4.50	.56983	.13540	.08177	12.502	1.6559	4.41874
5.00	.55555	.11111	.06667	18.634	1.6667	5.25000
6.00	.53633	.07849	.04669	38.946	1.6809	7.17499
7.00	.52437	.05826	.03448	75.414	1.6896	9.45000
8.00	.51646	.04491	.02649	136.62	1.6954	12.07500
9.00	.51098	.03565	.02098	233.88	1.6993	15.05003
10.00	.50702	.02897	.01702	381.62	1.7021	18.37500
∞	.48980	0	0	∞	1.7143	∞

Table 5 (Continued)
$\gamma = 1 \cdot 0$

$\gamma = 1 \cdot 0$

M	$\dfrac{T_0}{T_0{}^*} = \dfrac{T^*}{T}$	$\dfrac{p}{p^*}$	$\dfrac{p_0}{p_0{}^*}$	$\dfrac{V}{V^*}$	M	$\dfrac{T_0}{T_0{}^*} = \dfrac{T}{T^*}$	$\dfrac{p}{p^*}$	$\dfrac{p_0}{p_0{}^*}$	$\dfrac{V}{V^*}$
0	0	2.000	1.213	0	1.75	0.7422	0.4923	1.381	1.508
0.05	.00995	1.995	1.212	.00499	1.80	0.7209	0.4717	1.446	1.528
0.10	.03921	1.980	1.207	.01980	1.85	0.7000	0.4522	1.519	1.547
0.15	.08608	1.956	1.200	.04401	1.90	0.6795	0.4338	1.600	1.566
0.20	.14793	1.923	1.190	.07692	1.95	0.6595	0.4164	1.691	1.584
0.25	.2215	1.882	1.178	.1176	2.00	0.6400	0.4000	1.793	1.601
0.30	.3030	1.835	1.164	.1651	2.05	0.6211	0.3844	1.907	1.616
0.35	.3889	1.782	1.149	.2183	2.10	0.6027	0.3697	2.034	1.630
0.40	.4756	1.724	1.133	.2758	2.15	0.5849	0.3557	2.176	1.644
0.45	.5602	1.663	1.116	.3368	2.20	0.5677	0.3425	2.336	1.657
0.50	.6400	1.600	1.0997	.4000	2.25	0.5510	0.3299	2.515	1.670
0.55	.7132	1.536	1.0834	.4645	2.30	0.5348	0.3179	2.716	1.682
0.60	.7785	1.471	1.0679	.5294	2.35	0.5192	0.3066	2.942	1.693
0.65	.8352	1.406	1.0534	.5940	2.40	0.5042	0.2959	3.197	1.704
0.70	.8828	1.342	1.0402	.6577	2.45	0.4897	0.2857	3.484	1.714
0.75	.9216	1.280	1.0285	.7200	2.50	0.4757	0.2759	3.808	1.724
0.80	.9518	1.220	1.0186	.7805	2.55	0.4621	0.2666	4.175	1.733
0.85	.9740	1.161	1.0107	.8389	2.60	0.4490	0.2577	4.591	1.742
0.90	.9890	1.105	1.0048	.8950	2.65	0.4364	0.2493	5.064	1.751
0.95	.9974	1.0512	1.0012	.9488	2.70	0.4243	0.2413	5.602	1.759
1.00	1.0000	1.0000	1.0000	1.0000	2.75	0.4126	0.2336	6.215	1.766
1.05	.9976	.9512	1.0013	1.0488	2.80	0.4013	0.2262	6.916	1.774
1.10	.9910	.9049	1.0052	1.0951	2.85	0.3904	0.2192	7.719	1.781
1.15	.9807	.8611	1.0118	1.1389	2.90	0.3799	0.2125	8.640	1.787
1.20	.9675	.8197	1.0214	1.1802	2.95	0.3698	0.2061	9.699	1.794
1.25	.9518	.7805	1.0340	1.220	3.00	0.3600	0.2000	10.92	1.800
1.30	.9342	.7435	1.0498	1.257	3.50	0.2791	0.1509	41.85	1.849
1.35	.9151	.7086	1.0690	1.291	4.00	0.2215	0.1176	212.71	1.882
1.40	.8948	.6757	1.0919	1.324	4.50	0.1794	0.09412	1425	1.906
1.45	.8737	.6447	1.1187	1.355	5.00	0.1479	0.07692	12519	1.923
1.50	.8521	.6154	1.150	1.384	6.00	0.10519	0.05405	$215(10)^4$	1.946
1.55	.8301	.5878	1.186	1.412	7.00	0.07840	0.04000	$106(10)^7$	1.960
1.60	.8080	.5618	1.226	1.438	8.00	0.06059	0.03077	$147(10)^{10}$	1.969
1.65	.7859	.5373	1.271	1.463	9.00	0.04818	0.02439	$574(10)^{13}$	1.976
1.70	.7639	.5141	1.323	1.486	10.00	0.03921	0.01980	$623(10)^{17}$	1.980
					∞	0	0	∞	2.000

Table 5 (Continued)

$\gamma = 1{\cdot}1$

M	$\dfrac{T_0}{T_0^*}$	$\dfrac{T}{T^*}$	$\dfrac{p}{p^*}$	$\dfrac{p_0}{p_0^*}$	$\dfrac{V}{V^*}$	$\left(\dfrac{T_0}{T_0^*}\right)_{\text{isoth}}$
0	0	0	2.100	1.228	0	0.95652
0.05	.01044	0.01097	2.094	1.226	0.00524	0.95664
0.10	.04111	0.04315	2.077	1.221	0.02077	0.95700
0.15	.09009	0.09449	2.049	1.213	0.04611	0.95760
0.20	.15444	0.16184	2.011	1.203	0.08046	0.95843
0.25	.2305	0.2413	1.965	1.190	0.1228	0.95951
0.30	.3144	0.3286	1.911	1.174	0.1720	0.96083
0.35	.4020	0.4195	1.851	1.157	0.2267	0.96238
0.40	.4898	0.5102	1.786	1.140	0.2857	0.96417
0.45	.5746	0.5973	1.717	1.122	0.3478	0.96621
0.50	.6540	0.6782	1.647	1.1040	0.4118	0.96848
0.55	.7261	0.7510	1.576	1.0867	0.4766	0.97099
0.60	.7898	0.8147	1.504	1.0702	0.5416	0.97374
0.65	.8446	0.8684	1.434	1.0550	0.6057	0.97673
0.70	.8902	0.9123	1.365	1.0412	0.6686	0.97996
0.75	.9270	0.9467	1.297	1.0291	0.7297	0.98342
0.80	.9554	0.9720	1.232	1.0189	0.7887	0.98713
0.85	.9761	0.9892	1.170	1.0109	0.8453	0.99108
0.90	.9899	0.9989	1.111	1.0050	0.8995	0.99526
0.95	.9976	1.0023	1.0538	1.0013	0.9511	0.99968
1.00	1.0000	1.0000	1.0000	1.0000	1.0000	1.00435
1.05	.9979	0.9930	0.9490	1.0013	1.0463	1.00925
1.10	.9919	0.9821	0.9009	1.0051	1.0901	1.01439
1.15	.9827	0.9679	0.8555	1.0116	1.1314	1.01977
1.20	.9710	0.9511	0.8127	1.0209	1.1703	1.02539
1.25	.9572	0.9322	0.7724	1.0331	1.207	1.03125
1.30	.9418	0.9118	0.7345	1.0483	1.241	1.03735
1.35	.9251	0.8902	0.6989	1.0665	1.273	1.04368
1.40	.9074	0.8678	0.6654	1.0880	1.304	1.05026
1.45	.8892	0.8449	0.6339	1.1130	1.333	1.05708
1.50	.8706	0.8217	0.6043	1.141	1.360	1.06413
1.55	.8518	0.7984	0.5765	1.173	1.385	1.07142
1.60	.8329	0.7753	0.5503	1.210	1.409	1.07896
1.65	.8141	0.7524	0.5257	1.251	1.431	1.08673
1.70	.7955	0.7298	0.5025	1.297	1.452	1.09474

$\gamma = 1{\cdot}1$

M	$\dfrac{T_0}{T_0^*}$	$\dfrac{T}{T^*}$	$\dfrac{p}{p^*}$	$\dfrac{p_0}{p_0^*}$	$\dfrac{V}{V^*}$	$\left(\dfrac{T_0}{T_0^*}\right)_{\text{isoth}}$
1.75	.7771	.7076	.4807	1.347	1.472	1.10299
1.80	.7591	.6859	.4601	1.403	1.491	1.11148
1.85	.7415	.6648	.4407	1.465	1.508	1.12021
1.90	.7243	.6443	.4224	1.532	1.525	1.21917
1.95	.7076	.6243	.4052	1.607	1.541	1.13838
2.00	.6914	.6049	.3889	1.689	1.556	1.14783
2.05	.6756	.5862	.3735	1.780	1.570	1.15751
2.10	.6603	.5681	.3589	1.879	1.583	1.16743
2.15	.6456	.5506	.3451	1.987	1.595	1.17760
2.20	.6313	.5337	.3321	2.106	1.607	1.18800
2.25	.6175	.5174	.3197	2.237	1.618	1.19864
2.30	.6042	.5017	.3079	2.380	1.629	1.20952
2.35	.5914	.4866	.2968	2.537	1.639	1.22064
2.40	.5790	.4720	.2863	2.709	1.649	1.23200
2.45	.5671	.4580	.2763	2.897	1.658	1.24360
2.50	.5556	.4444	.2667	3.104	1.667	1.25543
2.55	.5445	.4314	.2576	3.332	1.675	1.26751
2.60	.5338	.4189	.2489	3.581	1.683	1.27983
2.65	.5235	.4068	.2406	3.855	1.690	1.29238
2.70	.5136	.3952	.2328	4.156	1.697	1.30517
2.75	.5041	.3840	.2253	4.487	1.704	1.31821
2.80	.4949	.3733	.2182	4.851	1.711	1.33148
2.85	.4860	.3629	.2114	5.251	1.717	1.34499
2.90	.4775	.3529	.2049	5.692	1.723	1.35874
2.95	.4693	.3433	.1986	6.176	1.729	1.37273
3.00	.4613	.3341	.1927	6.710	1.734	1.38696
3.50	.3960	.2578	.1451	16.26	1.777	1.54239
4.00	.3496	.2040	.1129	42.42	1.806	1.72174
4.50	.3160	.1648	.0902	115.70	1.827	1.92500
5.00	.2909	.1357	.0737	322.33	1.842	2.15217
6.00	.2568	.09631	.05172	2508	1.862	2.67826
7.00	.2356	.07169	.03825	18430	1.874	3.30000
8.00	.2215	.05536	.02941	$123(10)^3$	1.882	4.01739
9.00	.2116	.04400	.02331	$743(10)^3$	1.888	4.83044
10.00	.2045	.03579	.01892	$401(10)^4$	1.892	5.73913
∞	.1736	0	0	∞	1.909	∞

Table 5 (Continued)

$\gamma = 1\cdot2$

M	$\dfrac{T_0}{T_0{}^*}$	$\dfrac{T}{T^*}$	$\dfrac{p}{p^*}$	$\dfrac{p_0}{p_0{}^*}$	$\dfrac{V}{V^*}$	$\left(\dfrac{T_0}{T_0{}^*}\right)_{\text{isoth}}$
0	0	0	2.200	1.242	0	0.92308
0.05	.01094	.01203	2.193	1.239	.00548	0.92331
0.10	.04301	.04726	2.173	1.234	.02174	0.92400
0.15	.09408	.10325	2.141	1.226	.04820	0.92515
0.20	.16089	.17627	2.099	1.214	.08397	0.92677
0.25	.2395	.2618	2.047	1.199	.1279	0.92885
0.30	.3255	.3548	1.986	1.183	.1787	0.93138
0.35	.4147	.4507	1.918	1.165	.2350	0.93438
0.40	.5034	.5450	1.846	1.146	.2953	0.93785
0.45	.5884	.6343	1.770	1.127	.3584	0.94177
0.50	.6672	.7160	1.692	1.1078	.4231	0.94615
0.55	.7381	.7881	1.614	1.0895	.4884	0.95100
0.60	.8003	.8497	1.536	1.0722	.5531	0.95631
0.65	.8531	.9004	1.460	1.0563	.6168	0.96208
0.70	.8969	.9405	1.385	1.0420	.6788	0.96831
0.75	.9318	.9704	1.313	1.0296	.7388	0.97500
0.80	.9585	.9910	1.244	1.0191	.7964	0.98215
0.85	.9779	1.0032	1.178	1.0109	.8514	0.98977
0.90	.9907	1.0081	1.115	1.0049	.9037	0.99785
0.95	.9978	1.0067	1.0562	1.0012	.9532	1.00638
1.00	1.0000	1.0000	1.0000	1.0000	1.0000	1.01538
1.05	.9981	.9888	.9471	1.0013	1.0441	1.02485
1.10	.9927	.9741	.8972	1.0050	1.0856	1.03477
1.15	.9845	.9564	.8504	1.0114	1.1247	1.04515
1.20	.9740	.9365	.8065	1.0204	1.1613	1.05600
1.25	.9617	.9149	.7653	1.0322	1.196	1.06731
1.30	.9481	.8921	.7266	1.0467	1.228	1.07908
1.35	.9334	.8685	.6903	1.0640	1.258	1.09131
1.40	.9180	.8443	.6563	1.0843	1.286	1.10400
1.45	.9021	.8199	.6245	1.1077	1.313	1.11715
1.50	.8859	.7955	.5946	1.134	1.338	1.13077
1.55	.8695	.7712	.5666	1.164	1.361	1.14485
1.60	.8532	.7473	.5403	1.197	1.383	1.15938
1.65	.8370	.7237	.5156	1.234	1.404	1.17438
1.70	.8211	.7007	.4924	1.275	1.423	1.18985

$\gamma = 1\cdot2$

M	$\dfrac{T_0}{T_0{}^*}$	$\dfrac{T}{T^*}$	$\dfrac{p}{p^*}$	$\dfrac{p_0}{p_0{}^*}$	$\dfrac{V}{V^*}$	$\left(\dfrac{T_0}{T_0{}^*}\right)_{\text{isoth}}$
1.75	.8054	.6782	.4706	1.320	1.441	1.20577
1.80	.7900	.6563	.4501	1.369	1.458	1.22215
1.85	.7750	.6351	.4308	1.422	1.474	1.23900
1.90	.7604	.6146	.4126	1.480	1.490	1.25631
1.95	.7462	.5947	.3955	1.543	1.504	1.27408
2.00	.7325	.5755	.3793	1.612	1.517	1.29231
2.05	.7192	.5570	.3641	1.687	1.530	1.31100
2.10	.7063	.5391	.3497	1.767	1.542	1.33015
2.15	.6939	.5219	.3360	1.854	1.553	1.34977
2.20	.6819	.5054	.3231	1.948	1.564	1.36985
2.25	.6703	.4895	.3109	2.050	1.574	1.39039
2.30	.6591	.4742	.2994	2.159	1.584	1.41139
2.35	.6484	.4595	.2884	2.277	1.593	1.43285
2.40	.6381	.4453	.2780	2.405	1.602	1.45477
2.45	.6281	.4317	.2682	2.542	1.610	1.47715
2.50	.6185	.4187	.2588	2.690	1.618	1.50000
2.55	.6093	.4062	.2499	2.849	1.625	1.52331
2.60	.6004	.3941	.2414	3.021	1.632	1.54708
2.65	.5918	.3825	.2334	3.205	1.639	1.57131
2.70	.5836	.3713	.2257	3.403	1.645	1.59600
2.75	.5757	.3606	.2184	3.617	1.651	1.62115
2.80	.5681	.3503	.2114	3.847	1.657	1.64677
2.85	.5608	.3404	.2047	4.094	1.663	1.67285
2.90	.5537	.3309	.1983	4.359	1.668	1.69938
2.95	.5469	.3217	.1923	4.644	1.673	1.72638
3.00	.5404	.3128	.1864	4.951	1.678	1.75385
3.50	.4865	.2405	.1401	9.597	1.717	2.05385
4.00	.4486	.1898	.1089	18.99	1.743	2.40000
4.50	.4211	.1531	.08696	37.61	1.761	2.79230
5.00	.4006	.1259	.07097	73.64	1.774	3.23077
6.00	.3730	.08919	.04977	266.2	1.792	4.24615
7.00	.3557	.06632	.03679	875.9	1.803	5.44615
8.00	.3443	.05118	.02828	2621	1.810	6.83077
9.00	.3363	.04065	.02240	7181	1.815	8.40002
10.00	.3306	.03306	.01818	18182	1.818	10.15385
∞	0.3056	0	0	∞	1.833	∞

Table 5 (Continued)
$\gamma = 1\cdot3$

M	$\dfrac{T_0}{T_0{}^*}$	$\dfrac{T}{T^*}$	$\dfrac{p}{p^*}$	$\dfrac{p_0}{p_0{}^*}$	$\dfrac{V}{V^*}$	$\left(\dfrac{T_0}{T_0{}^*}\right)_{\text{isoth}}$
0	0	0	2.300	1.255	0	.89655
0.05	.01143	.01314	2.293	1.253	.00573	.89689
0.10	.04489	.05155	2.270	1.247	.02270	.89790
0.15	.09803	.11236	2.234	1.237	.05028	.89958
0.20	.16726	.19120	2.186	1.224	.08745	.90193
0.25	.2482	.2828	2.127	1.209	.1329	.90496
0.30	.3363	.3816	2.059	1.191	.1853	.90866
0.35	.4270	4822	1 984	1.172	.2430	.91303
0.40	.5165	.5800	1.904	1.152	.3046	.91807
0.45	.6015	.6713	1.821	1.131	.3687	.92378
0.50	.6796	.7533	1.736	1.1112	.4340	.93017
0.55	.7494	.8244	1.651	1.0919	.4994	.93723
0.60	.8099	.8837	1.567	1.0739	.5640	.94497
0.65	.8611	.9312	1.485	1.0574	.6272	.95337
0.70	.9029	.9673	1.405	1.0426	.6885	.96245
0.75	.9361	.9928	1.328	1.0299	.7473	.97220
0.80	.9614	1.0088	1.255	1.0193	.8035	.98262
0.85	.9795	1.0163	1.186	1.0109	.8569	.99372
0.90	.9914	1.0166	1.120	1.0049	.9075	1.00548
0.95	.9980	1.0108	1.0583	1.0012	.9552	1.01792
1.00	1.0000	1.0000	1.0000	1.0000	1.0000	1.03103
1.05	.9982	.9851	.9452	1.0012	1.0421	1.04482
1.10	.9933	.9669	.8939	1.0049	1.0816	1.05928
1.15	.9859	.9461	.8458	1.0111	1.1186	1.07441
1.20	.9765	.9235	.8008	1.0199	1.1532	1.09021
1.25	.9656	.8996	.7588	1.0312	1.186	1.10668
1.30	.9534	.8747	.7194	1.0451	1.216	1.12383
1.35	.9404	.8493	.6826	1.0617	1.244	1.14165
1.40	.9268	.8237	.6483	1.0809	1.270	1.16014
1.45	.9128	.7980	.6161	1.1028	1.295	1.17930
1.50	.8986	.7726	.5860	1.128	1.318	1.19914
1.55	.8843	.7475	.5578	1.155	1.340	1.21965
1.60	.8701	.7230	.5314	1.185	1.360	1.24083
1.65	.8560	.6990	.5067	1.219	1.379	1.26268
1.70	.8421	.6756	.4835	1.256	1.397	1.28521

$\gamma = 1\cdot3$

M	$\dfrac{T_0}{T_0{}^*}$	$\dfrac{T}{T^*}$	$\dfrac{p}{p^*}$	$\dfrac{p_0}{p_0{}^*}$	$\dfrac{V}{V^*}$	$\left(\dfrac{T_0}{T_0{}^*}\right)_{\text{isoth}}$
1.75	.8285	.6529	.4617	1.296	1.414	1.30841
1.80	.8153	.6309	.4413	1.340	1.430	1.33227
1.85	.8024	.6097	.4221	1.387	1.445	1.35682
1.90	.7898	.5892	.4040	1.438	1.459	1.38204
1.95	.7776	.5695	.3870	1.493	1.472	1.40792
2.00	.7659	.5505	.3710	1.552	1.484	1.43448
2.05	.7545	.5322	.3559	1.615	1.495	1.46172
2.10	.7435	.5146	.3416	1.683	1.506	1.48962
2.15	.7329	.4977	.3281	1.755	1.517	1.51820
2.20	.7227	.4815	.3154	1.832	1.527	1.54745
2.25	.7129	.4659	.3034	1.915	1.536	1.57737
2.30	.7034	.4510	.2920	2.003	1.545	1.60797
2.35	.6943	.4367	.2812	2.097	1.553	1.63923
2.40	.6855	.4229	.2710	2.197	1.561	1.67117
2.45	.6771	.4097	.2613	2.303	1.568	1.70378
2.50	.6690	.3971	.2521	2.416	1.575	1.73707
2.55	.6612	.3850	.2433	2.536	1.582	1.77103
2.60	.6537	.3733	.2350	2.664	1.588	1.80565
2.65	.6465	.3621	.2271	2.800	1.594	1.84096
2.70	.6396	.3513	.2195	2.944	1.600	1.87693
2.75	.6329	.3410	.2123	3.096	1.606	1.91358
2.80	.6265	.3311	.2055	3.258	1.611	1.95090
2.85	.6203	.3216	.1990	3.429	1.616	1.98889
2.90	.6144	.3124	.1928	3.611	1.621	2.02755
2.95	.6087	.3036	.1868	3.804	1.626	2.06689
3.00	.6032	.2952	.1811	4.007	1.630	2.10690
3.50	.5582	.2262	.1359	6.806	1.665	2.54397
4.00	.5265	.1781	.1055	11.57	1.688	3.04828
4.50	.5037	.1435	.08417	19.44	1.704	3.61982
5.00	.4867	.1178	.06866	32.06	1.716	4.25862
6.00	.4639	.08335	.04812	76.97	1.732	5.73793
7.00	.4496	.06192	.03555	191.3	1.742	7.48621
8.00	.4402	.04775	.02732	413.4	1.748	9.50345
9.00	.4336	.03792	.02164	833.4	1.753	11.78968
10.00	.4289	.03082	.01756	1582	1.756	14.34483
∞	.4083	0	0	∞	1.769	∞

Table 5 (Continued)
$\gamma = 1\cdot67$

M	$\dfrac{T_0}{T_0{}^*}$	$\dfrac{T}{T^*}$	$\dfrac{p}{p^*}$	$\dfrac{p_0}{p_0{}^*}$	$\dfrac{V}{V^*}$	$\left(\dfrac{T_0}{T_0{}^*}\right)_{\mathrm{isoth}}$
0	0	0	2.670	1.299	0	.83292
0.05	.01325	.01767	2.659	1.297	.00665	.83362
0.10	.05183	.06896	2.626	1.289	.02626	.83571
0.15	.11243	.1490	2.573	1.276	.05790	.83920
0.20	.19020	.2506	2.503	1.259	.10011	.84408
0.25	.2794	.3653	2.418	1.239	.1511	.85036
0.30	.3742	.4849	2.321	1.216	.2089	.85803
0.35	.4693	.6018	2.216	1.192	.2715	.86710
0.40	.5606	.7103	2.107	1.168	.3371	.87756
0.45	.6448	.8062	1.995	1.144	.4040	.88942
0.50	.7201	.8870	1.884	1.1202	.4709	.90267
0.55	.7853	.9519	1.774	1.0981	.5366	.91732
0.60	.8402	1.0010	1.667	1.0778	.6003	.93337
0.65	.8853	1.0354	1.565	1.0597	.6614	.95081
0.70	.9213	1.0565	1.468	1.0438	.7195	.96964
0.75	.9491	1.0662	1.377	1.0303	.7744	.98987
0.80	.9697	1.0660	1.291	1.0193	.8260	1.01150
0.85	.9842	1.0578	1.210	1.0108	.8742	1.03452
0.90	.9935	1.0432	1.135	1.0048	.9192	1.05893
0.95	.9985	1.0235	1.0649	1.0012	.9611	1.08474
1.00	1.0000	1.0000	1.0000	1.0000	1.0000	1.11195
1.05	.9987	0.9736	.9398	1.0012	1.0361	1.14055
1.10	.9952	0.9454	.8839	1.0046	1.0695	1.17054
1.15	.9899	0.9158	.8321	1.0103	1.1005	1.20193
1.20	.9833	0.8855	.7842	1.0181	1.1292	1.23472
1.25	.9757	0.8550	.7397	1.0280	1.156	1.26890
1.30	.9674	0.8246	.6985	1.0400	1.181	1.30447
1.35	.9586	0.7946	.6603	1.0540	1.204	1.34145
1.40	.9495	0.7652	.6249	1.0700	1.225	1.37981
1.45	.9403	0.7365	.5919	1.0880	1.245	1.41957
1.50	.9310	0.7087	.5612	1.108	1.263	1.46073
1.55	.9217	0.6818	.5327	1.130	1.280	1.50328
1.60	.9125	0.6559	.5062	1.154	1.296	1.54723
1.65	.9035	0.6309	.4814	1.179	1.311	1.59257
1.70	.8947	0.6069	.4583	1.206	1.324	1.63931

$\gamma = 1\cdot67$

M	$\dfrac{T_0}{T_0{}^*}$	$\dfrac{T}{T^*}$	$\dfrac{p}{p^*}$	$\dfrac{p_0}{p_0{}^*}$	$\dfrac{V}{V^*}$	$\left(\dfrac{T_0}{T_0{}^*}\right)_{\mathrm{isoth}}$
1.75	.8862	.5840	.4367	1.235	1.337	1.68744
1.80	.8779	.5620	.4165	1.266	1.349	1.73696
1.85	.8699	.5410	.3976	1.299	1.360	1.78789
1.90	.8621	.5209	.3799	1.334	1.371	1.84021
1.95	.8546	.5018	.3633	1.370	1.381	1.89392
2.00	.8474	.4835	.3477	1.408	1.391	1.94903
2.05	.8405	.4660	.3330	1.448	1.400	2.00553
2.10	.8338	.4493	.3192	1.490	1.408	2.05343
2.15	.8274	.4334	.3062	1.534	1.415	2.12273
2.20	.8213	.4183	.2940	1.580	1.423	2.18341
2.25	.8154	.4038	.2824	1.628	1.430	2.24550
2.30	.8097	.3899	.2715	1.678	1.436	2.30897
2.35	.8043	.3767	.2612	1.729	1.442	2.37385
2.40	.7991	.3641	.2514	1.783	1.448	2.44011
2.45	.7941	.3521	.2422	1.839	1.454	2.50778
2.50	.7893	.3406	.2334	1.897	1.459	2.57684
2.55	.7847	.3296	.2251	1.956	1.464	2.64729
2.60	.7803	.3191	.2173	2.018	1.469	2.71914
2.65	.7761	.3090	.2098	2.082	1.473	2.79239
2.70	.7721	.2994	.2027	2.148	1.477	2.86703
2.75	.7682	.2902	.1959	2.216	1.481	2.94307
2.80	.7644	.2814	.1895	2.287	1.485	3.02049
2.85	.7608	.2730	.1834	2.360	1.489	3.09932
2.90	.7574	.2649	.1775	2.435	1.493	3.17954
2.95	.7541	.2571	.1719	2.512	1.496	3.26115
3.00	.7509	.2497	.1666	2.587	1.499	3.34416
3.50	.7251	.1897	.1244	3.521	1.524	4.25100
4.00	.7072	.1484	.09632	4.716	1.541	5.29736
4.50	.6943	.1191	.07669	6.213	1.553	6.48321
5.00	.6848	.0975	.06246	8.044	1.561	7.80860
6.00	.6721	.06870	.04368	12.86	1.573	10.87789
7.00	.6642	.05092	.03224	19.44	1.580	14.50526
8.00	.6590	.03920	.02475	28.07	1.584	18.69067
9.00	.6553	.03110	.01959	39.05	1.587	23.43419
10.00	.6528	.02526	.01589	52.66	1.589	28.73566
∞	.6414	0	0	∞	1.599	∞

Oblique shocks: shock-wave angle versus flow-deflection angle.

(Perfect gas, $\gamma = 1.4$)

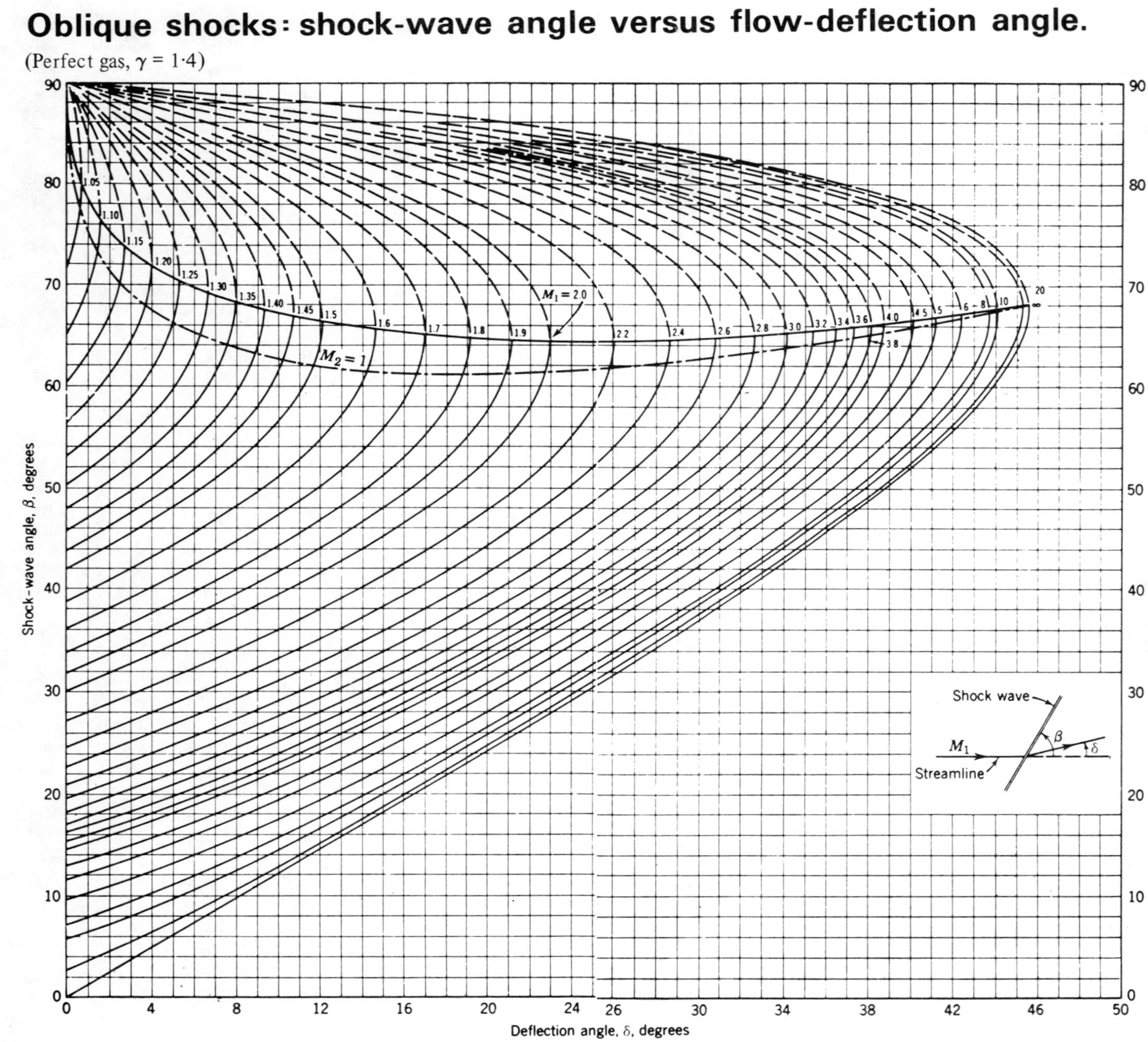

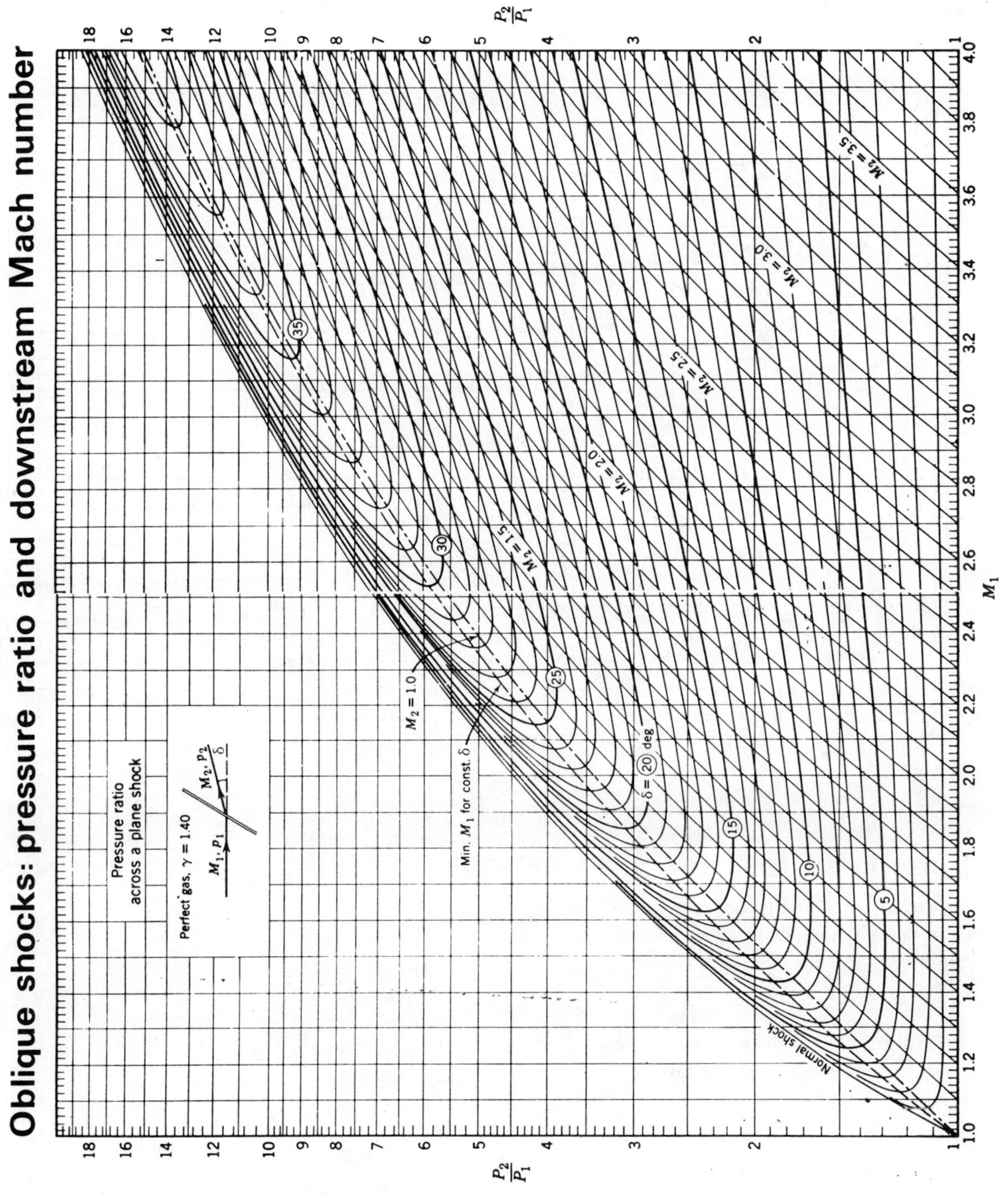

Oblique shocks: pressure ratio and downstream Mach number
Pressure ratio across a plane shock
Perfect gas, $\gamma = 1.40$
M_1, P_1
M_2, P_2
δ
$M_2 = 1.0$
Min. M_1 for const. δ
$M_2 = 1.5$
$M_2 = 2.0$
$M_2 = 2.5$
$M_2 = 3.0$
$M_2 = 3.5$
$\delta = 20$ deg
Normal shock
35
30
25
20
15
10
5
P_2/P_1
M_1

Coefficient of friction for pipes

The coefficient of friction f gives the head loss for an average flow velocity V in a pipe of radius r_0 and length L according to Darcy's equation:

$$h_L = f\frac{LV^2}{4r_0 g}$$

in which $f = 4C_f$ and C_f is the friction coefficient given by $\tau/\frac{1}{2}\rho V^2$ where τ is the shear stress at the wall. The curves show f as a function of Re for various values of r_0/ϵ, where ϵ is the effective surface roughness.

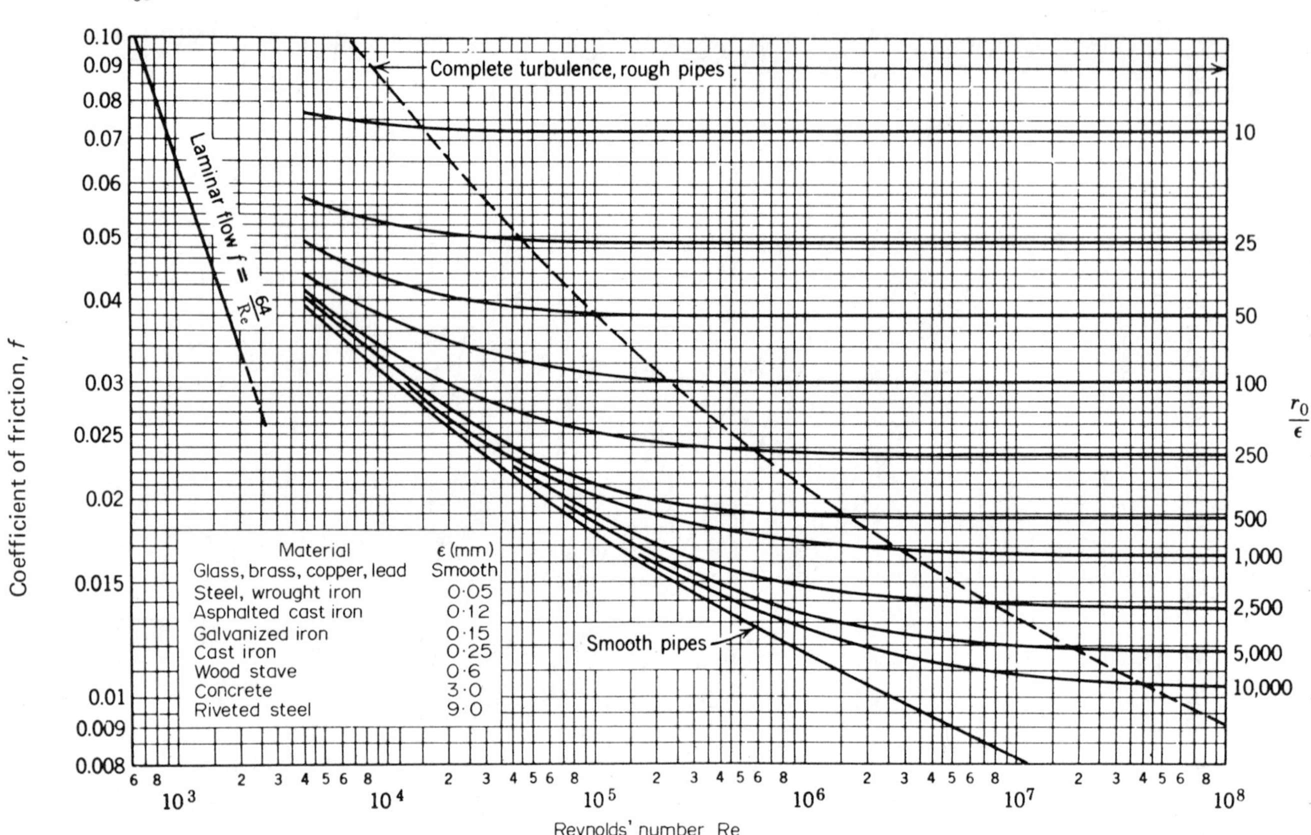

Coefficients of loss for pipe fittings

The loss of head incurred by fittings, valves or sudden contractions of area is given by the loss coefficient K_L according to the relation

$$h_L = K_L(V^2/2g)$$

where V is the average flow velocity. Values of K_L for fittings, valves and contractions of area ratio A_2/A_1 are given below

	K_L
Globe valve, fully open	10·0
Angle valve, fully open	5·0
Swing check valve, fully open	2·5
Gate valve, fully open	0·19
Three-quarters open	1·15
One-half open	5·6
One-quarter open	24·0

	K_L
Close return bend	2·2
Standard tee	1·8
Standard 90° elbow	0·9
Medium sweep 90° elbow	0·75
Long sweep 90° elbow	0·60
45° elbow	0·42
Rounded inlet	0·04
Re-entrant inlet	0·8
Sharp-edged inlet	0·5
Contraction, A_2/A_1 = 0·1	0·37
= 0·2	0·35
= 0·4	0·27
= 0·6	0·17
= 0·8	0·06
= 0·9	0·02

Boundary-layer friction and drag

In a two-dimensional constant-pressure boundary layer the local skin-friction coefficient C_f at distance x from the leading edge, and the friction drag coefficient C_F for unit width of a plate of length l (one surface) are as follows.

Laminar, $0 < \mathrm{Re} \lesssim 10^5$ (Blasius solution)

$$C_f = \tau / \tfrac{1}{2}\rho V^2 = 0 \cdot 664 \left(\frac{Vx}{\nu}\right)^{-1/2}$$

$$C_F = D / \tfrac{1}{2}\rho V^2 l = 1 \cdot 328 \left(\frac{Vl}{\nu}\right)^{-1/2}$$

Turbulent, $\mathrm{Re} \gtrsim 10^6$ ($\tfrac{1}{7}$th root velocity profile)

$$C_f = 0 \cdot 0576 \left(\frac{Vx}{\nu}\right)^{-1/5}$$

$$C_F = 0 \cdot 072 \left(\frac{Vl}{\nu}\right)^{-1/5}$$

or

$$C_F = 0 \cdot 455 \left\{\log_{10}\left(\frac{Vl}{\nu}\right)\right\}^{-2 \cdot 58}$$

which is the *Prandtl-Schlichting* formula.

In the above τ is the shear stress at the wall, V the velocity outside the boundary layer and D the drag.

Open-channel flow

The velocity V of uniform flow in an open channel of slope S may be estimated from the Chézy formula

$$V = C\sqrt{(RS)}$$

or from the Manning formula

$$V = \frac{0 \cdot 82}{n} R^{2/3} S^{1/2}$$

in which C is the Chézy coefficient given by $\sqrt{(8g/f)}$ or $\sqrt{(2g/C_f)}$, and R is the hydraulic radius (ratio of flow section to wetted perimeter) of the channel. Values of n are given below.

	$n\,(\mathrm{m}^{1/6})$
Smooth surface	0·008
Neat cement surface	0·009
Finished concrete, planed wood, or steel surface	0·010
Mortar, clay, or glazed brick surface	0·011
Vitrified clay surface	0·012
Brick surface lined with cement mortar	0·012
Unfinished cement surface	0·014
Rubble masonry or corrugated metal surface	0·016
Earth channel with gravel bottom	0·020
Earth channel with dense weed	0·030
Natural channel with clean bottom, brush on sides	0·040
Flood plain with dense brush	0·080

Elasticity and structures

In the following, u, v, w represent small displacements in the x, y, z directions (or as stated); σ, ϵ represent direct stress and strain, and τ, γ shear stress and strain; E, G, ν are Young's modulus, shear modulus and Poisson's ratio; l, m, n are direction cosines and ψ a stress function; ρ is mass density. M, T are moment and torque; I is second moment or product moment of area; ω is a rotation unless otherwise stated.

Two-dimensional stress and strain

Rectangular coordinates

Relations between strains and small displacements

$$\epsilon_{xx} = \frac{\partial u}{\partial x} \qquad \epsilon_{yy} = \frac{\partial v}{\partial y} \qquad \gamma_{xy} = \frac{\partial v}{\partial x} + \frac{\partial u}{\partial y}$$

$$2\omega_z = \frac{\partial v}{\partial x} - \frac{\partial u}{\partial y}$$

Transformation of strain

Axes Ox' and Oy' inclined at θ to axes Ox and Oy:

$$\epsilon_{x'x'} = \frac{\epsilon_{xx} + \epsilon_{yy}}{2} + \frac{\epsilon_{xx} - \epsilon_{yy}}{2} \cos 2\theta + \frac{\gamma_{xy}}{2} \sin 2\theta$$

$$\epsilon_{y'y'} = \frac{\epsilon_{xx} + \epsilon_{yy}}{2} - \frac{\epsilon_{xx} - \epsilon_{yy}}{2} \cos 2\theta - \frac{\gamma_{xy}}{2} \sin 2\theta$$

$$\gamma_{x'y'} = (\epsilon_{yy} - \epsilon_{xx}) \sin 2\theta + \gamma_{xy} \cos 2\theta$$

Principal strains

$$\epsilon_{max} \atop \epsilon_{min} = \frac{\epsilon_{xx} + \epsilon_{yy}}{2} \pm \left\{ \left(\frac{\epsilon_{xx} - \epsilon_{yy}}{2}\right)^2 + \left(\frac{\gamma_{xy}}{2}\right)^2 \right\}^{1/2}$$

The principal directions are given by $\tan 2\theta_{xp} = \dfrac{\gamma_{xy}}{\epsilon_x - \epsilon_y}$

Compatibility of strains

$$\frac{\partial^2 \epsilon_{xx}}{\partial y^2} + \frac{\partial^2 \epsilon_{yy}}{\partial x^2} = \frac{\partial^2 \gamma_{xy}}{\partial x \partial y}$$

Transformation of stress

Axes Ox' and Oy' inclined at θ to axes Ox and Oy:

$$\sigma_{x'x'} = \frac{\sigma_{xx} + \sigma_{yy}}{2} + \frac{\sigma_{xx} - \sigma_{yy}}{2} \cos 2\theta + \tau_{xy} \sin 2\theta$$

$$\sigma_{y'y'} = \frac{\sigma_{xx} + \sigma_{yy}}{2} - \frac{\sigma_{xx} - \sigma_{yy}}{2} \cos 2\theta - \tau_{xy} \sin 2\theta$$

$$\tau_{x'y'} = \frac{\sigma_{yy} - \sigma_{xx}}{2} \sin 2\theta + \tau_{xy} \cos 2\theta$$

Principal stress

$$\sigma_{max} \atop \sigma_{min} = \frac{\sigma_{xx} + \sigma_{yy}}{2} \pm \left\{ \left(\frac{\sigma_{xx} - \sigma_{yy}}{2}\right)^2 + \tau_{xy}{}^2 \right\}^{1/2}$$

The principal directions are given by $\tan 2\theta_{xp} = \dfrac{2\tau_{xy}}{\sigma_{xx} - \sigma_{yy}}$

Equilibrium equations

$$\frac{\partial \sigma_{xx}}{\partial x} + \frac{\partial \tau_{xy}}{\partial y} + X = 0$$

$$\frac{\partial \tau_{yx}}{\partial x} + \frac{\partial \sigma_{yy}}{\partial y} + Y = 0$$

(X and Y are body forces per unit volume.)

Boundary conditions

$$\sigma_{xx} l + \tau_{xy} m = \overline{X}$$

$$\tau_{yx} l + \sigma_{yy} m = \overline{Y}$$

($\overline{X}$ and $\overline{Y}$ are the surface forces per unit area at the boundary.)

Hooke's Law

$$\epsilon_{xx} = \frac{1}{E}(\sigma_{xx} - \nu\sigma_{yy})$$

$$\epsilon_{yy} = \frac{1}{E}(\sigma_{yy} - \nu\sigma_{xx});$$

also

$$\gamma_{xy} = \frac{\tau_{xy}}{G}$$

Plane stress $\qquad \sigma_{zz} = 0$

$$\epsilon_{xx} = \frac{1}{E}(\sigma_{xx} - \nu\sigma_{yy}) \qquad \sigma_{xx} = \frac{E}{(1-\nu^2)}(\epsilon_{xx} + \nu\epsilon_{yy})$$

Plane strain

$$\epsilon_{zz} = 0 \qquad \epsilon_{xx} = \frac{1-\nu^2}{E}\left(\sigma_{xx} - \frac{\nu}{1-\nu}\sigma_{yy}\right)$$

$$\sigma_{xx} = \frac{E}{(1+\nu)(1-2\nu)}\{(1-\nu)\epsilon_{xx} + \nu\epsilon_{yy}\}$$

Stress function: gravitational force

The stresses are

$$\sigma_{xx} = \frac{\partial^2 \psi}{\partial y^2} - \rho g y \qquad \sigma_{yy} = \frac{\partial^2 \psi}{\partial x^2} - \rho g y$$

$$\tau_{xy} = -\frac{\partial^2 \psi}{\partial x \partial y}$$

and the compatibility equation is

$$\frac{\partial^4 \psi}{\partial x^4} + \frac{2\partial^4 \psi}{\partial x^2 \partial y^2} + \frac{\partial^4 \psi}{\partial y^4} = 0$$

or

$$\nabla^4 \psi = 0$$

Polar coordinates

Relations between strains and small displacements

$$\epsilon_{rr} = \frac{\partial u}{\partial r} \qquad \epsilon_{\theta\theta} = \frac{u}{r} + \frac{1}{r}\frac{\partial v}{\partial \theta} \qquad \gamma_{r\theta} = \frac{\partial v}{\partial r} - \frac{v}{r} + \frac{1}{r}\frac{\partial u}{\partial \theta}$$

(u and v are displacements in the radial and tangential directions).

Equilibrium equations

$$\frac{\partial \sigma_r}{\partial r} + \frac{1}{r}\frac{\partial \tau_{r\theta}}{\partial \theta} + \frac{\sigma_r - \sigma_\theta}{r} + F_r = 0$$

$$\frac{1}{r}\frac{\partial \sigma_\theta}{\partial \theta} + \frac{\partial \tau_{r\theta}}{\partial r} + \frac{2\tau_{r\theta}}{r} + F_\theta = 0$$

where F_r and F_θ are the body forces per unit volume.

Stress function: body forces zero

The stresses are

$$\sigma_r = \frac{1}{r}\frac{\partial \psi}{\partial r} + \frac{1}{r^2}\frac{\partial^2 \psi}{\partial \theta^2}$$

$$\sigma_\theta = \frac{\partial^2 \psi}{\partial r^2}$$

$$\tau_{r\theta} = -\frac{\partial}{\partial r}\left(\frac{1}{r}\frac{\partial \psi}{\partial \theta}\right)$$

and the compatibility equation is

$$\left(\frac{\partial^2}{\partial r^2} + \frac{1}{r}\frac{\partial}{\partial r} + \frac{1}{r^2}\frac{\partial^2}{\partial \theta^2}\right)\left(\frac{\partial^2 \psi}{\partial r^2} + \frac{1}{r}\frac{\partial \psi}{\partial r} + \frac{1}{r^2}\frac{\partial^2 \psi}{\partial \theta^2}\right) = 0$$

Thick cylinder under uniform pressure

$$\sigma_r = A + \frac{B}{r^2} \qquad \sigma_\theta = A - \frac{B}{r^2}$$

where A, B are constants.

Rotating discs and cylinders (angular velocity ω)

$$F_r = \rho \omega^2 r$$

Stress function:

$$r\sigma_r = \psi \qquad \sigma_\theta = \frac{\partial \psi}{\partial r} + \rho \omega^2 r^2$$

Three-dimensional stress and strain

Rectangular coordinates

Relations between strains and small displacements

$$\epsilon_{xx} = \frac{\partial u}{\partial x} \qquad \epsilon_{yy} = \frac{\partial v}{\partial y} \qquad \epsilon_{zz} = \frac{\partial w}{\partial z}$$

$$\gamma_{xy} = \frac{\partial v}{\partial x} + \frac{\partial u}{\partial y} \qquad \gamma_{yz} = \frac{\partial w}{\partial y} + \frac{\partial v}{\partial z} \qquad \gamma_{zx} = \frac{\partial u}{\partial z} + \frac{\partial w}{\partial x}$$

$$2\omega_x = \frac{\partial w}{\partial y} - \frac{\partial v}{\partial z} \qquad 2\omega_y = \frac{\partial u}{\partial z} - \frac{\partial w}{\partial x}$$

$$2\omega_z = \frac{\partial v}{\partial x} - \frac{\partial u}{\partial y}$$

Transformation of strain

Original axes x, y, z; new axes x', y', z'

	x	y	z
x'	l_1	m_1	n_1
y'	l_2	m_2	n_2
z'	l_3	m_3	n_3

$$\epsilon_{x'x'} = \epsilon_{xx}l_1^2 + \epsilon_{yy}m_1^2 + \epsilon_{zz}n_1^2 + \gamma_{xy}l_1 m_1 + \gamma_{yz}m_1 n_1 + \gamma_{zx}n_1 l_1$$

etc.

$$\gamma_{y'z'} = 2\epsilon_{xx}l_2 l_3 + 2\epsilon_{yy}m_2 m_3 + 2\epsilon_{zz}n_2 n_3 + \gamma_{xy}(l_2 m_3 + m_2 l_3) + \gamma_{yz}(m_2 n_3 + n_2 m_3) + \gamma_{zx}(n_2 l_3 + l_2 n_3)$$

etc.

Compatibility of strains

$$\frac{\partial^2 \epsilon_{xx}}{\partial y^2} + \frac{\partial^2 \epsilon_{yy}}{\partial x^2} = \frac{\partial^2 \gamma_{xy}}{\partial x \partial y}$$

and two similar equations.

$$2\frac{\partial^2 \epsilon_{xx}}{\partial y \partial z} = \frac{\partial}{\partial x}\left(-\frac{\partial \gamma_{yz}}{\partial x} + \frac{\partial \gamma_{xz}}{\partial y} + \frac{\partial \gamma_{xy}}{\partial z}\right)$$

and two similar equations.

Principal stress

The direction cosines of a principal plane satisfy the equations

$$(\sigma_{xx} - \sigma)l + \tau_{xy}m + \tau_{xz}n = 0$$
$$\tau_{yx}l + (\sigma_{yy} - \sigma)m + \tau_{yz}n = 0$$
$$\tau_{zx}l + \tau_{zy}m + (\sigma_{zz} - \sigma)n = 0$$

The determinant of the coefficients vanishes and a cubic in σ is obtained. The direction cosines can be found from the above and

$$l^2 + m^2 + n^2 = 1$$

Equilibrium equations

$$\frac{\partial \sigma_{xx}}{\partial x} + \frac{\partial \tau_{xy}}{\partial y} + \frac{\partial \tau_{xz}}{\partial z} + X = 0$$

$$\frac{\partial \tau_{yx}}{\partial x} + \frac{\partial \sigma_{yy}}{\partial y} + \frac{\partial \tau_{yz}}{\partial z} + Y = 0$$

$$\frac{\partial \tau_{zx}}{\partial x} + \frac{\partial \tau_{zy}}{\partial y} + \frac{\partial \sigma_{zz}}{\partial z} + Z = 0$$

(X, Y and Z are body forces per unit volume).

Boundary conditions

$$\sigma_{xx}l + \tau_{xy}m + \tau_{xz}n = \overline{X}$$
$$\tau_{yx}l + \sigma_{yy}m + \tau_{yz}n = \overline{Y}$$
$$\tau_{zx}l + \tau_{zy}m + \sigma_{zz}n = \overline{Z}$$

($\overline{X}$, $\overline{Y}$ and $\overline{Z}$ are surface forces per unit area at the boundary).

Hooke's Law and relations between constants

$$\epsilon_{xx} = \frac{1}{E}\{\sigma_{xx} - \nu(\sigma_{yy} + \sigma_{zz})\}$$

etc.

$$\sigma_{xx} = \frac{\nu E}{(1+\nu)(1-2\nu)}\{\epsilon_{xx} + \epsilon_{yy} + \epsilon_{zz}\} + \frac{E}{1+\nu}\epsilon_{xx}$$

etc.

$$\gamma_{xy} = \frac{\tau_{xy}}{G}$$

etc.

$$G = \frac{E}{2(1+\nu)} \qquad K = \frac{E}{3(1-2\nu)}$$

Cylindrical coordinates

Relations between strains and small displacements

$$\epsilon_{rr} = \frac{\partial u}{\partial r} \qquad \epsilon_{\theta\theta} = \frac{u}{r} + \frac{1}{r}\frac{\partial v}{\partial \theta} \qquad \epsilon_{zz} = \frac{\partial w}{\partial z}$$

$$\gamma_{r\theta} = \frac{\partial v}{\partial r} - \frac{v}{r} + \frac{1}{r}\frac{\partial u}{\partial \theta} \qquad \gamma_{rz} = \frac{\partial w}{\partial r} + \frac{\partial u}{\partial z}$$

$$\gamma_{\theta z} = \frac{1}{r}\frac{\partial w}{\partial \theta} + \frac{\partial v}{\partial z}$$

Spherical coordinates

Relations between strains and small displacements

Displacements u_r, u_θ, u_ϕ.

$$\epsilon_{rr} = \frac{\partial u_r}{\partial r} \qquad \epsilon_{\theta\theta} = \frac{1}{r}\frac{\partial u_\theta}{\partial \theta} + \frac{u_r}{r}$$

$$\epsilon_{\phi\phi} = \frac{1}{r\sin\theta}\frac{\partial u_\phi}{\partial \phi} + \frac{u_\theta}{r}\cot\theta + \frac{u_r}{r}$$

$$\epsilon_{\theta\phi} = \frac{1}{r}\frac{\partial u_\phi}{\partial \theta} - \frac{u_\phi}{r}\cot\theta + \frac{1}{r\sin\theta}\frac{\partial u_\theta}{\partial \phi}$$

$$\epsilon_{\phi r} = \frac{1}{r\sin\theta}\frac{\partial u_r}{\partial \phi} + \frac{\partial u_\phi}{\partial r} - \frac{u_\phi}{r}$$

$$\epsilon_{r\theta} = \frac{\partial u_\theta}{\partial r} - \frac{u_\theta}{r} + \frac{1}{r}\frac{\partial u_r}{\partial \theta}$$

Bending of laterally loaded plates

Assume $\sigma_{zz} = 0$.

$$\sigma_{xx} = -\frac{Ez}{1-\nu^2}\left(\frac{\partial^2 w}{\partial x^2} + \nu\frac{\partial^2 w}{\partial y^2}\right)$$

$$\sigma_{yy} = -\frac{Ez}{1-\nu^2}\left(\frac{\partial^2 w}{\partial y^2} + \nu\frac{\partial^2 w}{\partial x^2}\right)$$

$$\tau_{xy} = -\frac{Ez}{1+\nu}\frac{\partial^2 w}{\partial x \partial y}$$

$$M_x = -D\left(\frac{\partial^2 w}{\partial x^2} + \nu\frac{\partial^2 w}{\partial y^2}\right) \quad M_y = -D\left(\frac{\partial^2 w}{\partial y^2} + \nu\frac{\partial^2 w}{\partial x^2}\right)$$

$$M_{xy} = -D(1-\nu)\frac{\partial^2 w}{\partial x \partial y}$$

where

$$D = \frac{Et^3}{12(1-\nu^2)}$$

and t is the plate thickness.

The differential equation for deflection is

$$\frac{\partial^4 w}{\partial x^4} + 2\frac{\partial^4 w}{\partial x^2 \partial y^2} + \frac{\partial^4 w}{\partial y^4} = \frac{p}{D}$$

where p is the load per unit area in the z direction.

Circular plates

$$M_r = -D\left[\frac{\partial^2 w}{\partial r^2} + \nu\left(\frac{1}{r}\frac{\partial w}{\partial r} + \frac{1}{r^2}\frac{\partial^2 w}{\partial \theta^2}\right)\right]$$

$$M_\theta = -D\left[\frac{1}{r}\frac{\partial w}{\partial r} + \frac{1}{r^2}\frac{\partial^2 w}{\partial \theta^2} + \nu\frac{\partial^2 w}{\partial r^2}\right]$$

$$M_{r\theta} = -(1-\nu)D\left(\frac{1}{r}\frac{\partial^2 w}{\partial r \partial \theta} - \frac{1}{r^2}\frac{dw}{d\theta}\right)$$

The differential equation for deflection is

$$\left(\frac{\partial^2}{\partial r^2} + \frac{1}{r}\frac{\partial}{\partial r} + \frac{1}{r^2}\frac{\partial^2}{\partial \theta^2}\right)^2 w = \frac{p}{D}$$

which for radial symmetry becomes

$$\frac{1}{r}\frac{d}{dr}\left\{r\frac{d}{dr}\left[\frac{1}{r}\frac{d}{dr}\left(r\frac{\partial w}{dr}\right)\right]\right\} = \frac{p}{D}$$

Torsion

The displacements for a rotation θz are

$$u = -\theta zy \qquad v = \theta zx \qquad w = \theta\phi(x, y)$$

where the function ϕ satisfies the equation

$$\frac{\partial^2 \phi}{\partial x^2} + \frac{\partial^2 \phi}{\partial y^2} = 0$$

The shear stresses are given by

$$\tau_{xz} = \frac{\partial \psi}{\partial y} \qquad \tau_{yz} = -\frac{\partial \psi}{\partial x}$$

where ψ is the stress function, and

$$\frac{\partial^2 \psi}{\partial x^2} + \frac{\partial^2 \psi}{\partial y^2} = -2G\theta$$

The torque is

$$T = 2\iint \psi \, dx \, dy$$

Yield criteria

Von Mises

$$(\sigma_1 - \sigma_2)^2 + (\sigma_2 - \sigma_3)^2 + (\sigma_3 - \sigma_1)^2 = 2\sigma_Y^2$$

where σ_1, σ_2 and σ_3 are principal stresses and σ_Y is the yield stress.

Tresca

If $\sigma_1 \geqslant \sigma_2 \geqslant \sigma_3$, then

$$\sigma_1 - \sigma_3 = \sigma_Y$$

Beams and structural members

In this section I_{xx} is denoted as I_x, etc.

Bending of straight beams of asymmetrical section

In a beam aligned with the x-axis, the bending stress is

$$\sigma_{xx} = \frac{(M_y I_z - M_z I_{yz})z + (M_z I_y - M_y I_{yz})y}{I_y I_z - I_{yz}^2}$$

Winkler theory for curved beams

$$\sigma_{xx} = \frac{M}{A R_0}\left[1 + \frac{R_0^2 y}{h^2(R_0 + y)}\right] \quad \text{where} \quad h^2 = \int_A \frac{1}{A}\frac{R_0 y^2}{R_0 + y}\, dA$$

R_0 is the original radius of curvature, and A the section area.

Deflection of beams

The curvature is

$$\frac{1}{R} = \frac{d^2 w/dx^2}{\left\{1 + \left(\frac{dw}{dx}\right)^2\right\}^{3/2}}$$

for displacement $w(x)$. The deflections δ_A for point load P or u.d.l. p are as shown.

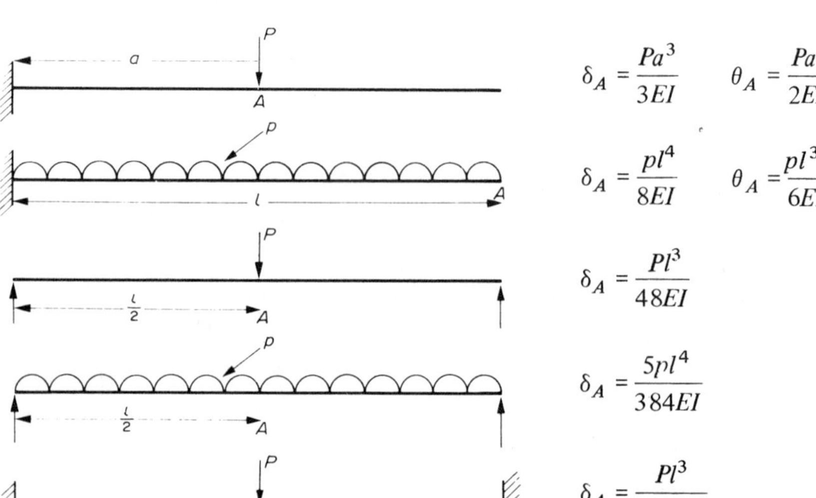

$$\delta_A = \frac{Pa^3}{3EI} \qquad \theta_A = \frac{Pa^2}{2EI}$$

$$\delta_A = \frac{pl^4}{8EI} \qquad \theta_A = \frac{pl^3}{6EI}$$

$$\delta_A = \frac{Pl^3}{48EI}$$

$$\delta_A = \frac{5pl^4}{384EI}$$

$$\delta_A = \frac{Pl^3}{192EI}$$

$$\delta_A = \frac{pl^4}{384EI}$$

Flexibility coefficients due to bending

If m_i, m_j are the moment distributions due to unit reactions at points i, j on a member of length l, then the flexibility coefficient is

$$f_{ij} = \int_0^l \frac{m_i m_j}{EI}\, ds$$

where s is the axial coordinate.

Product integrals $\displaystyle\int_0^l m_i m_j\, ds$

m_j \ m_i	rectangle a	triangle (right, high left) a	triangle (right, high right) a	parabola a	triangle (apex) a	trapezoid a, b
rectangle c	lac	$\frac{1}{2}lac$	$\frac{1}{2}lac$	$\frac{2}{3}lac$	$\frac{1}{2}lac$	$\frac{1}{2}l(a+b)c$
triangle (high left) c	$\frac{1}{2}lac$	$\frac{1}{3}lac$	$\frac{1}{6}lac$	$\frac{1}{3}lac$	$\frac{1}{4}lac$	$\frac{1}{6}l(2a+b)c$
triangle (high right) c	$\frac{1}{2}lac$	$\frac{1}{6}lac$	$\frac{1}{3}lac$	$\frac{1}{3}lac$	$\frac{1}{4}lac$	$\frac{1}{6}l(a+2b)c$
parabola c	$\frac{2}{3}lac$	$\frac{1}{3}lac$	$\frac{1}{3}lac$	$\frac{8}{15}lac$	$\frac{5}{12}lac$	$\frac{1}{3}l(a+b)c$
triangle (apex) c	$\frac{1}{2}lac$	$\frac{1}{4}lac$	$\frac{1}{4}lac$	$\frac{5}{12}lac$	$\frac{1}{3}lac$	$\frac{1}{4}l(a+b)c$
trapezoid c, d	$\frac{1}{2}la(c+d)$	$\frac{1}{6}la(2c+d)$	$\frac{1}{6}la(c+2d)$	$\frac{1}{3}la(c+d)$	$\frac{1}{4}la(c+d)$	$\frac{1}{6}l\{a(2c+d)+b(2d+c)\}$

Flexibility matrix: one-dimensional member

Length L, area A

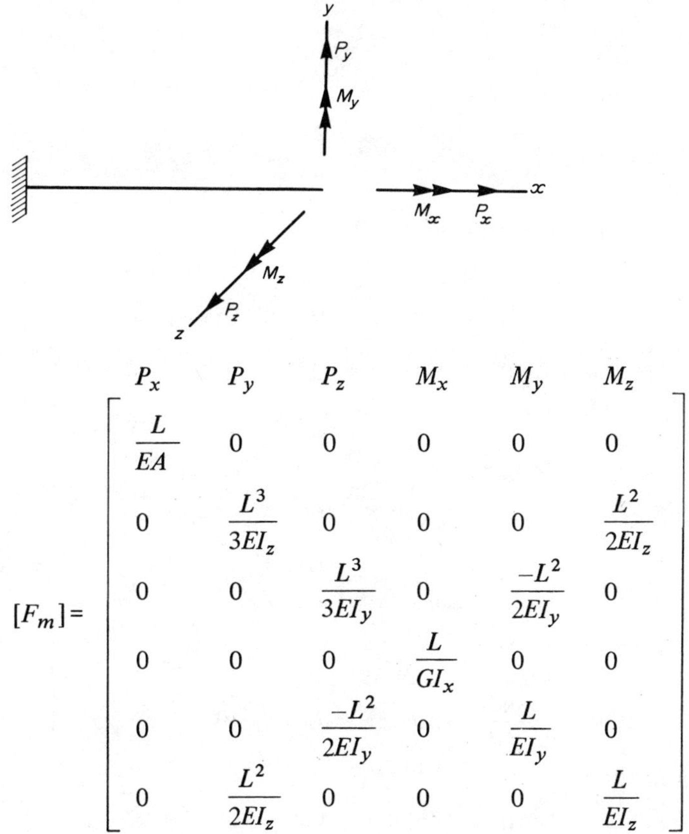

$$[F_m] = \begin{array}{c} \\ \\ \end{array} \begin{array}{cccccc} P_x & P_y & P_z & M_x & M_y & M_z \\ \dfrac{L}{EA} & 0 & 0 & 0 & 0 & 0 \\[2ex] 0 & \dfrac{L^3}{3EI_z} & 0 & 0 & 0 & \dfrac{L^2}{2EI_z} \\[2ex] 0 & 0 & \dfrac{L^3}{3EI_y} & 0 & \dfrac{-L^2}{2EI_y} & 0 \\[2ex] 0 & 0 & 0 & \dfrac{L}{GI_x} & 0 & 0 \\[2ex] 0 & 0 & \dfrac{-L^2}{2EI_y} & 0 & \dfrac{L}{EI_y} & 0 \\[2ex] 0 & \dfrac{L^2}{2EI_z} & 0 & 0 & 0 & \dfrac{L}{EI_z} \end{array}$$

Stiffness matrix: one-dimensional member

Length L, area A.

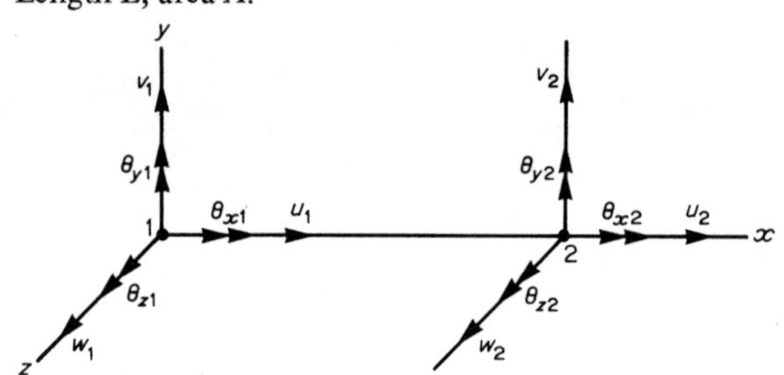

$$[K_m] =$$

	u_1	v_1	w_1	θ_{x1}	θ_{y1}	θ_{z1}	u_2	v_2	w_2	θ_{x2}	θ_{y2}	θ_{z2}
u_1	$\dfrac{AE}{L}$	0	0	0	0	0	$-\dfrac{AE}{L}$	0	0	0	0	0
v_1	0	$\dfrac{12EI_z}{L^3}$	0	0	0	$\dfrac{6EI_z}{L^2}$	0	$-\dfrac{12EI_z}{L^3}$	0	0	0	$\dfrac{6EI_z}{L^2}$
w_1	0	0	$\dfrac{12EI_y}{L^3}$	0	$-\dfrac{6EI_y}{L^2}$	0	0	0	$-\dfrac{12EI_y}{L^3}$	0	$-\dfrac{6EI_y}{L^2}$	0
θ_{x1}	0	0	0	$\dfrac{GI_x}{L}$	0	0	0	0	0	$-\dfrac{GI_x}{L}$	0	0
θ_{y1}	0	0	$-\dfrac{6EI_y}{L^2}$	0	$\dfrac{4EI_y}{L}$	0	0	0	$\dfrac{6EI_y}{L^2}$	0	$\dfrac{2EI_y}{L}$	0
θ_{z1}	0	$\dfrac{6EI_z}{L^2}$	0	0	0	$\dfrac{4EI_z}{L}$	0	$-\dfrac{6EI_z}{L^2}$	0	0	0	$\dfrac{2EI_z}{L}$
u_2	$-\dfrac{AE}{L}$	0	0	0	0	0	$\dfrac{AE}{L}$	0	0	0	0	0
v_2	0	$-\dfrac{12EI_z}{L^3}$	0	0	0	$-\dfrac{6EI_z}{L^2}$	0	$\dfrac{12EI_z}{L^3}$	0	0	0	$-\dfrac{6EI_z}{L^2}$
w_2	0	0	$-\dfrac{12EI_y}{L^3}$	0	$\dfrac{6EI_y}{L^2}$	0	0	0	$\dfrac{12EI_y}{L^3}$	0	$\dfrac{6EI_y}{L^2}$	0
θ_{x2}	0	0	0	$-\dfrac{GI_x}{L}$	0	0	0	0	0	$\dfrac{GI_x}{L}$	0	0
θ_{y2}	0	0	$-\dfrac{6EI_y}{L^2}$	0	$\dfrac{2EI_y}{L}$	0	0	0	$\dfrac{6EI_y}{L^2}$	0	$\dfrac{4EI_y}{L}$	0
θ_{z2}	0	$\dfrac{6EI_z}{L^2}$	0	0	0	$\dfrac{2EI_z}{L}$	0	$-\dfrac{6EI_z}{L^2}$	0	0	0	$\dfrac{4EI_z}{L}$

Slope-deflection equation for uniform section

$$M_{AB} = \frac{2EI}{l}\left(2\theta_A + \theta_B - \frac{3\Delta}{l}\right) \pm \text{F.E.M.}$$

Moment distribution: stiffness and carry-over for uniform section

$$M_{AB} = s\theta_A \qquad M_{BA} = sc\theta_A \qquad \text{Sidesway } \Delta$$

$$\theta_B = \Delta = 0: \qquad s = \frac{4EI}{l} \quad c = \tfrac{1}{2}$$

$$M_{BA} = \Delta = 0: \qquad s = \frac{3EI}{l} \qquad \text{(Pinned end)}$$

$$\theta_B = -\theta_A \quad \Delta = 0: \qquad s = \frac{2EI}{l} \qquad \text{(Symmetry)}$$

$$\theta_B = \theta_A \quad \Delta = 0: \qquad s = \frac{6EI}{l} \qquad \text{(Skew symmetry)}$$

$$\theta_A = \theta_B = 0: M_{AB} = M_{BA} = -\frac{6EI\Delta}{l^2} \qquad \text{(Sidesway)}$$

Fixed end moments

General case

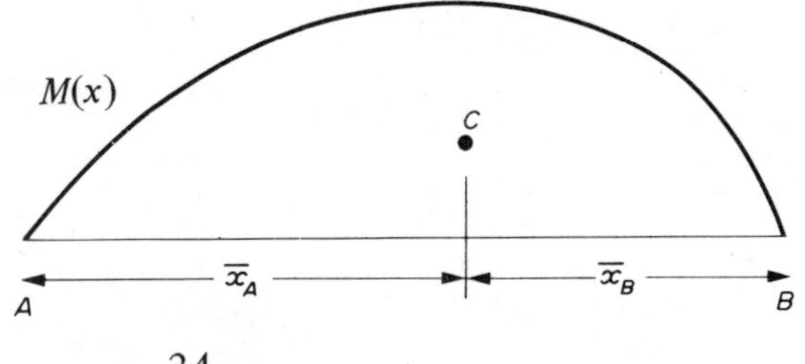

$$M_{AB} = \frac{2A}{l^2}(2\bar{x}_B - \bar{x}_A)$$

$$M_{BA} = \frac{2A}{l^2}(2\bar{x}_A - x_B)$$

where A is the area of the B.M. diagram and C its centroid.

Uniformly distributed load p

$$M_{AB} = M_{BA} = \frac{pl^2}{12}$$

Concentrated load

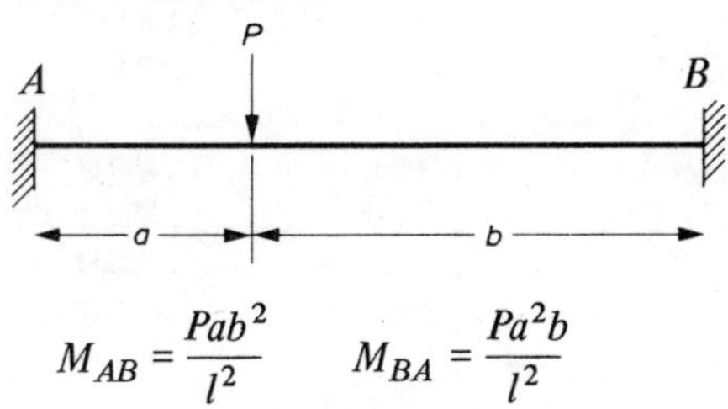

$$M_{AB} = \frac{Pab^2}{l^2} \qquad M_{BA} = \frac{Pa^2b}{l^2}$$

Elastic centre and column analogy

The elastic centre is such that

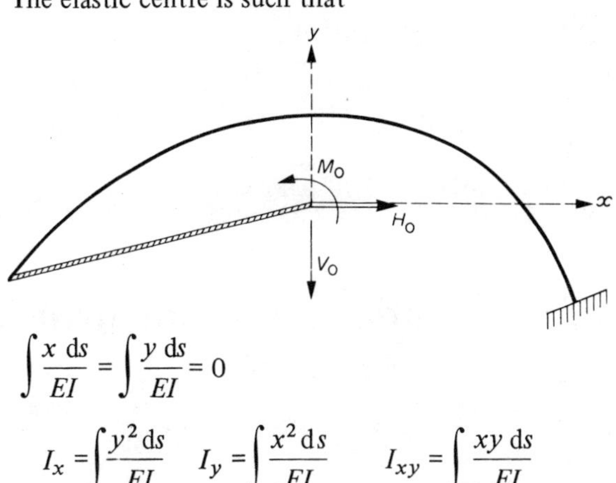

$$\int \frac{x\,ds}{EI} = \int \frac{y\,ds}{EI} = 0$$

$$I_x = \int \frac{y^2\,ds}{EI} \quad I_y = \int \frac{x^2\,ds}{EI} \quad I_{xy} = \int \frac{xy\,ds}{EI}$$

$$A = \int \frac{ds}{EI} \qquad P = \int \frac{M_s\,ds}{EI} \qquad M_x = \int \frac{M_s y\,ds}{EI}$$

$$M_y = \int \frac{M_s x\,ds}{EI}$$

where M_s is the statically determinate bending moment

$$V_0 = \frac{M_y - M_x\dfrac{I_{xy}}{I_x}}{I_y - \dfrac{I_{xy}^2}{I_x}} \qquad H_0 = \frac{M_x - M_y\dfrac{I_{xy}}{I_y}}{I_x = \dfrac{I_{xy}^2}{I_y}} \qquad M_0 = \frac{P}{A}$$

Stability

Euler critical loads

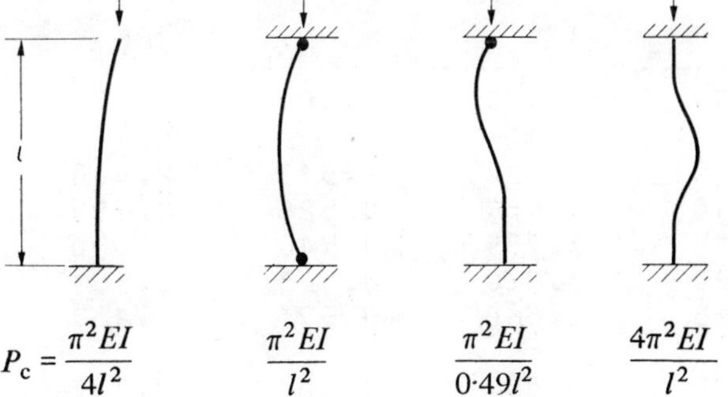

$$P_c = \frac{\pi^2 EI}{4l^2} \qquad \frac{\pi^2 EI}{l^2} \qquad \frac{\pi^2 EI}{0.49l^2} \qquad \frac{4\pi^2 EI}{l^2}$$

Energy methods

Rayleigh

$$P_c = \frac{\displaystyle\int_0^l EI\left(\frac{d^2y}{dx^2}\right)^2 dx}{\displaystyle\int_0^l \left(\frac{dy}{dx}\right)^2 dx}$$

Timoshenko

$$P_c = \frac{\int_0^l \left(\frac{dy}{dx}\right)^2 dx}{\int_0^l \frac{y^2}{EI}\, dx}$$

Stability functions for uniform sections

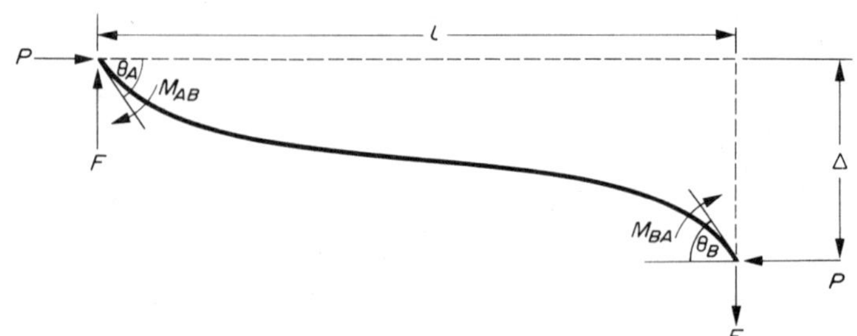

With

$$k = \frac{EI}{l}, \qquad P_E = \frac{\pi^2 EI}{l^2}, \qquad \rho = \frac{P}{P_E} \qquad \text{and} \qquad \alpha = \frac{\pi}{2}\sqrt{\rho}$$

When P is compressive

$$s = \frac{\alpha^2}{1 - \alpha \cot \alpha} + \alpha \cot \alpha$$

$$c = \frac{1}{s}\left\{\frac{\alpha^2}{1 - \alpha \cot \alpha} - \alpha \cot \alpha\right\}$$

When P is tensile

$$s = \frac{\alpha^2}{\alpha \coth \alpha - 1} + \alpha \coth \alpha$$

$$c = \frac{1}{s}\left\{\frac{\alpha^2}{\alpha \coth \alpha - 1} - \alpha \coth \alpha\right\}$$

The tables enable the following to be found

$$\theta_B = \Delta = 0: \qquad M_{AB} = sk\theta_A,\ M_{BA} = csk\theta_A$$
$$M_{BA} = \Delta = 0: \qquad M_{AB} = s''k\theta_A$$
$$\theta_A = \theta_B = 0: \qquad M_{AB} = M_{BA} = -s(1+c)\frac{k\Delta}{l} = -\frac{mFl}{2}$$
$$\theta_B = F = 0: \qquad M_{AB} = nk\theta_A,\ M_{BA} = -ok\theta_A$$

Axial compression

ρ	s	c	s''	$s(1+c)$	m	n	o
0.00	4.000	0.500	3.000	6.000	1.000	1.000	1.000
0.05	3.934	0.513	2.900	5.950	1.043	0.830	1.087
0.10	3.867	0.526	2.797	5.901	1.091	0.647	1.186
0.15	3.799	0.540	2.690	5.850	1.145	0.450	1.297
0.20	3.730	0.555	2.581	5.800	1.205	0.235	1.425
0.25	3.660	0.571	2.467	5.749	1.273	-0.000	1.571
0.30	3.589	0.588	2.350	5.697	1.351	-0.260	1.740
0.35	3.517	0.605	2.228	5.646	1.441	-0.550	1.938
0.40	3.444	0.624	2.102	5.594	1.545	-0.878	2.172
0.45	3.370	0.644	1.971	5.541	1.669	-1.254	2.452
0.50	3.294	0.666	1.834	5.488	1.817	-1.691	2.792
0.55	3.218	0.689	1.691	5.435	1.998	-2.210	3.212
0.60	3.140	0.714	1.541	5.381	2.223	-2.842	3.741
0.65	3.061	0.740	1.385	5.327	2.514	-3.634	4.429
0.70	2.981	0.769	1.220	5.272	2.900	-4.665	5.354
0.75	2.899	0.800	1.046	5.217	3.441	-6.078	6.659

Axial compression—*continued*

ρ	s	c	s''	$s(1+c)$	m	n	o
0.80	2.816	0.833	0.862	5.162	4.253	−8.159	8.630
0.85	2.731	0.869	0.667	5.106	5.604	−11.576	11.932
0.90	2.645	0.909	0.460	5.049	8.307	−18.327	18.568
0.95	2.557	0.952	0.238	4.992	16.414	−38.414	38.536
1.00	2.467	1.000	−0.000	4.935	INF.	INF.	INF.
1.05	2.376	1.053	−0.256	4.877	−16.007	41.409	−41.534
1.10	2.283	1.111	−0.534	4.818	−7.901	21.319	−21.572
1.15	2.187	1.176	−0.837	4.760	−5.199	14.559	−14.944
1.20	2.090	1.249	−1.169	4.700	−3.847	11.131	−11.651
1.25	1.991	1.331	−1.536	4.640	−3.036	9.034	−9.693
1.30	1.889	1.424	−1.944	4.580	−2.495	7.601	−8.403
1.35	1.785	1.532	−2.402	4.518	−2.108	6.547	−7.496
1.40	1.678	1.656	−2.922	4.457	−1.818	5.729	−6.829
1.45	1.569	1.801	−3.519	4.395	−1.592	5.066	−6.323
1.50	1.457	1.973	−4.215	4.332	−1.411	4.512	−5.930
1.55	1.342	2.180	−5.038	4.268	−1.262	4.036	−5.620
1.60	1.224	2.435	−6.032	4.204	−1.139	3.618	−5.374
1.65	1.103	2.754	−7.261	4.139	−1.034	3.243	−5.177
1.70	0.978	3.166	−8.825	4.074	−0.944	2.901	−5.019
1.75	0.849	3.719	−10.897	4.008	−0.866	2.585	−4.894
1.80	0.717	4.497	−13.783	3.941	−0.798	2.289	−4.796
1.85	0.580	5.674	−18.108	3.874	−0.737	2.008	−4.721
1.90	0.439	7.661	−25.353	3.806	−0.683	1.740	−4.667
1.95	0.294	11.727	−40.088	3.737	−0.635	1.480	−4.630
2.00	0.143	24.686	−86.873	3.668	−0.591	1.227	−4.609
2.05	−0.014	−266.942	963.875	3.597	−0.552	0.979	−4.603
2.10	−0.176	−21.071	77.827	3.526	−0.516	0.734	−4.611
2.15	−0.344	−11.037	41.581	3.454	−0.483	0.490	−4.632
2.20	−0.519	−7.510	28.780	3.382	−0.452	0.246	−4.666
2.25	−0.702	−5.712	22.206	3.308	−0.424	−0.000	−4.712
2.30	−0.893	−4.623	18.184	3.234	−0.398	−0.248	−4.771
2.35	−1.092	−3.893	15.456	3.159	−0.374	−0.501	−4.842
2.40	−1.301	−3.370	13.472	3.083	−0.352	−0.758	−4.926
2.45	−1.520	−2.978	11.957	3.006	−0.331	−1.022	−5.023
2.50	−1.750	−2.673	10.754	2.928	−0.311	−1.294	−5.133
2.55	−1.993	−2.430	9.771	2.849	−0.293	−1.576	−5.258
2.60	−2.249	−2.231	8.947	2.769	−0.275	−1.868	−5.399
2.65	−2.521	−2.066	8.243	2.688	−0.259	−2.173	−5.557
2.70	−2.809	−1.928	7.631	2.606	−0.243	−2.492	−5.732
2.75	−3.116	−1.810	7.090	2.523	−0.228	−2.828	−5.928
2.80	−3.445	−1.708	6.606	2.439	−0.214	−3.183	−6.146
2.85	−3.797	−1.620	6.168	2.354	−0.201	−3.561	−6.388
2.90	−4.177	−1.543	5.767	2.268	−0.188	−3.963	−6.658
2.95	−4.587	−1.475	5.397	2.180	−0.176	−4.395	−6.959
3.00	−5.032	−1.416	5.053	2.092	−0.165	−4.860	−7.296

Axial compression–*continued*

ρ	s	c	s''	$s(1+c)$	m	n	o
3.05	−5.518	−1.363	4.730	2.002	−0.153	−5.365	−7.674
3.10	−6.052	−1.316	4.424	1.911	−0.143	−5.916	−8.099
3.15	−6.642	−1.274	4.134	1.818	−0.132	−6.521	−8.580
3.20	−7.297	−1.236	3.856	1.724	−0.123	−7.192	−9.127
3.25	−8.032	−1.203	3.588	1.629	−0.113	−7.940	−9.753
3.30	−8.863	−1.173	3.329	1.532	−0.104	−8.783	−10.475
3.35	−9.812	−1.146	3.077	1.434	−0.095	−9.744	−11.314
3.40	−10.908	−1.122	2.832	1.334	−0.086	−10.851	−12.300
3.45	−12.192	−1.101	2.591	1.233	−0.078	−12.144	−13.473
3.50	−13.719	−1.082	2.353	1.130	−0.070	−13.680	−14.889
3.55	−15.570	−1.066	2.119	1.025	−0.062	−15.539	−16.628
3.60	−17.867	−1.051	1.886	0.919	−0.055	−17.842	−18.811
3.65	−20.800	−1.039	1.654	0.811	−0.047	−20.781	−21.630
3.70	−24.686	−1.028	1.422	0.701	−0.040	−24.672	−25.401
3.75	−30.097	−1.020	1.191	0.590	−0.033	−30.087	−30.696
3.80	−38.176	−1.012	0.957	0.476	−0.026	−38.170	−38.658
3.85	−51.590	−1.007	0.723	0.360	−0.019	−51.587	−51.954
3.90	−78.342	−1.003	0.485	0.242	−0.013	−78.340	−78.585
3.95	−158.444	−1.001	0.245	0.122	−0.006	−158.443	−158.566
4.00	−INF.	−1.000	0.000	0.000	0.000	−INF.	−INF.

Axial tension

ρ	s	c	s''	$s(1+c)$	m	n	o
0.0	4.000	0.500	3.000	6.000	1.000	1.000	1.000
0.1	4.130	0.477	3.192	6.098	0.925	1.309	0.853
0.2	4.257	0.455	3.374	6.195	0.863	1.585	0.734
0.3	4.380	0.436	3.548	6.290	0.809	1.835	0.636
0.4	4.501	0.418	3.714	6.384	0.764	2.063	0.555
0.5	4.619	0.402	3.872	6.477	0.724	2.274	0.488
0.6	4.735	0.387	4.025	6.569	0.689	2.471	0.430
0.7	4.848	0.374	4.172	6.659	0.658	2.656	0.381
0.8	4.959	0.361	4.314	6.749	0.631	2.830	0.340
0.9	5.068	0.349	4.451	6.837	0.606	2.996	0.303
1.0	5.175	0.338	4.583	6.924	0.584	3.153	0.272
1.1	5.280	0.328	4.712	7.010	0.564	3.304	0.245
1.2	5.382	0.318	4.837	7.096	0.545	3.449	0.221
1.3	5.483	0.309	4.959	7.180	0.528	3.588	0.199
1.4	5.583	0.301	5.077	7.263	0.513	3.722	0.181
1.5	5.681	0.293	5.192	7.346	0.498	3.851	0.164
1.6	5.777	0.286	5.305	7.427	0.485	3.977	0.149
1.7	5.871	0.279	5.415	7.508	0.472	4.098	0.136
1.8	5.965	0.272	5.523	7.588	0.461	4.217	0.125
1.9	6.056	0.266	5.628	7.667	0.450	4.332	0.114
2.0	6.147	0.260	5.731	7.745	0.440	4.444	0.105

Axial tension *—continued*

ρ	s	c	s''	$s(1+c)$	m	n	o
2.1	6.236	0.254	5.832	7.822	0.430	4.554	0.096
2.2	6.324	0.249	5.932	7.899	0.421	4.661	0.088
2.3	6.411	0.244	6.029	7.975	0.413	4.765	0.081
2.4	6.496	0.239	6.125	8.050	0.405	4.868	0.075
2.5	6.581	0.235	6.219	8.125	0.397	4.968	0.069
2.6	6.664	0.230	6.311	8.198	0.390	5.066	0.064
2.7	6.747	0.226	6.402	8.272	0.383	5.163	0.059
2.8	6.828	0.222	6.491	8.344	0.377	5.257	0.055
2.9	6.908	0.218	6.579	8.416	0.370	5.350	0.051
3.0	6.988	0.215	6.666	8.487	0.364	5.442	0.047
3.1	7.066	0.211	6.752	8.557	0.359	5.532	0.044
3.2	7.144	0.208	6.836	8.627	0.353	5.620	0.041
3.3	7.221	0.204	6.919	8.697	0.348	5.707	0.038
3.4	7.297	0.201	7.001	8.766	0.343	5.793	0.035
3.5	7.372	0.198	7.082	8.834	0.338	5.877	0.033
3.6	7.446	0.195	7.162	8.901	0.334	5.961	0.031
3.7	7.520	0.193	7.241	8.969	0.329	6.043	0.029
3.8	7.593	0.190	7.319	9.035	0.325	6.124	0.027
3.9	7.665	0.187	7.396	9.101	0.321	6.204	0.025
4.0	7.737	0.185	7.472	9.167	0.317	6.283	0.023
4.1	7.808	0.182	7.548	9.232	0.313	6.361	0.022
4.2	7.878	0.180	7.622	9.296	0.310	6.438	0.021
4.3	7.947	0.178	7.696	9.360	0.306	6.515	0.019
4.4	8.016	0.176	7.769	9.424	0.303	6.590	0.018
4.5	8.084	0.174	7.841	9.487	0.299	6.664	0.017
4.6	8.152	0.171	7.912	9.550	0.296	6.738	0.016
4.7	8.219	0.170	7.983	9.612	0.293	6.811	0.015
4.8	8.286	0.168	8.053	9.674	0.290	6.883	0.014
4.9	8.352	0.166	8.122	9.735	0.287	6.954	0.013
5.0	8.417	0.164	8.191	9.796	0.284	7.025	0.012
5.1	8.482	0.162	8.259	9.857	0.281	7.095	0.012
5.2	8.546	0.160	8.326	9.917	0.279	7.164	0.011
5.3	8.610	0.159	8.393	9.977	0.276	7.233	0.010
5.4	8.673	0.157	8.459	10.036	0.274	7.300	0.010
5.5	8.736	0.156	8.525	10.095	0.271	7.368	0.009
5.6	8.799	0.154	8.590	10.154	0.269	7.434	0.009
5.7	8.861	0.153	8.654	10.212	0.266	7.500	0.008
5.8	8.922	0.151	8.718	10.270	0.264	7.566	0.008
5.9	8.983	0.150	8.782	10.328	0.262	7.631	0.007
6.0	9.044	0.148	8.845	10.385	0.260	7.695	0.007
6.1	9.104	0.147	8.907	10.442	0.258	7.759	0.007
6.2	9.163	0.146	8.969	10.498	0.255	7.823	0.006
6.3	9.223	0.144	9.031	10.554	0.253	7.885	0.006
6.4	9.282	0.143	9.092	10.610	0.251	7.948	0.006
6.5	9.340	0.142	9.152	10.666	0.250	8.010	0.005

Axial tension *—continued*

ρ	s	c	s''	$s(1+c)$	m	n	o
6.6	9.398	0.141	9.212	10.721	0.248	8.071	0.005
6.7	9.456	0.140	9.272	10.776	0.246	8.132	0.005
6.8	9.514	0.138	9.331	10.830	0.244	8.192	0.005
6.9	9.571	0.137	9.390	10.885	0.242	8.252	0.004
7.0	9.627	0.136	9.449	10.939	0.241	8.312	0.004
7.1	9.684	0.135	9.507	10.992	0.239	8.371	0.004
7.2	9.740	0.134	9.564	11.046	0.237	8.430	0.004
7.3	9.795	0.133	9.622	11.099	0.236	8.488	0.003
7.4	9.850	0.132	9.679	11.151	0.234	8.546	0.003
7.5	9.905	0.131	9.735	11.204	0.232	8.604	0.003
7.6	9.960	0.130	9.791	11.256	0.231	8.661	0.003
7.7	10.014	0.129	9.847	11.308	0.229	8.718	0.003
7.8	10.068	0.128	9.903	11.360	0.228	8.774	0.003
7.9	10.122	0.127	9.958	11.411	0.226	8.830	0.003
8.0	10.175	0.126	10.013	11.463	0.225	8.886	0.002

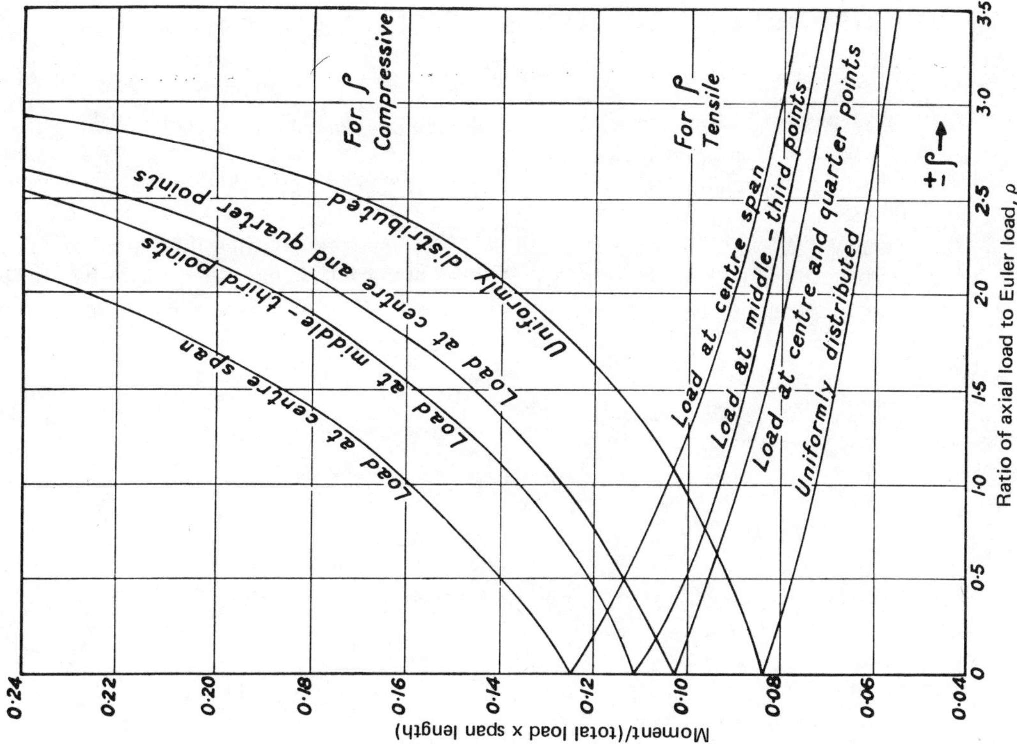

Fixed-end moments at the ends of an axially-loaded uniform beam due to four kinds of symmetrical load.

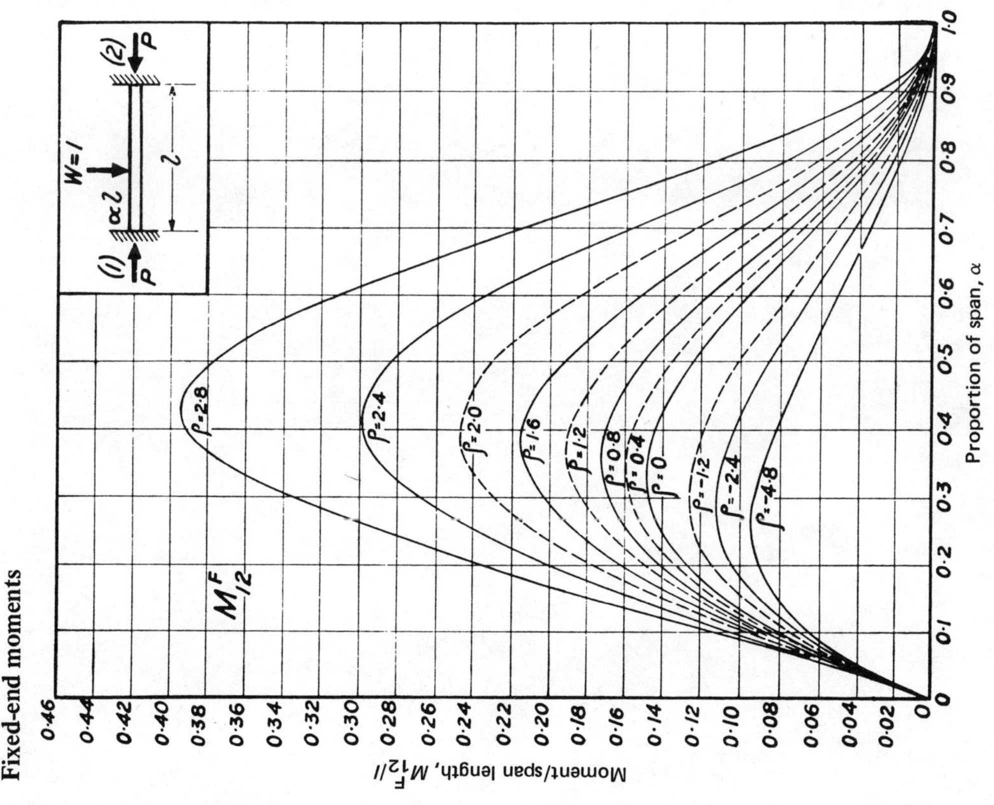

Influence lines for the fixed-end moments at the ends of an axially loaded uniform beam. ρ is the ratio of the axial load to the Euler load ($\pi^2 EI/l^2$)

Dimensions and properties of British Standard sections to B S 4

Notes

In the dimensions and properties tables for Universal Beams, Universal Columns, Joists and Universal Bearing Piles: one hole is deducted from each flange under 300 mm wide and two holes from each flange 300 mm and over, in calculating the Net Moment of Inertia* about $x - x$.

In the dimensions and properties tables for channels: one hole is deducted from each flange in calculating the Net Moment of Inertia about $x - x$.

In the tables giving plastic moduli, $n = \dfrac{\sigma}{\sigma_Y}$ where σ is the mean axial stress and σ_Y the yield stress. The formula for the lower values of n is to be used for values of n below the change values and indicates that the neutral axis is in the web. For n above the change values the higher values of n should be used, the neutral axis is then in the tension flange.

* The following tables refer to second moment of area as Moment of Inertia, and to centroid as Centre of Gravity.

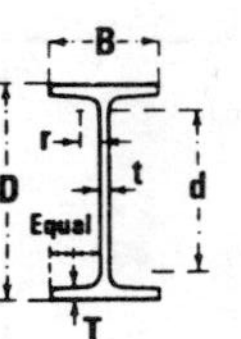

Universal beams

Dimensions and properties

Serial Size	Mass per metre	Depth of Section D	Width of Section B	Thickness Web t	Thickness Flange T	Root Radius r	Depth between Fillets d	Area of Section	Moment of Inertia Axis x–x Gross	Moment of Inertia Axis x–x Net	Moment of Inertia Axis y–y	Radius of Gyration Axis x–x	Radius of Gyration Axis y–y	Elastic Modulus Axis x–x	Elastic Modulus Axis y–y	Ratio D/T
mm	kg	mm	mm	mm	mm	mm	mm	cm²	cm⁴	cm⁴	cm⁴	cm	cm	cm³	cm³	
914 × 419	388	920.5	420.5	21.5	36.6	24.1	791.5	493.9	717325	639177	42481	38.1	9.27	15586	2021	25.2
	343	911.4	418.5	19.4	32.0	24.1	791.5	436.9	623866	555835	36251	37.8	9.11	13691	1733	28.5
914 × 305	289	926.6	307.8	19.6	32.0	19.1	819.2	368.5	503781	469903	14793	37.0	6.34	10874	961.3	29.0
	253	918.5	305.5	17.3	27.9	19.1	819.2	322.5	435796	406504	12512	36.8	6.23	9490	819.2	32.9
	224	910.3	304.1	15.9	23.9	19.1	819.2	284.9	375111	350209	10425	36.3	6.05	8241	685.6	38.1
	201	903.0	303.4	15.2	20.2	19.1	819.2	256.1	324715	303783	8632	35.6	5.81	7192	569.1	44.7
838 × 292	226	850.9	293.8	16.1	26.8	17.8	756.4	288.4	339130	315153	10661	34.3	6.08	7971	725.9	31.8
	194	840.7	292.4	14.7	21.7	17.8	756.4	246.9	278833	259625	8384	33.6	5.83	6633	573.6	38.7
	176	834.9	291.6	14.0	18.8	17.8	756.4	223.8	245412	228867	7111	33.1	5.64	5879	487.6	44.4
762 × 267	197	769.6	268.0	15.6	25.4	16.5	681.2	250.5	239464	221138	7699	30.9	5.54	6223	574.6	30.3
	173	762.0	266.7	14.3	21.6	16.5	681.2	220.2	204747	189341	6376	30.5	5.38	5374	478.1	35.3
	147	753.9	265.3	12.9	17.5	16.5	681.2	187.8	168535	156213	5002	30.0	5.16	4471	377.1	43.1
686 × 254	170	692.9	255.8	14.5	23.7	15.2	610.6	216.3	169843	156106	6225	28.0	5.36	4902	486.8	29.2
	152	687.6	254.5	13.2	21.0	15.2	610.6	193.6	150015	137965	5391	27.8	5.28	4364	423.7	32.7
	140	683.5	253.7	12.4	19.0	15.2	610.6	178.4	135972	125156	4789	27.6	5.18	3979	377.5	36.0
	125	677.9	253.0	11.7	16.2	15.2	610.6	159.4	117700	108580	3992	27.2	5.00	3472	315.5	41.8
610 × 305	238	633.0	311.5	18.6	31.4	16.5	531.6	303.5	207252	192203	14973	26.1	7.02	6549	961.3	20.2
	179	617.5	307.0	14.1	23.6	16.5	531.6	227.7	151312	140269	10571	25.8	6.81	4901	688.6	26.2
	149	609.6	304.8	11.9	19.7	16.5	531.6	189.9	124341	115233	8471	25.6	6.68	4079	555.9	30.9
610 × 229	140	617.0	230.1	13.1	22.1	12.7	543.1	178.2	111673	101699	4253	25.0	4.88	3620	369.6	27.9
	125	611.9	229.0	11.9	19.6	12.7	543.1	159.4	98408	89675	3676	24.8	4.80	3217	321.1	31.2
	113	607.3	228.2	11.2	17.3	12.7	543.1	144.3	87260	79645	3184	24.6	4.70	2874	279.1	35.1
	101	602.2	227.6	10.6	14.8	12.7	543.1	129.0	75549	69132	2658	24.2	4.54	2509	233.6	40.7
610 × 178	91	602.5	178.4	10.6	15.0	12.7	547.1	115.9	63970	57238	1427	23.5	3.51	2124	160.0	40.2
	82	598.2	177.8	10.1	12.8	12.7	547.1	104.4	55779	50076	1203	23.1	3.39	1865	135.3	46.7
533 × 330	212	545.1	333.6	16.7	27.8	16.5	450.1	269.6	141682	121777	16064	22.9	7.72	5199	963.2	19.6
	189	539.5	331.7	14.9	25.0	16.5	450.1	241.2	125618	107882	14093	22.8	7.64	4657	849.6	21.6
	167	533.4	330.2	13.4	22.0	16.5	450.1	212.7	109109	93647	12057	22.6	7.53	4091	730.3	24.2
533 × 210	122	544.6	211.9	12.8	21.3	12.7	472.7	155.6	76078	68719	3208	22.1	4.54	2794	302.8	25.6
	109	539.5	210.7	11.6	18.8	12.7	472.7	138.4	66610	60218	2755	21.9	4.46	2469	261.5	28.7
	101	536.7	210.1	10.9	17.4	12.7	472.7	129.1	61530	55671	2512	21.8	4.41	2293	239.2	30.8
	92	533.1	209.3	10.2	15.6	12.7	472.7	117.6	55225	50040	2212	21.7	4.34	2072	211.3	34.2
	82	528.3	208.7	9.6	13.2	12.7	472.7	104.3	47363	43062	1826	21.3	4.18	1793	175.0	40.0
533 × 165	73	528.8	165.6	9.3	13.5	12.7	476.5	93.0	40414	35752	1027	20.8	3.32	1528	124.1	39.2
	66	524.8	165.1	8.8	11.5	12.7	476.5	83.6	35083	31144	863	20.5	3.21	1337	104.5	45.6
457 × 191	98	467.4	192.8	11.4	19.6	10.2	404.4	125.2	45653	40469	2216	19.1	4.21	1954	229.9	23.8
	89	463.6	192.0	10.6	17.7	10.2	404.4	113.8	40956	36313	1960	19.0	4.15	1767	204.2	26.2
	82	460.2	191.3	9.9	16.0	10.2	404.4	104.4	37039	32869	1746	18.8	4.09	1610	182.6	28.8
	74	457.2	190.5	9.1	14.5	10.2	404.4	94.9	33324	29570	1547	18.7	4.04	1458	162.4	31.5
	67	453.6	189.9	8.5	12.7	10.2	404.4	85.4	29337	26072	1328	18.5	3.95	1293	139.9	35.7

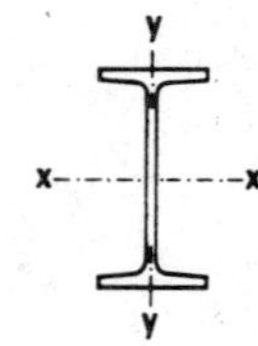

Universal beams

Dimensions and properties

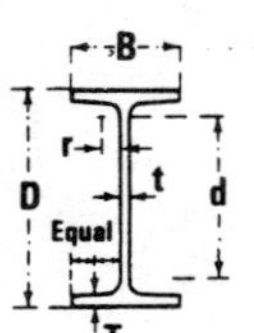

Serial Size	Mass per metre	Depth of Section D	Width of Section B	Thickness Web t	Thickness Flange T	Root Radius r	Depth between Fillets d	Area of Section	Moment of Inertia Axis x–x Gross	Moment of Inertia Axis x–x Net	Moment of Inertia Axis y–y	Radius of Gyration Axis x–x	Radius of Gyration Axis y–y	Elastic Modulus Axis x–x	Elastic Modulus Axis y–y	Ratio D/T
mm	kg	mm	mm	mm	mm	mm	mm	cm²	cm⁴	cm⁴	cm⁴	cm	cm	cm³	cm³	
457 × 152	82	465.1	153.5	10.7	18.9	10.2	404.4	104.4	36160	32058	1093	18.6	3.24	1555	142.5	24.6
	74	461.3	152.7	9.9	17.0	10.2	404.4	94.9	32380	28731	963	18.5	3.18	1404	126.1	27.1
	67	457.2	151.9	9.1	15.0	10.2	404.4	85.3	28522	25342	829	18.3	3.12	1248	109.1	30.5
	60	454.7	152.9	8.0	13.3	10.2	407.7	75.9	25464	22613	794	18.3	3.23	1120	104.0	34.2
	52	449.8	152.4	7.6	10.9	10.2	407.7	66.5	21345	19034	645	17.9	3.11	949.0	84.61	41.3
406 × 178	74	412.8	179.7	9.7	16.0	10.2	357.4	94.9	27279	23981	1448	17.0	3.91	1322	161.2	25.8
	67	409.4	178.8	8.8	14.3	10.2	357.4	85.4	24279	21357	1269	16.9	3.85	1186	141.9	28.6
	60	406.4	177.8	7.8	12.8	10.2	357.4	76.1	21520	18928	1108	16.8	3.82	1059	124.7	31.8
	54	402.6	177.6	7.6	10.9	10.2	357.4	68.3	18576	16389	922	16.5	3.67	922.8	103.8	36.9
406 × 152	74	416.3	153.7	10.1	18.1	10.2	357.4	94.8	26938	23811	1047	16.9	3.32	1294	136.2	23.0
	67	412.2	152.9	9.3	16.0	10.2	357.4	85.3	23798	21069	908	16.7	3.26	1155	118.8	25.8
	60	407.9	152.2	8.6	13.9	10.2	357.4	75.8	20619	18283	768	16.5	3.18	1011	100.9	29.3
406 × 140	46	402.3	142.4	6.9	11.2	10.2	357.4	58.9	15603	13699	500	16.3	2.92	775.6	70.26	35.9
	39	397.3	141.8	6.3	8.6	10.2	357.4	49.3	12408	10963	373	15.9	2.75	624.7	52.61	46.2
381 × 152	67	388.6	154.3	9.7	16.3	10.2	333.2	85.4	21276	18817	947	15.8	3.33	1095	122.7	23.8
	60	384.8	153.4	8.7	14.4	10.2	333.2	75.9	18632	16489	814	15.7	3.27	968.4	106.2	26.7
	52	381.0	152.4	7.8	12.4	10.2	333.2	66.4	16046	14226	685	15.5	3.21	842.3	89.96	30.7
356 × 171	67	364.0	173.2	9.1	15.7	10.2	309.1	85.3	19483	17002	1278	15.1	3.87	1071	147.6	23.2
	57	358.6	172.1	8.0	13.0	10.2	309.1	72.1	16038	14018	1026	14.9	3.77	894.3	119.2	27.6
	51	355.6	171.5	7.3	11.5	10.2	309.1	64.5	14118	12349	885	14.8	3.71	794.0	103.3	30.9
	45	352.0	171.0	6.9	9.7	10.2	309.1	56.9	12052	10578	730	14.6	3.58	684.7	85.39	36.3
356 × 127	39	352.8	126.0	6.5	10.7	10.2	309.1	49.3	10054	8688	333	14.3	2.60	570.0	52.87	33.0
	33	348.5	125.4	5.9	8.5	10.2	309.1	41.7	8167	7099	257	14.0	2.48	468.7	40.99	41.0
305 × 165	54	310.9	166.8	7.7	13.7	8.9	262.6	68.3	11686	10119	988	13.1	3.80	751.8	118.5	22.7
	46	307.1	165.7	6.7	11.8	8.9	262.6	58.8	9924	8596	825	13.0	3.74	646.4	99.54	26.0
	40	303.8	165.1	6.1	10.2	8.9	262.6	51.4	8500	7368	691	12.9	3.67	559.6	83.71	29.8
305 × 127	48	310.4	125.2	8.9	14.0	8.9	262.6	60.8	9485	8137	438	12.5	2.68	611.1	69.94	22.2
	42	306.6	124.3	8.0	12.1	8.9	262.6	53.1	8124	6978	367	12.4	2.63	530.0	58.99	25.3
	37	303.8	123.5	7.2	10.7	8.9	262.6	47.4	7143	6142	316	12.3	2.58	470.3	51.11	28.4
305 × 102	33	312.7	102.4	6.6	10.8	7.6	275.3	41.8	6482	5792	189	12.5	2.13	414.6	37.00	29.0
	28	308.9	101.9	6.1	8.9	7.6	275.3	36.3	5415	4855	153	12.2	2.05	350.7	30.01	34.7
	25	304.8	101.6	5.8	6.8	7.6	275.3	31.4	4381	3959	116	11.8	1.92	287.5	22.85	44.8
254 × 146	43	259.6	147.3	7.3	12.7	7.6	216.2	55.0	6546	5683	633	10.9	3.39	504.3	85.97	20.4
	37	256.0	146.4	6.4	10.9	7.6	216.2	47.4	5544	4814	528	10.8	3.34	433.1	72.11	23.5
	31	251.5	146.1	6.1	8.6	7.6	216.2	39.9	4427	3859	406	10.5	3.19	352.1	55.53	29.2
254 × 102	28	260.4	102.1	6.4	10.0	7.6	224.5	36.2	4004	3565	174	10.5	2.19	307.6	34.13	26.0
	25	257.0	101.9	6.1	8.4	7.6	224.5	32.1	3404	3041	144	10.3	2.11	264.9	28.23	30.6
	22	254.0	101.6	5.8	6.8	7.6	224.5	28.4	2863	2572	116	10.0	2.02	225.4	22.84	37.4
203 × 133	30	206.8	133.8	6.3	9.6	7.6	169.9	38.0	2880	2469	354	8.71	3.05	278.5	52.85	21.5
	25	203.2	133.4	5.8	7.8	7.6	169.9	32.3	2348	2020	280	8.53	2.94	231.1	41.92	26.1

Universal beams

Plastic moduli—minor axis

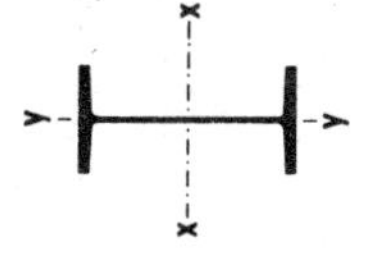

Serial Size (mm)	Mass per metre (kg)	Plastic Modulus Axis y–y (cm³)	Reduced Values of Plastic Modulus under Axial Load		
			Lower Values of n (cm³)	Change formula at n =	Higher Values of n (cm³)
914 × 419	388	3206	3206 − 662.5n^2	0.401	7744(1−n)(0.267+n)
	343	2756	2756 − 523.7n^2	0.405	6823(1−n)(0.253+n)
914 × 305	289	1552	1552 − 366.3n^2	0.493	4978(1−n)(0.087+n)
	253	1322	1322 − 283.0n^2	0.493	4320(1−n)(0.079+n)
	224	1112	1112 − 222.9n^2	0.508	3895(1−n)(0.042+n)
	201	932.2	932.2 − 181.6n^2	0.536	3667(1−n)(n−0.019)
838 × 292	226	1166	1166 − 244.4n^2	0.475	3588(1−n)(0.115+n)
	194	929.4	929.4 − 181.2n^2	0.501	3182(1−n)(0.056+n)
	176	796.6	796.6 − 150.0n^2	0.522	2982(1−n)(0.008+n)
762 × 267	197	924.8	924.8 − 203.9n^2	0.479	2863(1−n)(0.110+n)
	173	773.4	773.4 − 159.1n^2	0.495	2566(1−n)(0.072+n)
	147	615.2	615.2 − 117.0n^2	0.518	2260(1−n)(0.018+n)
686 × 254	170	780.8	780.8 − 168.9n^2	0.464	2302(1−n)(0.139+n)
	152	680.5	680.5 − 136.3n^2	0.469	2065(1−n)(0.124+n)
	140	608.2	608.2 − 116.4n^2	0.475	1901(1−n)(0.108+n)
	125	512.5	512.5 − 93.71n^2	0.498	1753(1−n)(0.058+n)
610 × 305	238	1522	1522 − 363.9n^2	0.388	3454(1−n)(0.306+n)
	179	1092	1092 − 209.8n^2	0.382	2522(1−n)(0.299+n)
	149	884.1	884.1 − 147.8n^2	0.382	2076(1−n)(0.290+n)
610 × 229	140	591.0	591.0 − 128.7n^2	0.454	1677(1−n)(0.163+n)
	125	514.2	514.2 − 103.8n^2	0.457	1495(1−n)(0.150+n)
	113	448.7	448.7 − 85.72n^2	0.471	1385(1−n)(0.116+n)
	101	378.6	378.6 − 69.12n^2	0.495	1281(1−n)(0.064+n)
610 × 178	91	256.2	256.2 − 55.70n^2	0.551	1051(1−n)(n−0.044)
	82	217.5	217.5 − 45.56n^2	0.579	1014(1−n)(n−0.105)
533 × 330	212	1518	1518 − 333.2n^2	0.338	3034(1−n)(0.399+n)
	189	1340	1340 − 269.6n^2	0.333	2678(1−n)(0.401+n)
	167	1156	1156 − 212.0n^2	0.336	2352(1−n)(0.389+n)
533 × 210	122	484.0	484.0 − 111.2n^2	0.448	1334(1−n)(0.179+n)
	109	418.5	418.5 − 88.81n^2	0.452	1187(1−n)(0.164+n)
	101	383.4	383.4 − 77.69n^2	0.453	1099(1−n)(0.158+n)
	92	339.6	339.6 − 64.87n^2	0.462	1015(1−n)(0.134+n)
	82	283.5	283.5 − 51.47n^2	0.486	930.8(1−n)(0.081+n)
533 × 165	73	197.3	197.3 − 40.93n^2	0.529	744.4(1−n)(0.001+n)
	66	166.9	166.9 − 33.27n^2	0.553	701.1(1−n)(n−0.053)
457 × 191	98	365.8	365.8 − 83.80n^2	0.426	938.9(1−n)(0.225+n)
	89	325.4	325.4 − 69.86n^2	0.432	861.2(1−n)(0.207+n)
	82	291.5	291.5 − 59.25n^2	0.436	789.9(1−n)(0.193+n)
	74	259.6	259.6 − 49.24n^2	0.439	716.5(1−n)(0.183+n)
	67	224.7	224.7 − 40.15n^2	0.452	654.2(1−n)(0.152+n)

Plastic moduli—major axis

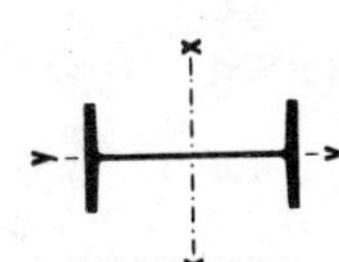

Serial Size (mm)	Mass per metre (kg)	Plastic Modulus Axis x–x (cm³)	Reduced Values of Plastic Modulus under Axial Load		
			Lower Values of n (cm³)	Change formula at n =	Higher Values of n (cm³)
914 × 419	388	17628	17628 − 28366n^2	0.368	1453(1−n)(14.65+n)
	343	15445	15445 − 24603n^2	0.375	1176(1−n)(15.93+n)
914 × 305	289	12566	12566 − 17318n^2	0.459	1088(1−n)(14.70+n)
	253	10930	10930 − 15026n^2	0.463	845.2(1−n)(16.52+n)
	224	9505	9505 − 12764n^2	0.481	661.2(1−n)(18.61+n)
	201	8345	8345 − 10787n^2	0.512	526.6(1−n)(20.96+n)
838 × 292	226	9144	9144 − 12919n^2	0.443	731.9(1−n)(15.77+n)
	194	7635	7635 − 10363n^2	0.473	548.4(1−n)(17.92+n)
	176	6795	6795 − 8947n^2	0.497	453.5(1−n)(19.60+n)
762 × 267	197	7156	7156 − 10057n^2	0.446	600.8(1−n)(15.05+n)
	173	6186	6186 − 8478n^2	0.465	473.6(1−n)(16.71+n)
	147	5163	5163 − 6836n^2	0.492	342.9(1−n)(19.65+n)
686 × 254	170	5616	5616 − 8069n^2	0.432	455.6(1−n)(15.45+n)
	152	4989	4989 − 7100n^2	0.440	360.1(1−n)(17.48+n)
	140	4552	4552 − 6417n^2	0.446	334.8(1−n)(17.21+n)
	125	3987	3987 − 5430n^2	0.471	270.0(1−n)(19.01+n)
610 × 305	238	7447	7447 − 12384n^2	0.349	745.2(1−n)(11.89+n)
	179	5512	5512 − 9190n^2	0.352	433.3(1−n)(15.22+n)
	149	4562	4562 − 7573n^2	0.357	299.7(1−n)(18.31+n)
610 × 229	140	4141	4141 − 6063n^2	0.420	350.6(1−n)(14.68+n)
	125	3672	3672 − 5338n^2	0.426	290.6(1−n)(15.78+n)
	113	3283	3283 − 4648n^2	0.444	225.3(1−n)(18.45+n)
	101	2877	2877 − 3927n^2	0.471	176.8(1−n)(20.97+n)
610 × 178	91	2484	2484 − 3166n^2	0.520	205.1(1−n)(16.02+n)
	82	2194	2194 − 2699n^2	0.556	138.3(1−n)(21.58+n)
533 × 330	212	5849	5849 − 10877n^2	0.302	556.1(1−n)(12.21+n)
	189	5212	5212 − 9760n^2	0.301	449.2(1−n)(13.48+n)
	167	4560	4560 − 8440n^2	0.308	347.3(1−n)(15.33+n)
533 × 210	122	3198	3198 − 4731n^2	0.413	280.1(1−n)(14.13+n)
	109	2820	2820 − 4131n^2	0.421	222.4(1−n)(15.79+n)
	101	2616	2616 − 3825n^2	0.422	205.7(1−n)(15.85+n)
	92	2362	2362 − 3390n^2	0.436	160.0(1−n)(18.59+n)
	82	2051	2051 − 2832n^2	0.463	122.8(1−n)(21.43+n)
533 × 165	73	1776	1776 − 2327n^2	0.498	143.7(1−n)(16.12+n)
	66	1562	1562 − 1984n^2	0.524	118.6(1−n)(17.49+n)
457 × 191	98	2229	2229 − 3436n^2	0.389	208.9(1−n)(13.00+n)
	89	2012	2012 − 3055n^2	0.399	166.7(1−n)(14.83+n)
	82	1830	1830 − 2754n^2	0.405	144.7(1−n)(15.61+n)
	74	1654	1654 − 2474n^2	0.410	119.1(1−n)(17.21+n)
	67	1469	1469 − 2143n^2	0.426	95.25(1−n)(19.32+n)

Universal beams

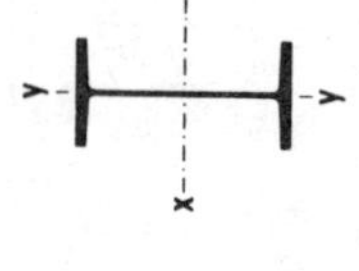

Plastic moduli—minor axis

Serial Size (mm)	Mass per metre (kg)	Plastic Modulus Axis y–y (cm³)	Reduced Values of Plastic Modulus under Axial Load — Lower Values of n (cm³)	Change formula at $n =$	Reduced Values of Plastic Modulus under Axial Load — Higher Values of n (cm³)
457 × 152	82	229.2	229.2 −58.56n^2	0.477	682.0(1−n)(0.128+n)
	74	203.0	203.0 −48.80n^2	0.481	621.8(1−n)(0.113+n)
	67	176.1	176.1 −39.81n^2	0.488	558.9(1−n)(0.094+n)
	60	162.9	162.9 −31.70n^2	0.479	515.3(1−n)(0.101+n)
	52	133.2	133.2 −24.58n^2	0.514	484.6(1−n)(0.024+n)
406 × 178	74	256.6	256.6 −54.50n^2	0.422	659.7(1−n)(0.226+n)
	67	226.2	226.2 −44.53n^2	0.422	588.1(1−n)(0.220+n)
	60	198.8	198.8 −35.62n^2	0.417	515.0(1−n)(0.224+n)
	54	167.2	167.2 −28.99n^2	0.448	482.2(1−n)(0.158+n)
406 × 152	74	217.5	217.5 −54.02n^2	0.443	581.7(1−n)(0.196+n)
	67	190.1	190.1 −44.18n^2	0.449	525.1(1−n)(0.177+n)
	60	162.1	162.1 −35.25n^2	0.463	474.4(1−n)(0.144+n)
406 × 140	46	113.1	113.1 −21.53n^2	0.472	349.3(1−n)(0.115+n)
	39	85.84	85.84 −15.30n^2	0.508	306.3(1−n)(0.036+n)
381 × 152	67	196.1	196.1 −46.94n^2	0.441	524.8(1−n)(0.196+n)
	60	169.8	169.8 −37.45n^2	0.441	460.9(1−n)(0.190+n)
	52	144.1	144.1 −28.93n^2	0.448	406.2(1−n)(0.169+n)
356 × 171	67	233.7	233.7 −50.01n^2	0.388	540.7(1−n)(0.295+n)
	57	189.5	189.5 −36.22n^2	0.398	459.2(1−n)(0.267+n)
	51	164.8	164.8 −29.24n^2	0.403	408.9(1−n)(0.253+n)
	45	137.4	137.4 −22.97n^2	0.427	371.6(1−n)(0.199+n)
356 × 127	39	85.07	85.07 −17.23n^2	0.465	254.3(1−n)(0.133+n)
	33	66.63	66.63 −12.50n^2	0.493	222.7(1−n)(0.070+n)
305 × 165	54	186.9	186.9 −37.52n^2	0.351	391.6(1−n)(0.366+n)
	46	157.4	157.4 −28.18n^2	0.350	333.7(1−n)(0.360+n)
	40	133.0	133.0 −21.77n^2	0.360	293.7(1−n)(0.333+n)
305 × 127	48	112.4	112.4 −29.73n^2	0.455	307.7(1−n)(0.179+n)
	42	94.91	94.91 −23.00n^2	0.462	271.3(1−n)(0.155+n)
	37	82.33	82.33 −18.49n^2	0.462	238.7(1−n)(0.149+n)
305 × 102	33	59.10	59.10 −13.94n^2	0.494	190.6(1−n)(0.084+n)
	28	48.18	48.18 −10.65n^2	0.519	173.1(1−n)(0.025+n)
	25	37.24	37.24 −8.070n^2	0.564	161.2(1−n)(n−0.071)
254 × 146	43	135.4	135.4 −29.18n^2	0.344	276.3(1−n)(0.384+n)
	37	113.9	113.9 −21.94n^2	0.346	236.6(1−n)(0.373+n)
	31	88.78	88.78 −15.86n^2	0.384	208.1(1−n)(0.290+n)
254 × 102	28	54.09	54.09 −12.56n^2	0.461	155.4(1−n)(0.153+n)
	25	45.07	45.07 −10.05n^2	0.488	143.4(1−n)(0.093+n)
	22	36.80	36.80 −7.939n^2	0.519	132.6(1−n)(0.025+n)
203 × 133	30	83.70	83.70 −17.41n^2	0.343	171.1(1−n)(0.383+n)
	25	67.04	67.04 −12.81n^2.	0.365	147.4(1−n)(0.333+n)

Plastic moduli—major axis

Serial Size (mm)	Mass per metre (kg)	Plastic Modulus Axis x–x (cm³)	Reduced Values of Plastic Modulus under Axial Load — Lower Values of n (cm³)	Change formula at $n =$	Reduced Values of Plastic Modulus under Axial Load — Higher Values of n (cm³)
457 × 152	82	1797	1797 −2545n^2	0.438	174.5(1−n)(12.91+n)
	74	1620	1620 −2274n^2	0.446	146.3(1−n)(13.96+n)
	67	1439	1439 −2000n^2	0.453	128.0(1−n)(14.24+n)
	60	1284	1284 −1802n^2	0.449	99.17(1−n)(16.41+n)
	52	1094	1094 −1455n^2	0.490	67.64(1−n)(21.11+n)
406 × 178	74	1502	1502 −2319n^2	0.390	122.5(1−n)(14.98+n)
	67	1343	1343 −2072n^2	0.392	102.0(1−n)(16.13+n)
	60	1195	1195 −1856n^2	0.389	82.74(1−n)(17.69+n)
	54	1046	1046 −1536n^2	0.423	66.91(1−n)(19.55+n)
406 × 152	74	1486	1486 −2226n^2	0.403	151.0(1−n)(12.07+n)
	67	1323	1323 −1958n^2	0.412	125.0(1−n)(13.07+n)
	60	1158	1158 −1672n^2	0.429	99.66(1−n)(14.52+n)
406 × 140	46	886.3	886.3 −1255n^2	0.444	63.01(1−n)(17.79+n)
	39	718.7	718.7 −964.8n^2	0.482	47.94(1−n)(19.43+n)
381 × 152	67	1254	1254 −1880n^2	0.404	117.3(1−n)(13.15+n)
	60	1106	1106 −1656n^2	0.406	98.55(1−n)(13.82+n)
	52	959.0	959.0 −1413n^2	0.418	71.08(1−n)(16.80+n)
356 × 171	67	1210	1210 −2000n^2	0.355	104.4(1−n)(13.88+n)
	57	1007	1007 −1624n^2	0.370	73.11(1−n)(16.68+n)
	51	892.9	892.9 −1424n^2	0.376	61.31(1−n)(17.70+n)
	45	771.7	771.7 −1172n^2	0.404	45.34(1−n)(21.08+n)
356 × 127	39	651.8	651.8 −935.0n^2	0.434	51.70(1−n)(15.82+n)
	33	537.9	537.9 −738.3n^2	0.465	37.78(1−n)(18.25+n)
305 × 165	54	843.4	843.4 −1515n^2	0.318	72.92(1−n)(13.56+n)
	46	721.3	721.3 −1292n^2	0.321	54.56(1−n)(15.56+n)
	40	623.1	623.1 −1084n^2	0.335	40.91(1−n)(18.10+n)
305 × 127	48	704.9	704.9 −1037n^2	0.413	73.97(1−n)(11.75+n)
	42	609.2	609.2 −881.4n^2	0.426	54.65(1−n)(13.90+n)
	37	539.3	539.3 −780.0n^2	0.427	47.17(1−n)(14.26+n)
305 × 102	33	479.6	479.6 −660.3n^2	0.459	43.31(1−n)(14.07+n)
	28	406.9	406.9 −539.5n^2	0.489	32.50(1−n)(16.24+n)
	25	337.5	337.5 −424.1n^2	0.532	28.77(1−n)(15.62+n)
254 × 146	43	567.4	567.4 −1038n^2	0.309	53.24(1−n)(12.42+n)
	37	484.5	484.5 −877.7n^2	0.315	38.90(1−n)(14.60+n)
	31	394.8	394.8 −653.9n^2	0.357	27.99(1−n)(16.94+n)
254 × 102	28	353.1	353.1 −511.2n^2	0.427	30.32(1−n)(14.53+n)
	25	305.3	305.3 −423.6n^2	0.455	25.38(1−n)(15.28+n)
	22	261.5	261.5 −347.7n^2	0.485	22.53(1−n)(15.01+n)
203 × 133	30	312.6	312.6 −571.6n^2	0.310	27.53(1−n)(13.26+n)
	25	259.1	259.1 −448.7n^2	0.334	20.90(1−n)(14.69+n)

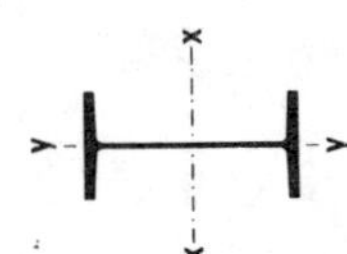

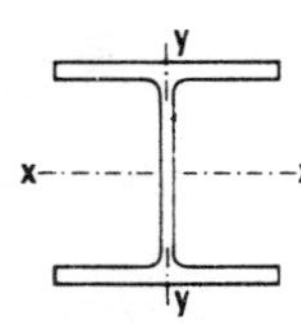

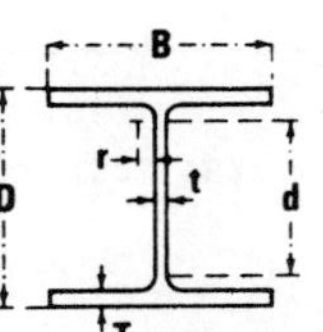

Universal columns
Parallel flanges

Dimensions and properties

Serial Size	Mass per metre	Depth of Section D	Width of Section B	Thickness Web t	Thickness Flange T	Root Radius r	Depth between Fillets d	Area of Section
mm	kg	mm	mm	mm	mm	mm	mm	cm²
356 × 406	634	474.7	424.1	47.6	77.0	15.2	290.1	808.1
	551	455.7	418.5	42.0	67.5	15.2	290.1	701.8
	467	436.6	412.4	35.9	58.0	15.2	290.1	595.5
	393	419.1	407.0	30.6	49.2	15.2	290.1	500.9
	340	406.4	403.0	26.5	42.9	15.2	290.1	432.7
	287	393.7	399.0	22.6	36.5	15.2	290.1	366.0
	235	381.0	395.0	18.5	30.2	15.2	290.1	299.8
Column Core	477	427.0	424.4	48.0	53.2	15.2	290.1	607.2
356 × 368	202	374.7	374.4	16.8	27.0	15.2	290.1	257.9
	177	368.3	372.1	14.5	23.8	15.2	290.1	225.7
	153	362.0	370.2	12.6	20.7	15.2	290.1	195.2
	129	355.6	368.3	10.7	17.5	15.2	290.1	164.9
305 × 305	283	365.3	321.8	26.9	44.1	15.2	246.6	360.4
	240	352.6	317.9	23.0	37.7	15.2	246.6	305.6
	198	339.9	314.1	19.2	31.4	15.2	246.6	252.3
	158	327.2	310.6	15.7	25.0	15.2	246.6	201.2
	137	320.5	308.7	13.8	21.7	15.2	246.6	174.6
	118	314.5	306.8	11.9	18.7	15.2	246.6	149.8
	97	307.8	304.8	9.9	15.4	15.2	246.6	123.3
254 × 254	167	289.1	264.5	19.2	31.7	12.7	200.2	212.4
	132	276.4	261.0	15.6	25.1	12.7	200.2	167.7
	107	266.7	258.3	13.0	20.5	12.7	200.2	136.6
	89	260.4	255.9	10.5	17.3	12.7	200.2	114.0
	73	254.0	254.0	8.6	14.2	12.7	200.2	92.9
203 × 203	86	222.3	208.8	13.0	20.5	10.2	160.8	110.1
	71	215.9	206.2	10.3	17.3	10.2	160.8	91.1
	60	209.6	205.2	9.3	14.2	10.2	160.8	75.8
	52	206.2	203.9	8.0	12.5	10.2	160.8	66.4
	46	203.2	203.2	7.3	11.0	10.2	160.8	58.8
152 × 152	37	161.8	154.4	8.1	11.5	7.6	123.4	47.4
	30	157.5	152.9	6.6	9.4	7.6	123.4	38.2
	23	152.4	152.4	6.1	6.8	7.6	123.4	29.8

Serial Size	Moment of Inertia Axis x–x Gross	Moment of Inertia Axis x–x Net	Moment of Inertia Axis y–y	Radius of Gyration Axis x–x	Radius of Gyration Axis y–y	Elastic Modulus Axis x–x	Elastic Modulus Axis y–y	Ratio D/T
mm	cm⁴	cm⁴	cm⁴	cm	cm	cm³	cm³	
356 × 406	275140	243076	98211	18.5	11.0	11592	4632	6.2
	227023	200312	82665	18.0	10.9	9964	3951	6.8
	183118	161331	67905	17.5	10.7	8388	3293	7.5
	146765	129159	55410	17.1	10.5	7004	2723	8.5
	122474	107667	46816	16.8	10.4	6027	2324	9.5
	99994	87843	38714	16.5	10.3	5080	1940	10.8
	79110	69424	31008	16.2	10.2	4153	1570	12.6
Column Core	172391	152936	68057	16.8	10.6	8075	3207	8.0
356 × 368	66307	57806	23632	16.0	9.57	3540	1262	13.9
	57153	49798	20470	15.9	9.52	3104	1100	15.5
	48525	42250	17470	15.8	9.46	2681	943.8	17.5
	40246	35040	14555	15.6	9.39	2264	790.4	20.3
305 × 305	78777	72827	24545	14.8	8.25	4314	1525	8.3
	64177	59295	20239	14.5	8.14	3641	1273	9.4
	50832	46935	16230	14.2	8.02	2991	1034	10.8
	38740	35766	12524	13.9	7.89	2368	806.3	13.1
	32838	30314	10672	13.7	7.82	2049	691.4	14.8
	27601	25472	9006	13.6	7.75	1755	587.0	16.8
	22202	20488	7268	13.4	7.68	1442	476.9	20.0
254 × 254	29914	27171	9796	11.9	6.79	2070	740.6	9.1
	22416	20350	7444	11.6	6.66	1622	570.4	11.0
	17510	15890	5901	11.3	6.57	1313	456.9	13.0
	14307	12976	4849	11.2	6.52	1099	378.9	15.1
	11360	10297	3873	11.1	6.46	894.5	305.0	17.9
203 × 203	9462	8374	3119	9.27	5.32	851.5	298.7	10.8
	7647	6758	2536	9.16	5.28	708.4	246.0	12.5
	6088	5383	2041	8.96	5.19	581.1	199.0	14.8
	5263	4653	1770	8.90	5.16	510.4	173.6	16.5
	4564	4035	1539	8.81	5.11	449.2	151.5	18.5
152 × 152	2218	1932	709	6.84	3.87	274.2	91.78	14.1
	1742	1515	558	6.75	3.82	221.2	73.06	16.8
	1263	1104	403	6.51	3.68	165.7	52.95	22.4

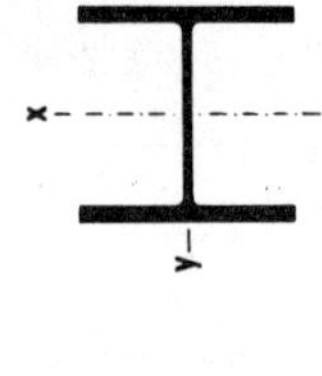

Universal columns
Parallel flanges

Plastic moduli—minor axis

Serial Size (mm)	Mass per metre (kg)	Plastic Modulus Axis y–y (cm³)	Reduced Values of Plastic Modulus under Axial Load		
			Lower Values of n (cm³)	Change formula at n =	Higher Values of n (cm³)
356 × 406	634	7114	$7114 - 3439n^2$	0.280	$10520(1-n)(0.624+n)$
	551	6058	$6058 - 2702n^2$	0.273	$9046(1-n)(0.618+n)$
	467	5038	$5038 - 2031n^2$	0.263	$7571(1-n)(0.615+n)$
	393	4157	$4157 - 1496n^2$	0.256	$6303(1-n)(0.610+n)$
	340	3541	$3541 - 1152n^2$	0.249	$5387(1-n)(0.609+n)$
	287	2952	$2952 - 850.5n^2$	0.243	$4519(1-n)(0.605+n)$
	235	2384	$2384 - 589.8n^2$	0.235	$3656(1-n)(0.606+n)$
Column Core	477	4979	$4979 - 2159n^2$	0.338	$8585(1-n)(0.495+n)$
356 × 368	202	1917	$1917 - 443.8n^2$	0.244	$3022(1-n)(0.584+n)$
	177	1668	$1668 - 345.9n^2$	0.237	$2615(1-n)(0.589+n)$
	153	1430	$1430 - 263.2n^2$	0.234	$2249(1-n)(0.587+n)$
	129	1196	$1196 - 191.2n^2$	0.231	$1889(1-n)(0.585+n)$
305 × 305	283	2337	$2337 - 888.9n^2$	0.273	$3626(1-n)(0.588+n)$
	240	1947	$1947 - 662.4n^2$	0.265	$3042(1-n)(0.585+n)$
	198	1576	$1576 - 468.3n^2$	0.259	$2485(1-n)(0.580+n)$
	158	1228	$1228 - 309.4n^2$	0.255	$1965(1-n)(0.570+n)$
	137	1052	$1052 - 237.8n^2$	0.253	$1697(1-n)(0.564+n)$
	118	891.7	$891.7 - 178.4n^2$	0.250	$1446(1-n)(0.562+n)$
	97	723.5	$723.5 - 123.4n^2$	0.247	$1182(1-n)(0.557+n)$
254 × 254	167	1132	$1132 - 390.0n^2$	0.261	$1749(1-n)(0.594+n)$
	132	869.9	$869.9 - 254.4n^2$	0.257	$1370(1-n)(0.581+n)$
	107	695.5	$695.5 - 175.0n^2$	0.254	$1110(1-n)(0.573+n)$
	89	575.4	$575.4 - 124.7n^2$	0.240	$904.9(1-n)(0.586+n)$
	73	462.4	$462.4 - 84.89n^2$	0.235	$730.3(1-n)(0.584+n)$
203 × 203	86	455.9	$455.9 - 136.3n^2$	0.263	$724.0(1-n)(0.574+n)$
	71	374.2	$374.2 - 96.02n^2$	0.244	$582.9(1-n)(0.592+n)$
	60	302.8	$302.8 - 68.60n^2$	0.257	$492.8(1-n)(0.558+n)$
	52	263.7	$263.7 - 53.53n^2$	0.248	$425.5(1-n)(0.566+n)$
	46	230.0	$230.0 - 42.59n^2$	0.252	$377.7(1-n)(0.553+n)$
152 × 152	37	140.1	$140.1 - 34.70n^2$	0.277	$236.1(1-n)(0.528+n)$
	30	111.2	$111.2 - 23.22n^2$	0.272	$189.2(1-n)(0.523+n)$
	23	80.87	$80.87 - 14.55n^2$	0.312	$154.7(1-n)(0.434+n)$

Plastic moduli—major axis

Serial Size (mm)	Mass per metre (kg)	Plastic Modulus Axis x–x (cm³)	Reduced Values of Plastic Modulus under Axial Load		
			Lower Values of n (cm³)	Change formula at n =	Higher Values of n (cm³)
356 × 406	634	14247	$14247 - 34295n^2$	0.189	$3850(1-n)(3.981+n)$
	551	12078	$12078 - 29319n^2$	0.192	$2946(1-n)(4.428+n)$
	467	10009	$10009 - 24694n^2$	0.193	$2151(1-n)(5.042+n)$
	393	8229	$8229 - 20495n^2$	0.196	$1542(1-n)(5.806+n)$
	340	6994	$6994 - 17659n^2$	0.196	$1164(1-n)(6.553+n)$
	287	5818	$5818 - 14816n^2$	0.198	$840.4(1-n)(7.573+n)$
	235	4689	$4689 - 12147n^2$	0.197	$571.6(1-n)(8.991+n)$
Column Core	477	9700	$9700 - 19203n^2$	0.253	$2174(1-n)(4.963+n)$
356 × 368	202	3976	$3976 - 9899n^2$	0.209	$444.6(1-n)(9.869+n)$
	177	3457	$3457 - 8785n^2$	0.206	$342.5(1-n)(11.14+n)$
	153	2964	$2964 - 7561n^2$	0.207	$257.9(1-n)(12.70+n)$
	129	2482	$2482 - 6355n^2$	0.208	$184.4(1-n)(14.91+n)$
305 × 305	283	5101	$5101 - 12071n^2$	0.206	$1012(1-n)(5.506+n)$
	240	4245	$4245 - 10154n^2$	0.208	$736.3(1-n)(6.319+n)$
	198	3436	$3436 - 8290n^2$	0.210	$507.9(1-n)(7.443+n)$
	158	2680	$2680 - 6448n^2$	0.215	$329.0(1-n)(9.006+n)$
	137	2298	$2298 - 5523n^2$	0.218	$248.7(1-n)(10.25+n)$
	118	1953	$1953 - 4714n^2$	0.219	$185.5(1-n)(11.70+n)$
	97	1589	$1589 - 3837n^2$	0.222	$125.5(1-n)(14.12+n)$
254 × 254	167	2417	$2417 - 5873n^2$	0.203	$427.0(1-n)(6.189+n)$
	132	1861	$1861 - 4507n^2$	0.210	$270.9(1-n)(7.554+n)$
	107	1485	$1485 - 3591n^2$	0.214	$180.4(1-n)(9.103+n)$
	89	1228	$1228 - 3092n^2$	0.207	$128.7(1-n)(10.53+n)$
	73	988.5	$988.5 - 2507n^2$	0.208	$86.22(1-n)(12.68+n)$
203 × 203	86	978.8	$978.8 - 2330n^2$	0.214	$145.1(1-n)(7.432+n)$
	71	802.4	$802.4 - 2013n^2$	0.204	$100.7(1-n)(8.763+n)$
	60	652.0	$652.0 - 1546n^2$	0.222	$70.26(1-n)(10.31+n)$
	52	568.1	$568.1 - 1380n^2$	0.218	$54.26(1-n)(11.63+n)$
	46	497.4	$497.4 - 1186n^2$	0.224	$43.12(1-n)(12.86+n)$
152 × 152	37	310.1	$310.1 - 693.2n^2$	0.236	$36.73(1-n)(9.438+n)$
	30	247.1	$247.1 - 554.0n^2$	0.239	$23.72(1-n)(11.70+n)$
	23	184.3	$184.3 - 363.5n^2$	0.283	$14.67(1-n)(14.47+n)$

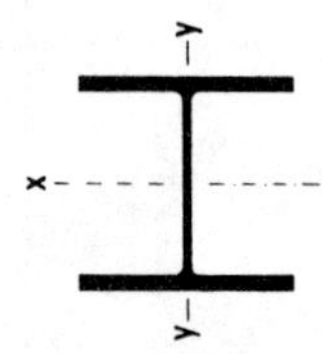

Joists

Dimensions and properties

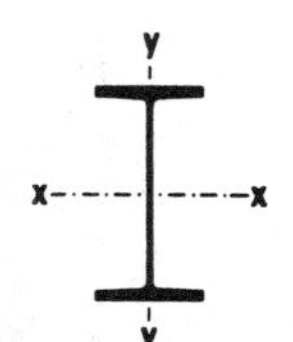

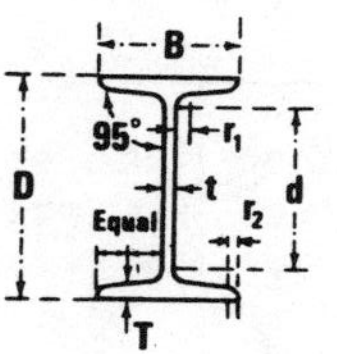

Nominal Size	Mass per metre	Depth of Section D	Width of Section B	Thickness		Radius		Depth between Fillets d	Area of Section
				Web t	Flange T	Root r_1	Toe r_2		
mm	kg	mm	mm	mm	mm	mm	mm	mm	cm²
203 × 102	25.33	203.2	101.6	5.8	10.4	9.4	3.2	161.0	32.3
178 × 102	21.54	177.8	101.6	5.3	9.0	9.4	3.2	138.2	27.4
152 × 89	17.09	152.4	88.9	4.9	8.3	7.9	2.4	117.9	21.8
127 × 76	13.36	127.0	76.2	4.5	7.6	7.9	2.4	94.2	17.0
102 × 64	9.65	101.6	63.5	4.1	6.6	6.9	2.4	73.2	12.3
76 × 51	6.67	76.2	50.8	3.8	5.6	6.9	2.4	50.3	8.49

Nominal Size	Moment of Inertia			Radius of Gyration		Elastic Modulus		Ratio $\frac{D}{T}$
	Axis x–x		Axis y–y	Axis x–x	Axis y–y	Axis x–x	Axis y–y	
	Gross	Net						
mm	cm⁴	cm⁴	cm⁴	cm	cm	cm³	cm³	
203 × 102	2294	2023	162.6	8.43	2.25	225.8	32.02	19.5
178 × 102	1519	1340	139.2	7.44	2.25	170.9	27.41	19.8
152 × 89	881.1	762.1	85.98	6.36	1.99	115.6	19.34	18.4
127 × 76	475.9	399.8	50.18	5.29	1.72	74.94	13.17	16.7
102 × 64	217.6	181.9	25.30	4.21	1.43	42.84	7.97	15.4
76 × 51	82.58	68.85	11.11	3.12	1.14	21.67	4.37	13.6

Joists

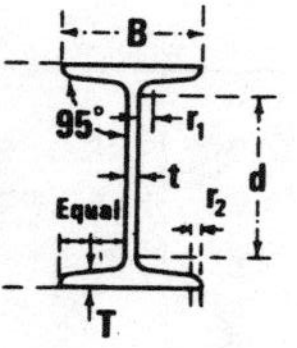

Plastic moduli—major axis

Nominal Size	Mass per metre	Plastic Modulus Axis x–x	Reduced Values of Plastic Modulus under Axial Load		
			Lower Values of n	Change formula at n =	Higher Values of n
mm	kg	cm³	cm³		cm³
203 × 102	25.33	256.3	$256.3 - 448.7n^2$	0.324	$27.23(1-n)(11.04+n)$
178 × 102	21.54	193.0	$193.0 - 355.2n^2$	0.305	$19.78(1-n)(11.33+n)$
152 × 89	17.09	131.0	$131.0 - 241.7n^2$	0.302	$14.08(1-n)(10.78+n)$
127 × 76	13.36	85.23	$85.23 - 161.0n^2$	0.292	$10.02(1-n)(9.786+n)$
102 × 64	9.65	48.98	$48.98 - 92.11n^2$	0.292	$6.148(1-n)(9.155+n)$
76 × 51	6.67	25.07	$25.07 - 47.43n^2$	0.284	$3.752(1-n)(7.624+n)$

Plastic moduli—minor axis

Nominal Size	Mass per metre	Plastic Modulus Axis y–y	Reduced Values of Plastic Modulus under Axial Load		
			Lower Values of n	Change formula at n =	Higher Values of n
mm	kg	cm³	cm³		cm³
203 × 102	25.33	51.79	$51.79 - 12.81n^2$	0.365	$109.6(1-n)(0.355+n)$
178 × 102	21.54	44.48	$44.48 - 10.59n^2$	0.343	$89.21(1-n)(0.395+n)$
152 × 89	17.09	31.29	$31.29 - 7.772n^2$	0.343	$62.26(1-n)(0.400+n)$
127 × 76	13.36	21.29	$21.29 - 5.704n^2$	0.336	$41.04(1-n)(0.422+n)$
102 × 64	9.65	12.91	$12.91 - 3.717n^2$	0.339	$24.75(1-n)(0.424+n)$
76 × 51	6.67	7.14	$7.142 - 2.366n^2$	0.341	$13.36(1-n)(0.439+n)$

Channels

Dimensions and properties

Nominal Size	Mass per metre	Depth of Section D	Width of Section B	Thickness Web t	Thickness Flange T	Radius Root r_1	Radius Toe r_2	Depth between Fillets d	Ratio $\frac{D}{T}$	Area of Section
mm	in kg	mm	mm	mm	mm	mm	mm	mm		cm²
432 × 102	65.54	431.8	101.6	12.2	16.8	15.2	4.8	362.5	25.7	83.49
381 × 102	55.10	381.0	101.6	10.4	16.3	15.2	4.8	312.4	23.4	70.19
305 × 102	46.18	304.8	101.6	10.2	14.8	15.2	4.8	239.3	20.6	58.83
305 × 89	41.69	304.8	88.9	10.2	13.7	13.7	3.2	245.4	22.2	53.11
254 × 89	35.74	254.0	88.9	9.1	13.6	13.7	3.2	194.8	18.7	45.52
254 × 76	28.29	254.0	76.2	8.1	10.9	12.2	3.2	203.7	23.3	36.03
229 × 89	32.76	228.6	88.9	8.6	13.3	13.7	3.2	169.9	17.2	41.73
229 × 76	26.06	228.6	76.2	7.6	11.2	12.2	3.2	178.1	20.4	33.20
203 × 89	29.78	203.2	88.9	8.1	12.9	13.7	3.2	145.3	15.8	37.94
203 × 76	23.82	203.2	76.2	7.1	11.2	12.2	3.2	152.4	18.1	30.34
178 × 89	26.81	177.8	88.9	7.6	12.3	13.7	3.2	120.9	14.5	34.15
178 × 76	20.84	177.8	76.2	6.6	10.3	12.2	3.2	128.8	17.3	26.54
152 × 89	23.84	152.4	88.9	7.1	11.6	13.7	3.2	97.0	13.1	30.36
152 × 76	17.88	152.4	76.2	6.4	9.0	12.2	2.4	105.9	16.9	22.77
127 × 64	14.90	127.0	63.5	6.4	9.2	10.7	2.4	84.1	13.8	18.98
102 × 51	10.42	101.6	50.8	6.1	7.6	9.1	2.4	65.8	13.4	13.28
76 × 38	6.70	76.2	38.1	5.1	6.8	7.6	2.4	45.7	11.2	8.53

Nominal Size	Dimension p	Moment of Inertia Axis x—x Gross	Moment of Inertia Axis x—x Net	Moment of Inertia Axis y—y	Radius of Gyration Axis x—x	Radius of Gyration Axis y—y	Elastic Modulus Axis x—x	Elastic Modulus Axis y—y
mm	cm	cm⁴	cm⁴	cm⁴	cm	cm	cm³	cm³
432 × 102	2.32	21399	17602	628.6	16.0	2.74	991.1	80.15
381 × 102	2.52	14894	12060	579.8	14.6	2.87	781.8	75.87
305 × 102	2.66	8214	6587	499.5	11.8	2.91	539.0	66.60
305 × 89	2.18	7061	5824	325.4	11.5	2.48	463.3	48.49
254 × 89	2.42	4448	3612	302.4	9.88	2.58	350.2	46.71
254 × 76	1.86	3367	2673	162.6	9.67	2.12	265.1	28.22
229 × 89	2.53	3387	2733	285.0	9.01	2.61	296.4	44.82
229 × 76	2.00	2610	2040	158.7	8.87	2.19	228.3	28.22
203 × 89	2.65	2491	1996	264.4	8.10	2.64	245.2	42.34
203 × 76	2.13	1950	1506	151.4	8.02	2.23	192.0	27.59
178 × 89	2.76	1753	1397	241.0	7.16	2.66	197.2	39.29
178 × 76	2.20	1337	1028	134.0	7.10	2.25	150.4	24.73
152 × 89	2.86	1166	923.7	215.1	6.20	2.66	153.0	35.70
152 × 76	2.21	851.6	654.3	113.8	6.12	2.24	111.8	21.05
127 × 64	1.94	482.6	367.5	67.24	5.04	1.88	75.99	15.25
102 × 51	1.51	207.7	167.9	29.10	3.96	1.48	40.89	8.16
76 × 38	1.19	74.14	54.52	10.66	2.95	1.12	19.46	4.07

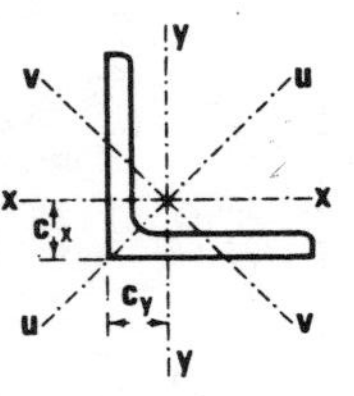

Equal angles

Dimensions and properties

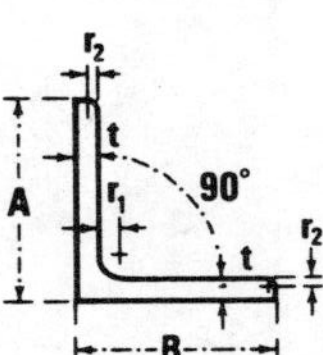

Nominal Size	Leg Lengths A×B	Actual Thickness	Mass per metre	Radii Root r₁	Radii Toe r₂	Area of Section	Centre of Gravity Cx	Centre of Gravity Cy	Moment of Inertia Axis x–x	Moment of Inertia Axis y–y	Moment of Inertia Axis u–u Max.	Moment of Inertia Axis v–v Min.	Radius of Gyration Axis x–x	Radius of Gyration Axis y–y	Radius of Gyration Axis u–u Max.	Radius of Gyration Axis v–v Min.	Elastic Modulus Axis x–x	Elastic Modulus Axis y–y
mm	mm	mm	kg	mm	mm	cm²	cm	cm	cm⁴	cm⁴	cm⁴	cm⁴	cm	cm	cm	cm	cm³	cm³
203 × 203	203.2 × 203.2	25.3	76.00	15.2	4.8	96.81	5.99	5.99	3686	3686	5845	1527	6.17	6.17	7.77	3.97	257	257
		23.7	71.51	15.2	4.8	91.09	5.93	5.93	3491	3491	5540	1442	6.19	6.19	7.80	3.98	243	243
		22.1	67.05	15.2	4.8	85.42	5.87	5.87	3294	3294	5232	1357	6.21	6.21	7.83	3.99	228	228
		20.5	62.56	15.2	4.8	79.69	5.81	5.81	3094	3094	4916	1271	6.23	6.23	7.85	3.99	213	213
		18.9	57.95	15.2	4.8	73.82	5.75	5.75	2885	2885	4587	1183	6.25	6.25	7.88	4.00	198	198
		17.3	53.30	15.2	4.8	67.89	5.69	5.69	2671	2671	4248	1093	6.27	6.27	7.91	4.01	183	183
		15.8	48.68	15.2	4.8	62.02	5.63	5.63	2455	2455	3907	1004	6.29	6.29	7.94	4.02	167	167
152 × 152	152.4 × 152.4	22.1	49.32	12.2	4.8	62.83	4.60	4.60	1321	1321	2089	553	4.58	4.58	5.77	2.97	124	124
		20.5	46.03	12.2	4.8	58.63	4.54	4.54	1243	1243	1968	517	4.60	4.60	5.79	2.97	116	116
		19.0	42.75	12.2	4.8	54.45	4.49	4.49	1164	1164	1846	482	4.62	4.62	5.82	2.98	108	108
		17.3	39.32	12.2	4.8	50.09	4.42	4.42	1080	1080	1714	446	4.64	4.64	5.85	2.98	99.8	99.8
		15.8	36.07	12.2	4.8	45.95	4.37	4.37	999	999	1587	411	4.66	4.66	5.88	2.99	91.9	91.9
		14.2	32.62	12.2	4.8	41.55	4.31	4.31	911	911	1448	374	4.68	4.68	5.90	3.00	83.3	83.3
		12.6	29.07	12.2	4.8	37.03	4.24	4.24	819	819	1303	335	4.70	4.70	5.93	3.01	74.5	74.5
		11.0	25.60	12.2	4.8	32.61	4.18	4.18	727	727	1156	297	4.72	4.72	5.96	3.02	65.7	65.7
		9.4	22.02	12.2	4.8	28.06	4.11	4.11	631	631	1003	258	4.74	4.74	5.98	3.03	56.7	56.7
127 × 127	127.0 × 127.0	19.0	35.16	10.7	4.8	44.80	3.85	3.85	651	651	1028	273	3.81	3.81	4.79	2.47	73.5	73.5
		17.4	32.47	10.7	4.8	41.37	3.79	3.79	607	607	961	253	3.83	3.83	4.82	2.47	68.1	68.1
		15.8	29.66	10.7	4.8	37.78	3.73	3.73	560	560	888	232	3.85	3.85	4.85	2.48	62.4	62.4
		14.2	26.80	10.7	4.8	34.14	3.67	3.67	511	511	811	211	3.87	3.87	4.87	2.48	56.6	56.6
		12.6	23.99	10.7	4.8	30.56	3.61	3.61	462	462	734	190	3.89	3.89	4.90	2.49	50.8	50.8
		11.0	21.14	10.7	4.8	26.93	3.55	3.55	411	411	654	169	3.91	3.91	4.93	2.50	44.9	44.9
		9.5	18.30	10.7	4.8	23.31	3.49	3.49	359	359	571	147	3.93	3.93	4.95	2.51	39.0	39.0
102 × 102	101.6 × 101.6	19.0	27.57	9.1	4.8	35.12	3.22	3.22	317	317	497	136	3.00	3.00	3.76	1.97	45.6	45.6
		17.4	25.48	9.1	4.8	32.45	3.16	3.16	296	296	466	126	3.02	3.02	3.79	1.97	42.3	42.3
		15.8	23.37	9.1	4.8	29.78	3.10	3.10	275	275	434	116	3.04	3.04	3.82	1.97	38.9	38.9
		14.2	21.17	9.1	4.8	26.96	3.04	3.04	252	252	399	105	3.06	3.06	3.84	1.97	35.4	35.4
		12.6	18.91	9.1	4.8	24.09	2.98	2.98	228	228	361	94.3	3.07	3.07	3.87	1.98	31.7	31.7
		11.0	16.69	9.1	4.8	21.27	2.92	2.92	203	203	323	83.8	3.09	3.09	3.90	1.99	28.1	28.1
		9.4	14.44	9.1	4.8	18.39	2.86	2.86	178	178	283	73.1	3.11	3.11	3.92	1.99	24.4	24.4
		7.8	12.06	9.1	4.8	15.37	2.79	2.79	150	150	239	61.7	3.13	3.13	3.95	2.00	20.4	20.4

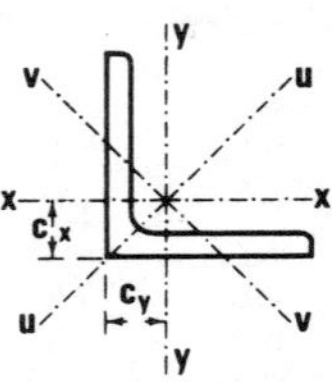

Equal angles

Dimensions and properties

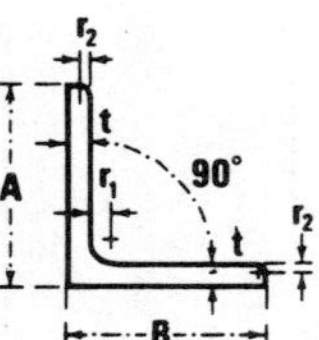

Nominal Size	Leg Lengths A×B	Actual Thickness	Mass per metre	Radii Root r_1	Radii Toe r_2	Area of Section	Centre of Gravity Cx	Centre of Gravity Cy	Moment of Inertia Axis x—x	Moment of Inertia Axis y—y	Moment of Inertia Axis u—u Max.	Moment of Inertia Axis v—v Min.	Radius of Gyration Axis x—x	Radius of Gyration Axis y—y	Radius of Gyration Axis u—u Max.	Radius of Gyration Axis v—v Min.	Elastic Modulus Axis x—x	Elastic Modulus Axis y—y
mm	mm	mm	kg	mm	mm	cm²	cm	cm	cm⁴	cm⁴	cm⁴	cm⁴	cm	cm	cm	cm	cm³	cm³
89 × 89	88.9 × 88.9	15.8	20.10	8.4	4.8	25.61	2.78	2.78	178	178	280	75.7	2.63	2.63	3.30	1.72	29.1	29.1
		14.2	18.31	8.4	4.8	23.32	2.72	2.72	164	164	259	69.1	2.65	2.65	3.33	1.72	26.6	26.6
		12.6	16.38	8.4	4.8	20.87	2.66	2.66	149	149	235	62.0	2.67	2.67	3.36	1.72	23.9	23.9
		11.0	14.44	8.4	4.8	18.40	2.60	2.60	133	133	211	55.0	2.69	2.69	3.38	1.73	21.1	21.1
		9.4	12.50	8.4	4.8	15.92	2.54	2.54	116	116	185	47.9	2.70	2.70	3.41	1.74	18.3	18.3
		7.9	10.58	8.4	4.8	13.47	2.48	2.48	99.8	99.8	159	41.0	2.72	2.72	3.43	1.74	15.6	15.6
		6.3	8.49	8.4	4.8	10.81	2.41	2.41	81.0	81.0	129	33.3	2.74	2.74	3.45	1.75	12.5	12.5
76 × 76	76.2 × 76.2	14.3	15.50	7.6	4.8	19.74	2.41	2.41	99.6	99.6	157	42.7	2.25	2.25	2.82	1.47	19.1	19.1
		12.6	13.85	7.6	4.8	17.64	2.35	2.35	90.4	90.4	143	38.2	2.26	2.26	2.84	1.47	17.1	17.1
		11.0	12.20	7.6	4.8	15.55	2.29	2.29	80.9	80.9	128	33.8	2.28	2.28	2.87	1.47	15.2	15.2
		9.4	10.57	7.6	4.8	13.47	2.23	2.23	71.1	71.1	113	29.5	2.30	2.30	2.89	1.48	13.2	13.2
		7.8	8.93	7.6	4.8	11.37	2.16	2.16	60.9	60.9	96.8	25.1	2.31	2.31	2.92	1.49	11.2	11.2
		6.2	7.16	7.6	4.8	9.12	2.10	2.10	49.6	49.6	78.8	20.3	2.33	2.33	2.94	1.49	8.97	8.97
64 × 64	63.5 × 63.5	12.5	11.31	6.9	2.4	14.41	2.03	2.03	50.4	50.4	78.9	21.8	1.87	1.87	2.34	1.23	11.7	11.7
		11.0	10.12	6.9	2.4	12.89	1.98	1.98	45.8	45.8	72.1	19.5	1.89	1.89	2.37	1.23	10.5	10.5
		9.4	8.78	6.9	2.4	11.18	1.92	1.92	40.5	40.5	64.0	17.0	1.90	1.90	2.39	1.23	9.15	9.15
		7.9	7.45	6.9	2.4	9.48	1.86	1.86	35.0	35.0	55.5	14.6	1.92	1.92	2.42	1.24	7.80	7.80
		6.2	5.96	6.9	2.4	7.59	1.80	1.80	28.6	28.6	45.4	11.8	1.94	1.94	2.45	1.25	6.28	6.28
57 × 57	57.2 × 57.2	9.3	7.74	6.6	2.4	9.86	1.76	1.76	28.6	28.6	45.0	12.1	1.70	1.70	2.14	1.11	7.22	7.22
		7.8	6.55	6.6	2.4	8.35	1.70	1.70	24.7	24.7	39.1	10.3	1.72	1.72	2.16	1.11	6.15	6.15
		6.2	5.35	6.6	2.4	6.82	1.64	1.64	20.6	20.6	32.6	8.53	1.74	1.74	2.19	1.12	5.05	5.05
		4.6	4.01	6.6	2.4	5.11	1.57	1.57	15.8	15.8	25.0	6.51	1.76	1.76	2.21	1.13	3.80	3.80
51 × 51	50.8 × 50.8	9.4	6.85	6.1	2.4	8.72	1.60	1.60	19.6	19.6	30.8	8.42	1.50	1.50	1.88	.98	5.64	5.64
		7.8	5.80	6.1	2.4	7.39	1.54	1.54	17.0	17.0	26.8	7.17	1.52	1.52	1.91	.98	4.81	4.81
		6.3	4.77	6.1	2.4	6.08	1.49	1.49	14.3	14.3	22.7	5.95	1.53	1.53	1.93	.99	3.98	3.98
		4.6	3.58	6.1	2.4	4.56	1.42	1.42	11.0	11.0	17.4	4.54	1.55	1.55	1.95	1.00	3.00	3.00

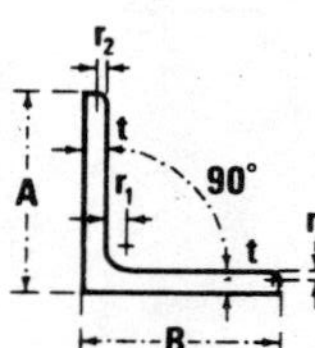

Equal angles

Dimensions and properties

Nominal Size	Leg Lengths A×B	Actual Thickness	Mass per metre	Radii Root r_1	Radii Toe r_2	Area of Section	Centre of Gravity Cx	Centre of Gravity Cy	Moment of Inertia Axis x–x	Axis y–y	Axis u–u Max.	Axis v–v Min.	Radius of Gyration Axis x–x	Axis y–y	Axis u–u Max.	Axis v–v Min.	Elastic Modulus Axis x–x	Axis y–y
mm	mm	mm	kg	mm	mm	cm²	cm	cm	cm⁴	cm⁴	cm⁴	cm⁴	cm	cm	cm	cm	cm³	cm³
45 × 45	44.5 × 44.5	7.9	5.06	5.8	2.4	6.45	1.39	1.39	11.1	11.1	17.5	4.75	1.31	1.31	1.65	.86	3.64	3.64
		6.1	4.02	5.8	2.4	5.12	1.32	1.32	9.09	9.09	14.4	3.80	1.33	1.33	1.68	.86	2.91	2.91
		4.7	3.13	5.8	2.4	3.99	1.26	1.26	7.24	7.24	11.5	3.00	1.35	1.35	1.70	.87	2.28	2.28
38 × 38	38.1 × 38.1	7.8	4.24	5.3	2.4	5.40	1.23	1.23	6.69	6.69	10.5	2.92	1.11	1.11	1.39	.73	2.59	2.59
		6.3	3.50	5.3	2.4	4.46	1.17	1.17	5.67	5.67	8.94	2.41	1.13	1.13	1.42	.73	2.15	2.15
		4.7	2.68	5.3	2.4	3.41	1.11	1.11	4.47	4.47	7.08	1.86	1.14	1.14	1.44	.74	1.66	1.66
32 × 32	31.8 × 31.8	6.2	2.83	5.1	2.4	3.61	1.01	1.01	3.10	3.10	4.87	1.34	.93	.93	1.16	.61	1.43	1.43
		4.6	2.16	5.1	2.4	2.75	0.95	0.95	2.45	2.45	3.87	1.03	.94	.94	1.19	.61	1.10	1.10
		3.1	1.49	5.1	2.4	1.90	0.88	0.88	1.74	1.74	2.75	.72	.96	.96	1.20	.62	.76	.76
25 × 25	25.4 × 25.4	6.4	2.23	4.6	2.4	2.84	0.85	0.85	1.50	1.50	2.33	.68	.73	.73	.90	.49	.89	.89
		4.7	1.72	4.6	2.4	2.19	0.79	0.79	1.20	1.20	1.89	.51	.74	.74	.93	.48	.69	.69
		3.1	1.19	4.6	2.4	1.52	0.73	0.73	.86	.86	1.37	.36	.75	.75	.95	.49	.48	.48

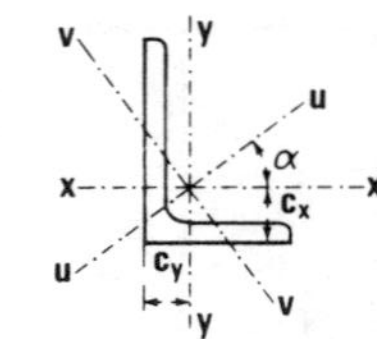

Unequal angles

Dimensions and properties

Nominal Size	Leg Lengths A×B	Actual Thickness	Mass per metre	Radii Root r_1	Radii Toe r_2	Area of Section	Centre of Gravity Cx	Centre of Gravity Cy	Moment of Inertia Axis x—x	Moment of Inertia Axis y—y	Moment of Inertia Axis u—u Max.	Moment of Inertia Axis v—v Min.	Radius of Gyration Axis x—x	Radius of Gyration Axis y—y	Radius of Gyration Axis u—u Max.	Radius of Gyration Axis v—v Min.	Angle Axis x—x to Axis u—u	Elastic Modulus Axis x—x	Elastic Modulus Axis y—y
mm	mm	mm	kg	mm	mm	cm²	cm	cm	cm⁴	cm⁴	cm⁴	cm⁴	cm	cm	cm	cm	tan α	cm³	cm³
229 × 102	228.6 × 101.6	22.1	53.77	13.0	4.8	68.49	8.73	2.41	3606	447	3747	306	7.26	2.56	7.40	2.11	.207	255	57.8
		20.6	50.21	13.0	4.8	63.97	8.67	2.35	3388	423	3523	287	7.28	2.57	7.42	2.12	.209	239	54.1
		18.9	46.45	13.0	4.8	59.17	8.60	2.29	3154	396	3283	267	7.30	2.59	7.45	2.13	.211	221	50.3
		17.4	42.87	13.0	4.8	54.61	8.54	2.23	2929	369	3051	248	7.32	2.60	7.47	2.13	.213	205	46.6
		15.8	39.20	13.0	4.8	49.93	8.47	2.17	2695	342	2808	229	7.35	2.62	7.50	2.14	.214	187	42.8
		14.2	35.43	13.0	4.8	45.13	8.40	2.10	2451	313	2556	208	7.37	2.63	7.53	2.15	.216	170	38.9
		12.6	31.56	13.0	4.8	40.20	8.33	2.04	2197	283	2292	187	7.39	2.65	7.55	2.16	.218	151	34.8
203 × 152	203.2 × 152.4	22.1	58.09	13.7	4.8	74.00	6.59	4.07	2992	1439	3648	783	6.36	4.41	7.02	3.25	.545	218	129
		20.5	54.22	13.7	4.8	69.07	6.53	4.01	2811	1355	3432	734	6.38	4.43	7.05	3.26	.547	204	121
		18.9	50.32	13.7	4.8	64.10	6.47	3.95	2625	1268	3209	684	6.40	4.45	7.08	3.27	.548	190	112
		17.3	46.30	13.7	4.8	58.99	6.41	3.89	2432	1177	2976	633	6.42	4.47	7.10	3.28	.550	175	104
		15.8	42.32	13.7	4.8	53.91	6.35	3.83	2237	1085	2740	582	6.44	4.49	7.13	3.29	.551	160	95.1
		14.2	38.29	13.7	4.8	48.78	6.29	3.77	2037	990	2497	530	6.46	4.51	7.15	3.30	.553	145	86.3
		12.6	34.10	13.7	4.8	43.44	6.22	3.70	1826	890	2240	476	6.48	4.53	7.18	3.31	.554	129	77.1
203 × 102	203.2 × 101.6	19.0	42.75	12.2	4.8	54.45	7.46	2.41	2277	386	2409	253	6.47	2.66	6.65	2.15	.256	177	49.7
		17.3	39.32	12.2	4.8	50.09	7.40	2.35	2109	359	2234	234	6.49	2.68	6.68	2.16	.259	163	46.0
		15.8	36.07	12.2	4.8	45.95	7.33	2.29	1947	333	2064	216	6.51	2.69	6.70	2.17	.260	150	42.4
		14.2	32.62	12.2	4.8	41.55	7.27	2.23	1773	305	1881	197	6.53	2.71	6.73	2.18	.262	136	38.5
		12.6	29.07	12.2	4.8	37.03	7.20	2.16	1591	276	1689	177	6.55	2.73	6.75	2.19	.264	121	34.5
178 × 89	177.8 × 88.9	15.8	31.30	11.2	4.8	39.87	6.49	2.08	1280	217	1355	142	5.67	2.33	5.83	1.89	.257	113	31.9
		14.2	28.28	11.2	4.8	36.02	6.43	2.01	1165	199	1235	129	5.69	2.35	5.86	1.89	.260	103	28.9
		12.6	25.31	11.2	4.8	32.24	6.36	1.95	1051	181	1115	117	5.71	2.37	5.88	1.90	.262	92.0	26.0
		11.1	22.36	11.2	4.8	28.48	6.29	1.89	935	162	993	104	5.73	2.38	5.90	1.91	.264	81.4	23.1
		9.4	19.22	11.2	4.8	24.48	6.22	1.83	810	141	861	90.3	5.75	2.40	5.93	1.92	.265	70.1	20.0
152 × 102	152.4 × 101.6	19.0	35.16	10.7	4.8	44.80	5.25	2.73	1015	358	1161	212	4.76	2.83	5.09	2.17	.427	102	48.2
		17.4	32.47	10.7	4.8	41.37	5.19	2.67	945	335	1083	196	4.78	2.84	5.12	2.18	.430	94.1	44.7
		15.8	29.66	10.7	4.8	37.78	5.13	2.61	871	309	1000	180	4.80	2.86	5.14	2.19	.432	86.1	41.0
		14.2	26.80	10.7	4.8	34.14	5.07	2.55	794	283	913	164	4.82	2.88	5.17	2.19	.435	78.0	37.2
		12.6	23.99	10.7	4.8	30.56	5.00	2.48	716	257	825	148	4.84	2.90	5.20	2.20	.437	70.0	33.4
		11.0	21.14	10.7	4.8	26.93	4.94	2.42	637	229	734	132	4.86	2.92	5.22	2.21	.439	61.8	29.6
		9.5	18.30	10.7	4.8	23.31	4.88	2.36	555	201	641	115	4.88	2.93	5.24	2.22	.441	53.6	25.7

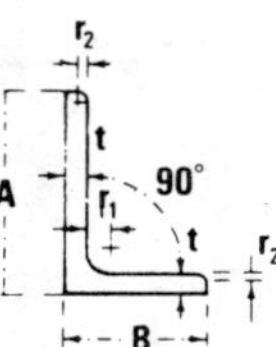

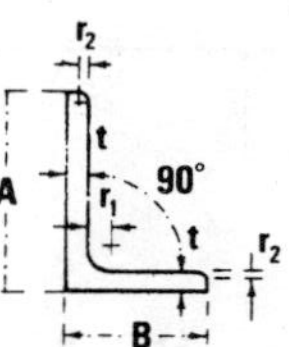

Unequal angles

Dimensions and properties

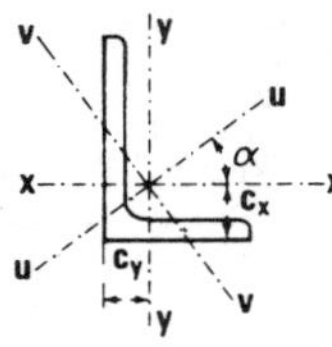

Nominal Size	Leg Lengths A×B	Actual Thickness	Mass per metre	Radii Root r_1	Radii Toe r_2	Area of Section	Centre of Gravity Cx	Centre of Gravity Cy	Moment of Inertia Axis x–x	Axis y–y	Axis u–u Max.	Axis v–v Min.	Radius of Gyration Axis x–x	Axis y–y	Axis u–u Max.	Axis v–v Min.	Angle Axis x–x to Axis u–u	Elastic Modulus Axis x–x	Axis y–y
mm	mm	mm	kg	mm	mm	cm²	cm	cm	cm⁴	cm⁴	cm⁴	cm⁴	cm	cm	cm	cm	tan α	cm³	cm³
152 × 89	152.4 × 88.9	15.7	27.99	10.4	4.8	35.66	5.37	2.22	828	208	907	129	4.82	2.42	5.04	1.90	.336	83.9	31.2
		14.2	25.46	10.4	4.8	32.43	5.31	2.16	759	192	833	118	4.84	2.43	5.07	1.91	.338	76.5	28.5
		12.6	22.77	10.4	4.8	29.00	5.25	2.10	685	174	752	107	4.86	2.45	5.09	1.92	.341	68.5	25.6
		11.1	20.12	10.4	4.8	25.63	5.19	2.04	610	156	671	95.2	4.88	2.47	5.12	1.93	.343	60.7	22.7
		9.4	17.26	10.4	4.8	21.99	5.12	1.97	528	136	581	82.6	4.90	2.48	5.14	1.94	.345	52.1	19.6
		7.8	14.44	10.4	4.8	18.40	5.04	1.91	445	115	490	69.9	4.92	2.50	5.16	1.95	.347	43.7	16.5
152 × 76	152.4 × 76.2	15.8	26.52	9.9	4.8	33.78	5.65	1.86	786	132	830	87.2	4.82	1.98	4.96	1.61	.253	81.9	22.9
		14.2	23.99	9.9	4.8	30.56	5.59	1.80	717	121	759	79.3	4.84	1.99	4.98	1.61	.256	74.3	20.8
		12.6	21.45	9.9	4.8	27.33	5.52	1.74	647	110	685	71.5	4.87	2.01	5.01	1.62	.259	66.6	18.7
		11.0	18.92	9.9	4.8	24.10	5.46	1.68	575	98.5	610	63.7	4.89	2.02	5.03	1.63	.261	58.8	16.6
		9.5	16.39	9.9	4.8	20.87	5.39	1.62	503	86.7	534	55.7	4.91	2.04	5.06	1.63	.263	51.0	14.4
		7.8	13.69	9.9	4.8	17.44	5.32	1.55	424	73.6	450	47.2	4.93	2.05	5.08	1.64	.265	42.7	12.1
127 × 89	127.0 × 88.9	15.8	24.86	9.7	4.8	31.67	4.29	2.40	496	198	580	114	3.96	2.50	4.28	1.90	.470	59.0	30.5
		14.2	22.64	9.7	4.8	28.84	4.24	2.34	456	183	534	104	3.97	2.52	4.30	1.90	.473	53.8	27.9
		12.6	20.26	9.7	4.8	25.81	4.17	2.28	412	166	484	93.9	3.99	2.54	4.33	1.91	.476	48.3	25.1
		11.1	17.89	9.7	4.8	22.79	4.11	2.22	367	149	432	83.6	4.01	2.55	4.35	1.92	.479	42.7	22.3
		9.4	15.35	9.7	4.8	19.56	4.04	2.16	318	129	375	72.5	4.03	2.57	4.38	1.93	.481	36.8	19.2
		7.9	12.94	9.7	4.8	16.48	3.98	2.10	271	110	319	61.7	4.05	2.59	4.40	1.94	.483	31.0	16.2
127 × 76	127.0 × 76.2	14.2	21.17	9.1	4.8	26.96	4.47	1.95	430	116	475	71.5	4.00	2.07	4.20	1.63	.351	52.3	20.4
		12.6	18.91	9.1	4.8	24.09	4.41	1.89	389	105	429	64.4	4.02	2.09	4.22	1.63	.355	46.9	18.3
		11.0	16.69	9.1	4.8	21.27	4.35	1.83	346	94.2	383	57.3	4.04	2.10	4.25	1.64	.358	41.5	16.3
		9.4	14.44	9.1	4.8	18.39	4.28	1.77	302	82.8	335	50.0	4.06	2.12	4.27	1.65	.360	35.9	14.1
		7.8	12.06	9.1	4.8	15.37	4.21	1.70	255	70.2	283	42.3	4.07	2.14	4.29	1.66	.362	30.1	11.9
102 × 89	101.6 × 88.9	15.8	21.75	8.9	4.8	27.71	3.27	2.64	262	186	357	91.5	3.08	2.59	3.59	1.82	.743	38.1	29.7
		14.2	19.67	8.9	4.8	25.06	3.21	2.58	240	170	328	83.0	3.10	2.61	3.62	1.82	.746	34.6	27.0
		12.6	17.72	8.9	4.8	22.57	3.15	2.52	219	156	299	75.0	3.11	2.63	3.64	1.82	.748	31.2	24.4
		11.0	15.62	8.9	4.8	19.90	3.09	2.46	195	139	268	66.6	3.13	2.64	3.67	1.83	.750	27.6	21.6
		9.5	13.55	8.9	4.8	17.27	3.03	2.40	171	122	235	58.2	3.15	2.66	3.69	1.84	.752	24.0	18.8
		7.8	11.31	8.9	4.8	14.41	2.96	2.33	145	103	199	49.1	3.17	2.68	3.72	1.85	.753	20.1	15.8

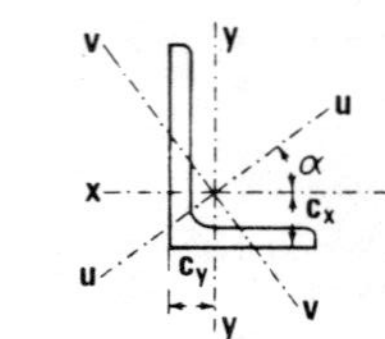

Unequal angles

Dimensions and properties

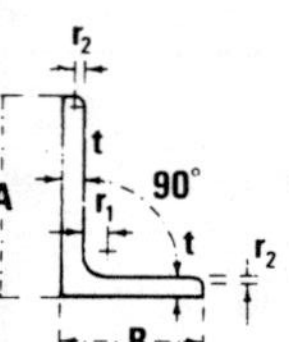

| Nominal Size | Leg Lengths A×B | Actual Thickness | Mass per metre | Radii | | Area of Section | Centre of Gravity | | Moment of Inertia | | | | Radius of Gyration | | | | Angle | Elastic Modulus | |
| | | | | Root r_1 | Toe r_2 | | Cx | Cy | Axis x—x | Axis y—y | Axis u—u Max. | Axis v—v Min. | Axis x—x | Axis y—y | Axis u—u Max. | Axis v—v Min. | Axis x—x to Axis u—u | Axis x—x | Axis y—y |
mm	mm	mm	kg	mm	mm	cm²	cm	cm	cm⁴	cm⁴	cm⁴	cm⁴	cm	cm	cm	cm	tan α	cm³	cm³
102 × 76	101.6 × 76.2	14.2	18.31	8.4	4.8	23.32	3.40	2.14	228	109	277	60.4	3.13	2.16	3.45	1.61	.537	33.8	19.9
		12.6	16.38	8.4	4.8	20.87	3.34	2.08	207	98.8	251	54.3	3.15	2.18	3.47	1.61	.540	30.3	17.8
		11.0	14.44	8.4	4.8	18.40	3.28	2.02	185	88.5	225	48.2	3.17	2.19	3.50	1.62	.544	26.8	15.8
		9.4	12.50	8.4	4.8	15.92	3.22	1.96	162	77.8	197	42.0	3.19	2.21	3.52	1.62	.547	23.3	13.7
		7.9	10.58	8.4	4.8	13.47	3.16	1.90	138	66.8	169	35.9	3.20	2.23	3.54	1.63	.549	19.7	11.7
102 × 64	101.6 × 63.5	11.0	13.40	8.1	4.8	17.07	3.51	1.62	174	51.9	194	31.4	3.19	1.74	3.37	1.36	.380	26.1	11.0
		9.5	11.61	8.1	4.8	14.79	3.45	1.56	152	45.8	171	27.4	3.21	1.76	3.40	1.36	.383	22.7	9.56
		7.8	9.69	8.1	4.8	12.35	3.38	1.49	129	38.9	145	23.2	3.23	1.78	3.42	1.37	.386	19.0	8.02
		6.3	7.89	8.1	4.8	10.05	3.31	1.43	106	32.2	119	19.1	3.25	1.79	3.44	1.38	.388	15.4	6.54
89 × 76	88.9 × 76.2	14.2	16.83	8.1	4.8	21.44	2.89	2.26	155	104	207	52.4	2.69	2.20	3.11	1.56	.710	25.9	19.4
		12.7	15.20	8.1	4.8	19.36	2.84	2.21	142	95.4	190	47.5	2.71	2.22	3.13	1.57	.713	23.4	17.6
		11.0	13.40	8.1	4.8	17.07	2.77	2.14	127	85.4	170	42.0	2.73	2.24	3.16	1.57	.715	20.7	15.6
		9.5	11.61	8.1	4.8	14.79	2.71	2.08	111	75.1	150	36.7	2.74	2.25	3.18	1.58	.718	18.0	13.6
		7.8	9.69	8.1	4.8	12.35	2.65	2.02	94.2	63.7	127	30.9	2.76	2.27	3.21	1.58	.720	15.1	11.4
		6.3	7.89	8.1	4.8	10.05	2.58	1.96	77.5	52.5	104	25.5	2.78	2.29	3.23	1.59	.721	12.3	9.27
89 × 64	88.9 × 63.5	11.0	12.20	7.6	4.8	15.55	2.97	1.71	118	49.8	140	28.2	2.76	1.79	3.00	1.35	.489	20.0	10.7
		9.4	10.57	7.6	4.8	13.47	2.91	1.65	104	43.9	123	24.6	2.78	1.80	3.02	1.35	.493	17.4	9.34
		7.8	8.93	7.6	4.8	11.37	2.85	1.59	88.8	37.7	106	21.0	2.79	1.82	3.05	1.36	.496	14.7	7.92
		6.2	7.16	7.6	4.8	9.12	2.78	1.53	72.1	30.7	85.8	17.0	2.81	1.83	3.07	1.37	.498	11.8	6.37
76 × 64	76.2 × 63.5	11.0	11.17	7.4	4.8	14.23	2.46	1.83	76.6	47.8	99.9	24.4	2.32	1.83	2.65	1.31	.669	14.8	10.6
		9.4	9.68	7.4	4.8	12.33	2.40	1.77	67.3	42.1	88.2	21.3	2.34	1.85	2.67	1.31	.673	12.9	9.20
		7.9	8.19	7.4	4.8	10.43	2.34	1.71	57.8	36.2	75.9	18.1	2.35	1.86	2.70	1.32	.676	10.9	7.81
		6.2	6.56	7.4	4.8	8.36	2.27	1.64	46.9	29.5	61.7	14.7	2.37	1.88	2.72	1.33	.678	8.77	6.27
76 × 51	76.2 × 50.8	11.0	10.12	6.9	2.4	12.89	2.68	1.42	71.7	25.1	81.7	15.1	2.36	1.40	2.52	1.08	.420	14.5	6.86
		9.4	8.78	6.9	2.4	11.18	2.62	1.36	63.2	22.3	72.3	13.2	2.38	1.41	2.54	1.09	.426	12.6	5.99
		7.9	7.45	6.9	2.4	9.48	2.56	1.30	54.5	19.3	62.5	11.3	2.40	1.43	2.57	1.09	.431	10.8	5.12
		6.2	5.96	6.9	2.4	7.59	2.49	1.24	44.4	15.9	51.1	9.20	2.42	1.45	2.59	1.10	.436	8.66	4.13
		4.7	4.62	6.9	2.4	5.88	2.43	1.18	34.9	12.6	40.2	7.26	2.44	1.46	2.62	1.11	.438	6.72	3.22

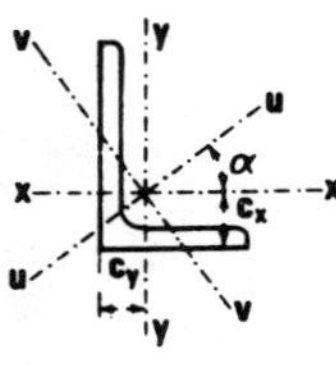

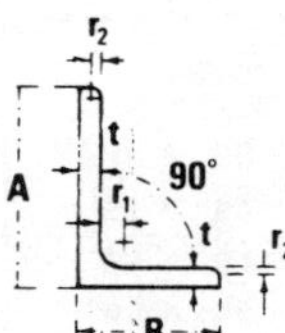

Unequal angles

Dimensions and properties

Nominal Size	Leg Lengths A×B	Actual Thickness	Mass per metre	Radii Root r_1	Radii Toe r_2	Area of Section	C.G. C_x	C.G. C_y	M.I. Axis x–x	M.I. Axis y–y	M.I. Axis u–u Max.	M.I. Axis v–v Min.	R.G. Axis x–x	R.G. Axis y–y	R.G. Axis u–u Max.	R.G. Axis v–v Min.	Angle Axis x–x to Axis u–u	E.M. Axis x–x	E.M. Axis y–y
mm	mm	mm	kg	mm	mm	cm²	cm	cm	cm⁴	cm⁴	cm⁴	cm⁴	cm	cm	cm	cm	tan α	cm³	cm³
64 × 51	63.5 × 50.8	9.3	7.74	6.6	2.4	9.86	2.09	1.46	37.2	20.9	47.0	11.1	1.94	1.46	2.18	1.06	.613	8.73	5.78
		7.8	6.55	6.6	2.4	8.35	2.03	1.40	32.1	18.1	40.7	9.50	1.96	1.47	2.21	1.07	.618	7.44	4.93
		6.2	5.35	6.6	2.4	6.82	1.97	1.35	26.7	15.1	34.0	7.86	1.98	1.49	2.23	1.07	.622	6.10	4.06
		4.6	4.01	6.6	2.4	5.11	1.90	1.28	20.4	11.6	26.0	6.01	2.00	1.51	2.26	1.08	.625	4.59	3.06
64 × 38	63.5 × 38.1	7.8	5.80	6.1	2.4	7.39	2.26	1.00	29.2	7.79	32.1	4.87	1.99	1.03	2.08	.81	.347	7.13	2.77
		6.3	4.77	6.1	2.4	6.08	2.20	0.94	24.4	6.59	27.0	4.05	2.00	1.04	2.11	.82	.353	5.89	2.30
		4.6	3.58	6.1	2.4	4.56	2.13	0.88	18.7	5.10	20.7	3.10	2.03	1.06	2.13	.82	.358	4.43	1.74
51 × 38	50.8 × 38.1	7.9	5.06	5.8	2.4	6.45	1.73	1.10	15.5	7.37	18.8	4.15	1.55	1.07	1.71	.80	.532	4.64	2.72
		6.1	4.02	5.8	2.4	5.12	1.66	1.03	12.7	6.04	15.4	3.32	1.57	1.09	1.73	.81	.540	3.70	2.17
		4.7	3.13	5.8	2.4	3.99	1.60	0.98	10.1	4.83	12.3	2.63	1.59	1.10	1.75	.81	.544	2.89	1.70

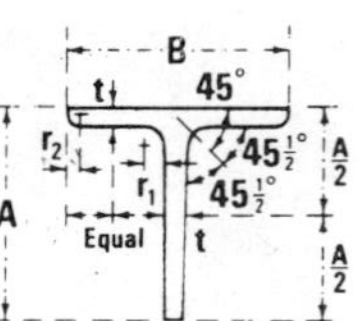
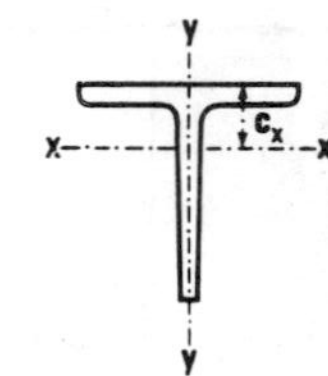

T–Bars

Dimensions and properties

| Designation | | Width of Section B | Depth of Section A | Thick–ness t | Radius | | Area of Section | Centre of Gravity c_x | Moment of Inertia | | Radius of Gyration | | Elastic Modulus | |
| Nominal Size | Mass per metre | | | | Root r_1 | Toe r_2 | | | Axis x–x | Axis y–y | Axis x–x | Axis y–y | Axis x–x | Axis y–y |
mm	kg	mm	mm	mm	mm	mm	cm²	cm	cm⁴	cm⁴	cm	cm	cm³	cm³
152 × 152	36	152.4	152.4	15.9	12.2	8.6	45.97	4.29	970.2	452.4	4.57	3.12	88.49	59.32
152 × 152	29	152.4	152.4	12.7	12.2	8.6	37.23	4.14	792.5	356.3	4.62	3.10	71.45	46.70
152 × 102	30	152.4	101.6	15.9	10.7	7.4	37.94	2.59	304.7	454.9	2.84	3.45	40.31	59.65
152 × 102	24	152.4	101.6	12.7	10.7	7.4	30.78	2.46	252.7	359.6	2.87	3.53	32.77	47.19
152 × 76	22	152.4	76.2	12.7	9.9	6.9	27.55	1.73	109.5	360.9	1.98	3.61	18.68	47.36
152 × 76	16	152.4	76.2	9.5	9.9	6.9	21.02	1.60	85.74	266.4	2.03	3.56	14.26	34.90
127 × 102	22	127.0	101.6	12.7	9.9	6.9	27.55	2.67	240.2	209.0	2.95	2.77	32.12	32.94
127 × 102	16	127.0	101.6	9.5	9.9	6.9	20.96	2.54	186.1	154.0	2.97	2.72	24.42	24.25
127 × 76	19	127.0	76.2	12.7	9.1	6.4	24.32	1.88	104.5	209.8	2.08	2.95	18.19	32.94
127 × 76	15	127.0	76.2	9.5	9.1	6.4	18.58	1.75	82.00	154.8	2.18	2.90	13.93	24.42
102 × 102	19	101.6	101.6	12.7	9.1	6.4	24.25	2.95	224.8	107.8	3.05	2.11	31.14	21.30
102 × 102	15	101.6	101.6	9.5	9.1	6.4	18.51	2.79	174.4	79.08	3.07	2.06	23.76	15.57
102 × 76	16	101.6	76.2	12.7	8.4	5.8	21.02	2.08	98.65	108.2	2.16	2.26	17.70	21.30
102 × 76	13	101.6	76.2	9.5	8.4	5.8	16.13	1.96	77.42	79.50	2.18	2.21	13.60	15.73
76 × 76	11	76.2	76.2	9.5	7.6	5.3	13.67	2.21	71.18	33.71	2.29	1.57	13.11	8.85
64 × 64	9	63.5	63.5	9.5	6.9	4.8	11.22	1.90	39.96	19.56	1.88	1.32	9.01	6.23
64 × 64	6	63.5	63.5	6.4	6.9	4.8	7.74	1.78	28.30	12.49	1.90	1.27	6.23	3.93
51 × 51	5	50.8	50.8	6.4	6.1	4.3	6.06	1.47	14.15	6.66	1.52	1.04	3.93	2.62
38 × 38	4	38.1	38.1	6.4	5.3	3.8	4.45	1.17	5.83	2.91	1.12	0.79	2.13	1.47

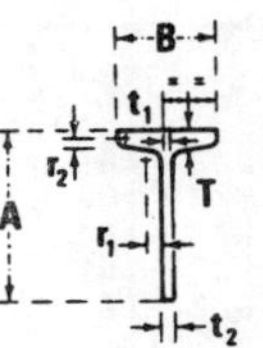

Long stalk T–Bars

Dimensions and properties

Designation		Width of Section	Depth of Section	Thickness			Radius		Area of Section	Centre of Gravity	Moment of Inertia		Radius of Gyration		Elastic Modulus	
Nominal Size	Mass per metre	B	A	T	t_1	t_2	Root r_1	Toe r_2		c_x	Axis x–x	Axis y–y	Axis x–x	Axis y–y	Axis x–x	Axis y–y
mm	kg	mm	mm	mm	mm	mm	mm	mm	cm²	cm	cm⁴	cm⁴	cm	cm	cm³	cm³
127 × 254	35.42	127.0	254.0	18.3	9.4	8.9	13.5	6.6	45.35	6.93	2811	273.0	7.85	2.46	153	42.9
102 × 203	25.02	101.6	203.2	16.3	8.4	7.9	12.2	7.6	31.93	5.84	1289	124.9	6.38	1.98	89.0	24.6
89 × 178	20.42	88.9	177.8	15.2	7.9	7.4	11.2	6.4	26.06	5.18	804.9	79.49	5.56	1.75	63.7	17.9
76 × 152	16.30	76.2	152.4	14.2	7.4	6.9	10.2	6.4	20.90	4.44	468.2	46.61	4.72	1.50	43.4	12.3
64 × 127	12.62	63.5	127.0	13.4	6.9	6.4	8.9	5.1	16.13	3.76	248.5	25.80	3.94	1.27	27.9	8.19
44 × 114	7.44	44.5	114.3	9.5	5.1	5.1	7.6	3.8	9.48	3.66	126.1	7.08	3.63	0.86	16.2	3.11
25 × 76	3.65	25.4	76.2	6.4	4.4	4.4	5.1	3.8	4.64	2.82	27.89	0.83	2.44	0.43	5.74	0.66

Structural tees
Cut from universal beams

Dimensions and properties

Serial Size	Mass per metre	Width of Section B	Depth of Section A	Thickness Web t	Thickness Flange T	Root Radius r	Slope inside Flange	Area of Section	Gravity Centre Distance C_x	Moment of Inertia Axis x–x	Moment of Inertia Axis y–y	Radius of Gyration Axis x–x	Radius of Gyration Axis y–y	Elastic Modulus Axis x–x C_x	Elastic Modulus Axis x–x E_x	Elastic Modulus Axis y–y	Cut from Universal Beam
mm	kg	mm	mm	mm	mm	mm	per cent	cm²	cm	cm⁴	cm⁴	cm	cm	cm³	cm³	cm³	mm×mm @ kg/m
305 × 457	127	305.5	459.2	17.3	27.9	19.1	5	161.2	12.03	32664	6256	14.2	6.23	2716	963.7	409.6	914 × 305 @ 253
305 × 457	112	304.1	455.2	15.9	23.9	19.1	5	142.5	12.16	29001	5212	14.3	6.05	2386	869.3	342.8	914 × 305 @ 224
305 × 457	101	303.4	451.5	15.2	20.2	19.1	5	128.0	12.56	26399	4316	14.4	5.81	2101	810.2	284.5	914 × 305 @ 201
292 × 419	113	293.8	425.5	16.1	26.8	17.8	5	144.2	10.84	24636	5331	13.1	6.08	2272	777.2	362.9	838 × 292 @ 226
292 × 419	97	292.4	420.4	14.7	21.7	17.8	5	123.4	11.11	21354	4192	13.2	5.83	1922	690.4	286.8	838 × 292 @ 194
292 × 419	88	291.6	417.4	14.0	18.8	17.8	5	111.9	11.39	19560	3555	13.2	5.64	1718	644.3	243.8	838 × 292 @ 176
267 × 381	99	268.0	384.8	15.6	25.4	16.5	5	125.3	9.91	17512	3850	11.8	5.54	1766	613.0	287.3	762 × 267 @ 197
267 × 381	87	266.7	381.0	14.3	21.6	16.5	5	110.1	10.01	15477	3188	11.9	5.38	1547	550.9	239.1	762 × 267 @ 173
267 × 381	74	265.3	376.9	12.9	17.5	16.5	5	93.9	10.20	13308	2501	11.9	5.16	1304	484.1	188.6	762 × 267 @ 147
254 × 343	85	255.8	346.5	14.5	23.7	15.2	5	108.2	8.69	12025	3113	10.5	5.36	1384	463.2	243.4	686 × 254 @ 170
254 × 343	76	254.5	343.8	13.2	21.0	15.2	5	96.8	8.61	10726	2695	10.5	5.28	1246	416.2	211.8	686 × 254 @ 152
254 × 343	70	253.7	341.8	12.4	19.0	15.2	5	89.2	8.66	9926	2395	10.5	5.18	1146	389.1	188.7	686 × 254 @ 140
254 × 343	63	253.0	339.0	11.7	16.2	15.2	5	79.7	8.88	8984	1996	10.6	5.00	1011	359.2	157.7	686 × 254 @ 125
305 × 305	119	311.5	316.5	18.6	31.4	16.5	5	151.8	7.12	12283	7487	9.00	7.02	1726	500.7	480.7	610 × 305 @ 238
305 × 305	90	307.0	308.7	14.1	23.6	16.5	5	113.8	6.66	8939	5285	8.86	6.81	1341	369.2	344.3	610 × 305 @ 179
305 × 305	75	304.8	304.8	11.9	19.7	16.5	5	94.9	6.45	7355	4236	8.80	6.68	1140	306.1	277.9	610 × 305 @ 149
229 × 305	70	230.1	308.5	13.1	22.1	12.7	5	89.1	7.62	7739	2126	9.32	4.88	1016	333.1	184.8	610 × 229 @ 140
229 × 305	63	229.0	305.9	11.9	19.6	12.7	5	79.7	7.56	6904	1838	9.31	4.80	913.7	299.7	160.5	610 × 229 @ 125
229 × 305	57	228.2	303.7	11.2	17.3	12.7	5	72.2	7.62	6288	1592	9.34	4.70	825.6	276.4	139.5	610 × 229 @ 113
229 × 305	51	227.6	301.1	10.6	14.8	12.7	5	64.5	7.82	5702	1329	9.40	4.54	729.6	255.7	116.8	610 × 229 @ 101
178 × 305	46	178.4	301.2	10.6	15.0	12.7	0	57.9	8.68	5351	713.5	9.61	3.51	616.3	249.6	80.0	610 × 178 @ 91
178 × 305	41	177.8	299.1	10.1	12.8	12.7	0	52.2	8.90	4848	601.3	9.64	3.39	544.7	230.8	67.6	610 × 178 @ 82
330 × 267	106	333.6	272.5	16.7	27.8	16.5	5	134.8	5.56	7381	8032	7.40	7.72	1329	340.2	481.6	533 × 330 @ 212
330 × 267	95	331.7	269.7	14.9	25.0	16.5	5	120.6	5.36	6484	7046	7.33	7.64	1209	300.0	424.8	533 × 330 @ 189
330 × 267	84	330.2	266.7	13.4	22.0	16.5	5	106.3	5.23	5678	6029	7.31	7.53	1085	264.9	365.2	533 × 330 @ 167
210 × 267	61	211.9	272.3	12.8	21.3	12.7	5	77.8	6.68	5178	1604	8.16	4.54	775.1	252.0	151.4	533 × 210 @ 122
210 × 267	55	210.7	269.7	11.6	18.8	12.7	5	69.2	6.61	4588	1377	8.14	4.46	694.5	225.3	130.7	533 × 210 @ 109
210 × 267	51	210.1	268.4	10.9	17.4	12.7	5	64.6	6.58	4277	1256	8.14	4.41	649.9	211.2	119.6	533 × 210 @ 101
210 × 267	46	209.3	266.6	10.2	15.6	12.7	5	58.8	6.58	3900	1106	8.14	4.34	593.0	194.2	105.7	533 × 210 @ 92
210 × 267	41	208.7	264.2	9.6	13.2	12.7	5	52.1	6.75	3511	912.8	8.21	4.18	520.3	178.5	87.5	533 × 210 @ 82
165 × 267	37	165.6	264.4	9.3	13.5	12.7	0	46.5	7.35	3258	513.6	8.37	3.32	443.0	170.7	62.0	533 × 165 @ 73
165 × 267	33	165.1	262.4	8.8	11.5	12.7	0	41.8	7.55	2949	431.5	8.40	3.21	390.6	157.8	52.3	533 × 165 @ 66
191 × 229	49	192.8	233.7	11.4	19.6	10.2	5	62.6	5.56	2976	1108	6.90	4.21	535.4	167.1	114.9	457 × 191 @ 98
191 × 229	45	192.0	231.8	10.6	17.7	10.2	5	56.9	5.50	2698	980.1	6.89	4.15	490.5	152.7	102.1	457 × 191 @ 89
191 × 229	41	191.3	230.1	9.9	16.0	10.2	5	52.2	5.49	2479	873.1	6.89	4.09	451.9	141.5	91.3	457 × 191 @ 82
191 × 229	37	190.5	228.6	9.1	14.5	10.2	5	47.4	5.43	2244	773.6	6.88	4.04	413.4	128.7	81.2	457 × 191 @ 74
191 × 229	34	189.9	226.8	8.5	12.7	10.2	5	42.7	5.48	2034	664.2	6.90	3.95	371.5	118.2	70.0	457 × 191 @ 67

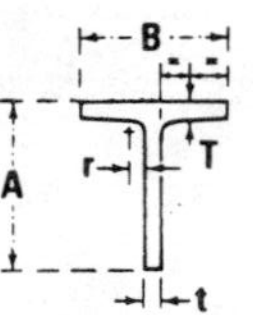

Structural tees
Cut from universal beams

Dimensions and properties

Cut from Universal Beam (mm×mm @ kg/m)	Elastic Modulus Axis y–y (cm³)	Elastic Modulus Axis x–x Ex (cm³)	Elastic Modulus Axis x–x Cx (cm³)	Radius of Gyration Axis y–y (cm)	Radius of Gyration Axis x–x (cm)	Moment of Inertia Axis x–x (cm⁴)	Moment of Inertia Axis y–y (cm⁴)	Gravity Centre Distance Cx (cm)
457 × 152 @ 82	71.2	151.3	431.8	3.24	7.07	2606	546.7	6.03
457 × 152 @ 74	63.0	138.4	394.3	3.18	7.06	2362	481.3	5.99
457 × 152 @ 67	54.6	126.0	354.7	3.12	7.06	2126	414.4	5.99
457 × 152 @ 60	52.0	110.6	321.4	3.23	7.02	1870	397.2	5.82
457 × 152 @ 52	42.3	101.3	276.3	3.11	7.08	1667	322.4	6.03
406 × 178 @ 74	80.6	110.9	365.2	3.91	6.08	1756	724.0	4.81
406 × 178 @ 67	71.0	99.9	331.7	3.85	6.07	1572	634.3	4.74
406 × 178 @ 60	62.3	88.0	299.2	3.82	6.03	1382	554.2	4.62
406 × 178 @ 54	51.9	83.6	265.6	3.67	6.12	1280	460.9	4.82
406 × 152 @ 74	68.1	116.3	354.3	3.32	6.20	1823	523.3	5.14
406 × 152 @ 67	59.4	105.6	320.8	3.26	6.20	1638	454.2	5.11
406 × 152 @ 60	50.5	95.7	285.5	3.18	6.21	1462	384.0	5.12
406 × 140 @ 46	35.1	75.0	223.1	2.92	6.19	1129	250.1	5.06
406 × 140 @ 39	26.3	66.3	182.8	2.75	6.26	966.4	186.5	5.29
381 × 152 @ 67	61.4	97.2	300.8	3.33	5.78	1427	473.4	4.75
381 × 152 @ 60	53.1	86.6	269.8	3.27	5.76	1261	407.0	4.67
381 × 152 @ 52	45.0	76.0	238.1	3.21	5.75	1097	342.7	4.61
356 × 171 @ 67	73.8	81.6	288.3	3.87	5.21	1157	639.2	4.01
356 × 171 @ 57	59.6	70.0	247.2	3.77	5.21	977.9	513.0	3.96
356 × 171 @ 51	51.6	63.3	222.9	3.71	5.21	876.6	442.7	3.93
356 × 171 @ 45	42.7	58.2	196.0	3.58	5.27	790.3	365.1	4.03
356 × 127 @ 39	26.4	54.4	162.6	2.60	5.40	718.5	166.5	4.42
356 × 127 @ 33	20.5	47.9	136.0	2.48	5.44	617.2	128.5	4.54
305 × 165 @ 54	59.3	51.5	198.9	3.80	4.31	635.8	494.1	3.20
305 × 165 @ 46	49.8	44.1	174.7	3.74	4.29	540.3	412.4	3.09
305 × 165 @ 40	41.9	39.2	154.7	3.67	4.30	475.4	345.5	3.07
305 × 127 @ 48	35.0	56.3	166.7	2.68	4.64	653.1	219.0	3.92
305 × 127 @ 42	29.5	49.4	147.1	2.63	4.62	567.2	183.3	3.86
305 × 127 @ 37	25.6	44.3	132.2	2.58	4.61	503.8	157.9	3.81
305 × 102 @ 33	18.5	42.4	117.3	2.13	4.83	486.5	94.68	4.15
305 × 102 @ 28	15.0	38.0	100.8	2.05	4.85	426.5	76.41	4.23
305 × 102 @ 25	11.4	34.9	83.8	1.92	4.89	375.4	58.05	4.48
254 × 146 @ 43	43.0	33.8	130.5	3.39	3.56	348.8	316.6	2.67
254 × 146 @ 37	36.1	29.0	114.8	3.34	3.54	296.4	263.8	2.58
254 × 146 @ 31	27.8	26.6	97.7	3.19	3.63	262.7	202.8	2.69
254 × 102 @ 28	17.1	28.6	85.6	2.19	3.93	278.7	87.12	3.26
254 × 102 @ 25	14.1	26.6	75.2	2.11	3.96	252.5	71.89	3.36
254 × 102 @ 22	11.4	24.7	65.1	2.02	4.00	227.1	58.00	3.49
203 × 133 @ 30	26.4	18.5	72.5	3.05	2.83	152.4	176.8	2.10
203 × 133 @ 25	21.0	16.6	62.7	2.94	2.88	133.5	139.8	2.13

Serial Size (mm)	Mass per metre (kg)	Width of Section B (mm)	Depth of Section A (mm)	Thickness Web t (mm)	Thickness Flange T (mm)	Root Radius r (mm)	Slope inside Flange (per cent)	Area of Section (cm²)
152 × 229	41	153.5	232.5	10.7	18.9	10.2	5	52.2
152 × 229	37	152.7	230.6	9.9	17.0	10.2	5	47.4
152 × 229	34	151.9	228.6	9.1	15.0	10.2	5	42.7
152 × 229	30	152.9	227.3	8.0	13.3	10.2	0	38.0
152 × 229	26	152.4	224.9	7.6	10.9	10.2	0	33.2
178 × 203	37	179.7	206.4	9.7	16.0	10.2	5	47.4
178 × 203	34	178.8	204.7	8.8	14.3	10.2	5	42.7
178 × 203	30	177.8	203.2	7.8	12.8	10.2	5	38.0
178 × 203	27	177.6	201.3	7.6	10.9	10.2	5	34.2
152 × 203	37	153.7	208.2	10.1	18.1	10.2	5	47.4
152 × 203	34	152.9	206.1	9.3	16.0	10.2	5	42.7
152 × 203	30	152.2	204.0	8.6	13.9	10.2	5	37.9
140 × 203	23	142.4	201.2	6.9	11.2	10.2	5	29.4
140 × 203	20	141.8	198.6	6.3	8.6	10.2	5	24.7
152 × 191	34	154.3	194.3	9.7	16.3	10.2	5	42.7
152 × 191	30	153.4	192.4	8.7	14.4	10.2	5	38.0
152 × 191	26	152.4	190.5	7.8	12.4	10.2	5	33.2
171 × 178	34	173.2	182.0	9.1	15.7	10.2	5	42.7
171 × 178	29	172.1	179.3	8.0	13.0	10.2	5	36.0
171 × 178	26	171.5	177.8	7.3	11.5	10.2	5	32.2
171 × 178	23	171.0	176.0	6.9	9.7	10.2	5	28.4
127 × 178	20	126.0	176.4	6.5	10.7	10.2	5	24.7
127 × 178	17	125.4	174.2	5.9	8.5	10.2	5	20.9
165 × 152	27	166.8	155.4	7.7	13.7	8.9	5	34.2
165 × 152	23	165.7	153.5	6.7	11.8	8.9	5	29.4
165 × 152	20	165.1	151.9	6.1	10.2	8.9	5	25.7
127 × 152	24	125.2	155.2	8.9	14.0	8.9	2	30.4
127 × 152	21	124.3	153.3	8.0	12.1	8.9	2	26.6
127 × 152	19	123.5	151.9	7.2	10.7	8.9	2	23.7
102 × 152	17	102.4	156.3	6.6	10.8	7.6	5	20.9
102 × 152	14	101.9	154.4	6.1	8.9	7.6	5	18.1
102 × 152	13	101.6	152.4	5.8	6.8	7.6	5	15.7
146 × 127	22	147.3	129.8	7.3	12.7	7.6	2	27.5
146 × 127	19	146.4	128.0	6.4	10.9	7.6	2	23.7
146 × 127	16	146.1	125.7	6.1	8.6	7.6	2	20.0
102 × 127	14	102.1	130.2	6.4	10.0	7.6	5	18.1
102 × 127	13	101.9	128.5	6.1	8.4	7.6	5	16.1
102 × 127	11	101.6	127.0	5.8	6.8	7.6	5	14.2
133 × 102	15	133.8	103.4	6.3	9.6	7.6	5	19.0
133 × 102	13	133.4	101.6	5.8	7.8	7.6	5	16.1

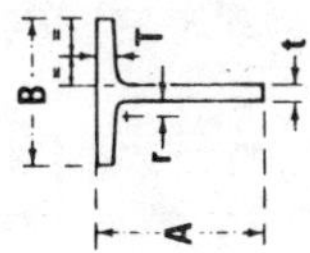

Structural tees
Cut from universal columns

Dimensions and properties

Serial Size	Mass per metre	Width of Section B	Depth of Section A	Thickness Web t	Thickness Flange T	Root Radius r	Slope inside Flange	Area of Section	Gravity Centre Distance Cx	Moment of Inertia Axis x–x	Moment of Inertia Axis y–y	Radius of Gyration Axis x–x	Radius of Gyration Axis y–y	Elastic Modulus Axis x–x Cx	Elastic Modulus Axis x–x Ex	Elastic Modulus Axis y–y	Cut from Universal Column
mm	kg	mm	mm	mm	mm	mm	per cent	cm²	cm	cm⁴	cm⁴	cm	cm	cm³	cm³	cm³	mm×mm @ kg/m
406 × 178	118	395.0	190.5	18.5	30.2	15.2	0	149.9	3.41	2886	15504	4.39	10.2	846.3	184.5	785.1	356 × 406 @ 235
368 × 178	101	374.4	187.3	16.8	27.0	15.2	0	129.0	3.32	2500	11816	4.40	9.57	754.1	162.1	631.2	356 × 368 @ 202
368 × 178	89	372.1	184.2	14.5	23.8	15.2	0	112.9	3.10	2099	10235	4.31	9.52	677.4	137.0	550.1	356 × 368 @ 177
368 × 178	77	370.2	181.0	12.6	20.7	15.2	0	97.6	2.92	1765	8735	4.25	9.46	605.6	116.3	471.9	356 × 368 @ 153
368 × 178	65	368.3	177.8	10.7	17.5	15.2	0	82.5	2.73	1451	7278	4.20	9.39	531.1	96.4	395.2	356 × 368 @ 129
305 × 152	79	310.6	163.6	15.7	25.0	15.2	0	100.6	3.04	1529	6262	3.90	7.89	502.9	114.9	403.2	305 × 305 @ 158
305 × 152	69	308.7	160.3	13.8	21.7	15.2	0	87.3	2.86	1291	5336	3.85	7.82	450.9	98.1	345.7	305 × 305 @ 137
305 × 152	59	306.8	157.2	11.9	18.7	15.2	0	74.9	2.69	1075	4503	3.79	7.75	399.9	82.5	293.5	305 × 305 @ 118
305 × 152	49	304.8	153.9	9.9	15.4	15.2	0	61.6	2.50	858.1	3634	3.73	7.68	343.1	66.6	238.5	305 × 305 @ 97
254 × 127	66	261.0	138.2	15.6	25.1	12.7	0	83.9	2.72	886.7	3722	3.25	6.66	325.6	79.9	285.2	254 × 254 @ 132
254 × 127	54	258.3	133.4	13.0	20.5	12.7	0	68.3	2.47	683.0	2951	3.16	6.57	277.0	62.8	228.5	254 × 254 @ 107
254 × 127	45	255.9	130.2	10.5	17.3	12.7	0	57.0	2.24	534.8	2424	3.06	6.52	238.7	49.6	189.5	254 × 254 @ 89
254 × 127	37	254.0	127.0	8.6	14.2	12.7	0	46.4	2.06	418.7	1936	3.00	6.46	203.7	39.3	152.5	254 × 254 @ 73
203 × 102	43	208.8	111.1	13.0	20.5	10.2	0	55.0	2.22	379.8	1560	2.63	5.32	171.0	42.7	149.4	203 × 203 @ 86
203 × 102	36	206.2	108.0	10.3	17.3	10.2	0	45.5	1.98	288.1	1268	2.52	5.28	145.3	32.7	123.0	203 × 203 @ 71
203 × 102	30	205.2	104.8	9.3	14.2	10.2	0	37.9	1.88	241.2	1021	2.52	5.19	128.3	28.1	99.5	203 × 203 @ 60
203 × 102	26	203.9	103.1	8.0	12.5	10.2	0	33.2	1.76	203.1	885.0	2.47	5.16	115.2	23.8	86.8	203 × 203 @ 52
203 × 102	23	203.2	101.6	7.3	11.0	10.2	0	29.4	1.71	179.6	769.4	2.47	5.11	105.3	21.2	75.7	203 × 203 @ 46
152 × 76	19	154.4	80.9	8.1	11.5	7.6	0	23.7	1.55	94.67	354.3	2.00	3.87	61.2	14.5	45.9	152 × 152 @ 37
152 × 76	15	152.9	78.7	6.6	9.4	7.6	0	19.1	1.41	72.79	279.2	1.95	3.82	51.5	11.3	36.5	152 × 152 @ 30
152 × 76	12	152.4	76.2	6.1	6.8	7.6	0	14.9	1.43	61.11	201.7	2.03	3.68	42.7	9.88	26.5	152 × 152 @ 23

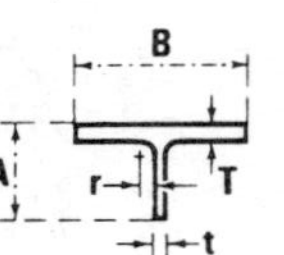

Universal bearing piles
Parallel flanges

Dimensions and properties

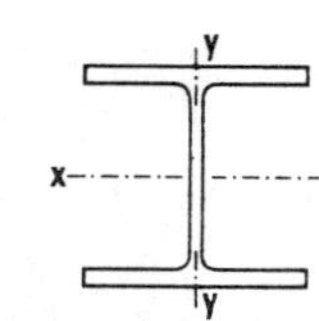

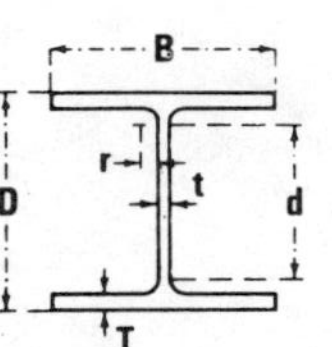

Serial Size	Mass per metre	Depth of Section D	Width of Section B	Thickness Web t	Thickness Flange T	Root Radius r	Depth between Fillets d	Area of Section	Moment of Inertia Axis x–x Gross	Moment of Inertia Axis x–x Net	Moment of Inertia Axis y–y	Radius of Gyration Axis x–x	Radius of Gyration Axis y–y	Elastic Modulus Axis x–x	Elastic Modulus Axis y–y	Ratio $\frac{D}{T}$
mm	kg	mm	mm	mm	mm	mm	mm	cm²	cm⁴	cm⁴	cm⁴	cm	cm	cm³	cm³	
356 × 368	174	361.5	378.1	20.4	20.4	15.2	290.1	222.2	51134	44954	18444	15.2	9.11	2829	975.7	17.7
	152	356.4	375.5	17.9	17.9	15.2	290.1	193.6	43916	38578	15799	15.1	9.03	2464	841.5	19.9
	133	351.9	373.3	15.6	15.6	15.2	290.1	169.0	37840	33248	13576	15.0	8.96	2150	727.4	22.6
	109	346.4	370.5	12.9	12.9	15.2	290.1	138.4	30515	26784	10901	14.8	8.87	1762	588.5	26.9
305 × 305	110	307.9	310.3	15.4	15.4	15.2	246.6	140.4	23580	21865	7689	13.0	7.40	1532	495.6	20.0
	79	299.2	306.0	11.1	11.1	15.2	246.6	100.4	16400	15201	5292	12.8	7.26	1096	345.9	27.0
254 × 254	85	254.3	259.7	14.3	14.3	12.7	200.2	108.1	12264	11192	4188	10.7	6.22	964.5	322.6	17.8
	63	246.9	256.0	10.6	10.6	12.7	200.2	79.7	8775	8005	2971	10.5	6.11	710.9	232.2	23.3
203 × 203	54	203.9	207.2	11.3	11.3	10.2	160.8	68.4	4987	4441	1683	8.54	4.96	489.2	162.4	18.0

Mechanics

Statics

Laws of Coulomb friction

(i) The friction force developed is independent of the magnitude of the area of contact.

(ii) The friction force is proportional to the normal force.

(iii) At low velocity of sliding the friction force is independent of the velocity.

Belt friction

The ratio of tensions at the ends of an arc of belt sustending angle θ is

$$T_1/T_2 = e^{\mu\theta}$$

where μ is the coefficient of friction.

The funicular curve

The funicular curve for a load p per unit length is given by

$$\frac{d^2 y}{dx^2} = \frac{p}{H}$$

where H is the polar distance.

Kinematics

In the following, v and a are velocity and acceleration, s is arc length and ω is angular velocity. Unit vectors are $\mathbf{i}, \mathbf{j}, \mathbf{k}$ for Cartesian, and $\mathbf{e}$ with the relevant subscript for other coordinates.

Rectangular coordinates

$$\mathbf{v} = v_x \mathbf{i} + v_y \mathbf{j} + v_z \mathbf{k}$$
$$\mathbf{a} = a_x \mathbf{i} + a_y \mathbf{j} + a_z \mathbf{k}$$

Normal and tangential components

$$\mathbf{v} = \frac{ds}{dt}\,\mathbf{e}_t$$

$$\mathbf{a} = \frac{d^2 s}{dt^2}\,\mathbf{e}_t - \frac{(ds/dt)^2}{R}\,\mathbf{e}_n$$

Cylindrical coordinates

$$\mathbf{v} = \dot{r}\mathbf{e}_r + r\dot{\theta}\mathbf{e}_\theta + \dot{z}\mathbf{e}_z$$
$$\mathbf{a} = (\ddot{r} - r\dot{\theta}^2)\mathbf{e}_r + (r\ddot{\theta} + 2\dot{r}\dot{\theta})\mathbf{e}_\theta + \ddot{z}\mathbf{e}_z$$

Spherical polar coordinates

$$\mathbf{v} = \dot{r}\mathbf{e}_r + r\dot{\theta}\mathbf{e}_\theta + r\dot{\phi}\sin\theta\,\mathbf{e}_\phi$$
$$\mathbf{a} = [\ddot{r} - r\dot{\theta}^2 - r\dot{\phi}^2\sin^2\theta]\mathbf{e}_r + [r\ddot{\theta} + 2\dot{r}\dot{\theta} - r\dot{\phi}^2\sin\theta\cos\theta]\mathbf{e}_\theta$$
$$+ [(r\ddot{\phi} + 2\dot{r}\dot{\phi})\sin\theta + 2r\dot{\phi}\dot{\theta}\cos\theta]\mathbf{e}_\phi$$

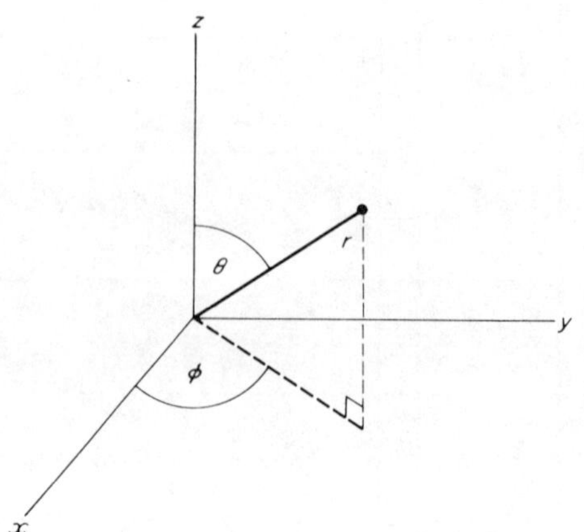

Motion referred to a moving coordinate system

$$\mathbf{r} = \mathbf{R} + \rho$$

$$\dot{\mathbf{r}} = \dot{\mathbf{R}} + \dot{\rho}_r + \omega \times \rho$$

$$\ddot{\mathbf{r}} = \ddot{\mathbf{R}} + \omega \times (\omega \times \rho) + \dot{\omega} \times \rho + \ddot{\rho}_r + 2\omega \times \dot{\rho}_r$$

$\dot{\rho}_r$ is the velocity of P measured relative to $O'xyz$, which has angular velocity ω relative to $OXYZ$.

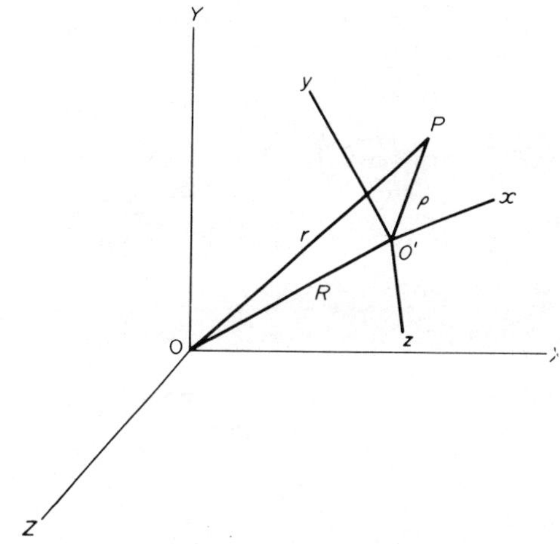

Dynamics

In the following, m is mass, F force, H momentum, M moment, V potential energy and T kinetic energy.

Newton's Laws

(1) Every body stays in a state of rest or uniform motion in a straight line unless it is acted on by a force which may change that state.

(2) The rate of change of momentum with respect to time is equal to the force producing it. The change takes place in the direction of the force.

(3) To every action there is an equal and opposite reaction.

Particle dynamics

Impulse and momentum

$$\int_{t_1}^{t_2} \mathbf{F}\, dt = m\mathbf{v}_2 - m\mathbf{v}_1.$$

Moment of momentum

$$\mathbf{r} \times \mathbf{F} = \frac{d}{dt}(\mathbf{r} \times m\mathbf{v})$$

where $\mathbf{r} \times m\mathbf{v} = \mathbf{H}_0$.

Conservation of momentum

If in a system of particles only mutual interactions are involved, the momentum of the system is constant.

Work and energy

$$\int_{r_1}^{r_2} \mathbf{F} \cdot d\mathbf{r} = \tfrac{1}{2}mv_2^2 - \tfrac{1}{2}mv_1^2$$

Potential energy

If $\mathbf{F} = \text{grad}\,\phi$,

$$\int_1^2 \mathbf{F} \cdot d\mathbf{r} = \phi_2 - \phi_1$$

and

$$-\int_1^2 \mathbf{F} \cdot d\mathbf{r} = V_2 - V_1,$$

the change in potential energy.
    For an inverse square law

$$\mathbf{F} = \frac{\mu m}{r^2}\,\mathbf{e}_r \qquad V = \frac{\mu m}{r}$$

where μ is a constant of the field of force acting on m.

Conservation of energy

For a conservative system

$$V + T = \text{constant}$$

Central force motion

Kepler's Laws

(1) Each planet has an elliptical orbit with the sun at a focus.
(2) The radius vector drawn from the sun to the planet sweeps out equal areas in equal times.
(3) The squares of the periods of the planets are proportional to the cubes of the semi-major axes of the elliptical orbits.

Inverse-square-law attraction

The equation of the orbit is

$$r = \frac{h^2}{\mu} + x\left(1 - \frac{h}{\mu}\dot{y}_0\right) + y\left(\frac{h}{\mu}\dot{x}_0\right)$$

where $h = \dfrac{H_0}{m}$ and the velocity of projection has components $\dot{x}_0$ and $\dot{y}_0$ at $x = x_0, y = 0$.

The energy per unit mass is $E = \frac{1}{2}v^2 - \dfrac{\mu}{r} = $ constant.

If $E < 0$, the path is elliptical, if $E = 0$, parabolic and if $E > 0$, hyperbolic.

Rigid-body dynamics

Moment of momentum about mass centre

$$H_x = I_{xx}\omega_x - I_{xy}\omega_y - I_{xz}\omega_z$$
$$H_y = -I_{yx}\omega_x + I_{yy}\omega_y - I_{yz}\omega_z$$
$$H_z = -I_{zx}\omega_x - I_{zy}\omega_y + I_{zz}\omega_z$$

where with mass density ρ

$$I_{xx} = \int_\tau \rho(y^2 + z^2)\,d\tau$$

etc. and

$$I_{xy} = \int_\tau \rho xy\,d\tau$$

etc., the integrals being taken over the volume τ of the body.

General equations of motion

$$M_x = \dot{H}_x - \omega_z H_y + \omega_y H_z$$
$$M_y = \dot{H}_y - \omega_x H_z + \omega_z H_x$$
$$M_z = \dot{H}_z - \omega_y H_x + \omega_x H_y.$$

Euler's equations

$$M_x = I_{xx}\dot{\omega}_x + (I_{zz} - I_{yy})\omega_y\omega_z$$
$$M_y = I_{yy}\dot{\omega}_y + (I_{xx} - I_{zz})\omega_x\omega_z$$
$$M_z = I_{zz}\dot{\omega}_z + (I_{yy} - I_{xx})\omega_x\omega_y$$

in which all moments of inertia are principal values.

Kinetic energy

$$T = \tfrac{1}{2}mv_C^2 + \tfrac{1}{2}\boldsymbol{\omega}.\mathbf{H}_C$$

where suffix C refers to mass centre.

Matrix notation for dynamics

Motion referred to moving coordinate system

$$\{\ddot{r}\} = \{\ddot{R}\} + [\omega]^2\{\rho\} + [\dot{\omega}]\{\rho\} + \{\ddot{\rho}\} + 2[\omega]\{\dot{\rho}\}$$

Moment of momentum about mass centre

$$\{H\} = [I]\{\omega\},$$

where

$$[I] = -\sum\left(m\begin{bmatrix} 0 & -z & y \\ z & 0 & -x \\ -y & x & 0 \end{bmatrix}^2\right)$$

General equations of motion

$$\{M\} = [I]\{\dot{\omega}\} + [\omega][I]\{\omega\}$$

Kinetic energy

$$T = \tfrac{1}{2}M\{v_C\}^T\{v_C\} + \tfrac{1}{2}\{\omega\}^T[I]\{\omega\}$$

where v_C is the velocity of the centre of mass.

Gyroscopic motion

ω is angular velocity of housing, Ω is angular velocity of rotor relative to housing.

$$\{H\} = [I]\{\omega\} + [I]\{\Omega\}$$

For principal moments of inertia:

$$\{\dot{H}\} + [\omega]\{H\} = [I]\{\dot{\omega}\} + [\omega][I]\{\omega\} + [I]\{\dot{\Omega}\}$$
$$+ [\omega][I]\{\Omega\}$$

Lagrange's equations

$$\frac{d}{dt}\left(\frac{\partial T}{\partial \dot{q}_j}\right) - \frac{\partial T}{\partial q_j} = Q_j$$

where q_j is a generalized coordinate and Q_j is a generalized force.

For a conservative system

$$\frac{d}{dt}\left(\frac{\partial L}{\partial \dot{q}_j}\right) - \frac{\partial L}{\partial q_j} = 0$$

where $L = (T - V)$.

Euler's differential equation

$$\frac{d}{dx}\left(\frac{\partial f}{\partial y'}\right) - \frac{\partial f}{\partial y} = 0$$

Hamilton's principle

$$\delta\int_{t_1}^{t_2}(T - V)\,dt = 0$$

Vibrations

In the following, k is spring stiffness, c a viscous damping constant, ω_n an undamped natural frequency and m, M masses.

Free vibration with viscous damping

For a mass m the undamped natural frequency is

$$\omega_n = \sqrt{(k/m)}$$

the critical damping constant is

$$c_c = 2\sqrt{(km)}$$

the damping ratio is $\zeta = c/c_c$ and the logarithmic decrement is

$$\delta = 2\pi\zeta/\sqrt{(1 - \zeta^2)} \approx 2\pi\zeta$$

Steady-state vibration with viscous damping

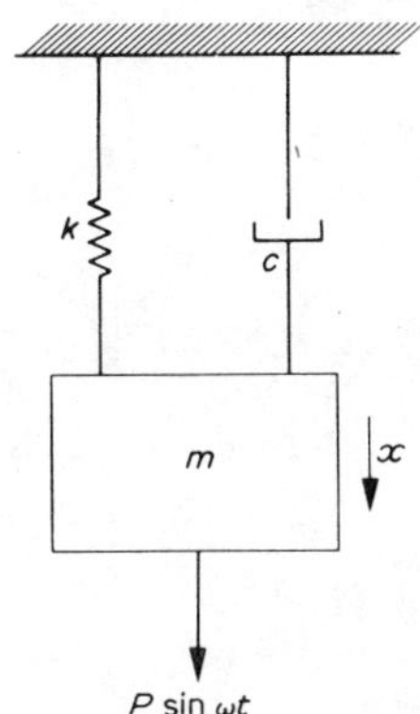

The ratio of peak amplitude X to the steady displacement $X_0 = P/k$ is

$$\frac{X}{X_0} = \frac{1}{[\{1 - (\omega/\omega_n)^2\}^2 + \{2\zeta\omega/\omega_n\}^2]^{1/2}}$$

and the phase angle ϕ is given by

$$\tan\phi = \frac{2\zeta\omega/\omega_n}{1 - (\omega/\omega_n)^2}$$

These relations yield the curves given below.

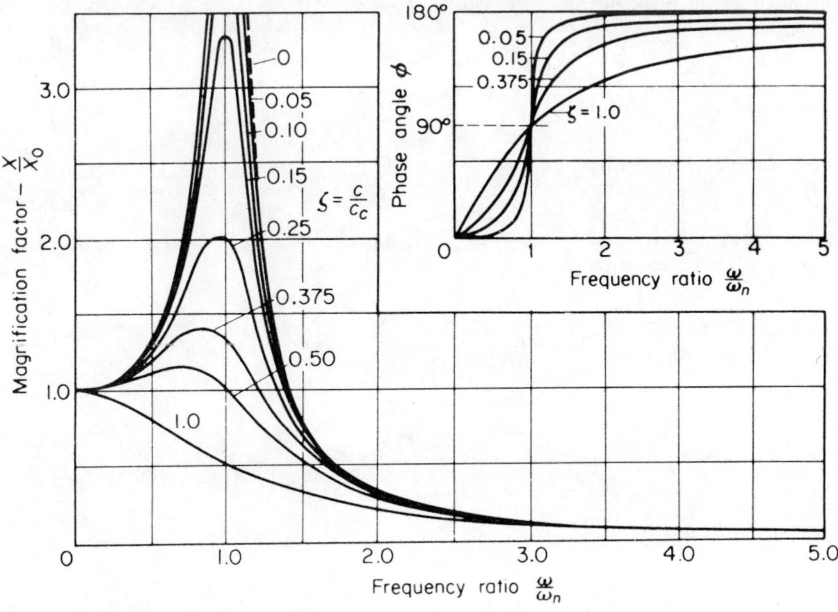

Rotating unbalance

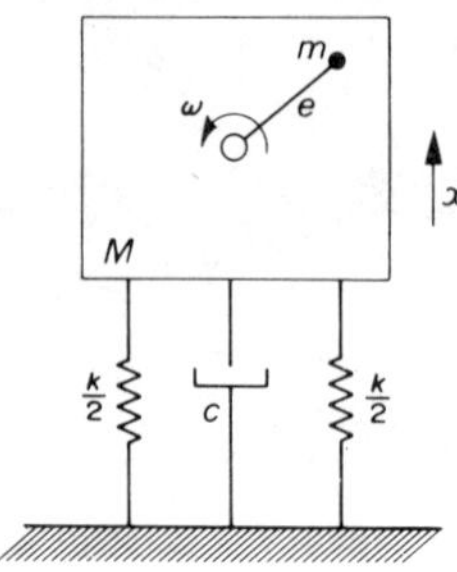

The peak amplitude X and the phase ϕ are given by

$$\frac{MX}{me} = \frac{(\omega/\omega_n)^2}{[\{1 - (\omega/\omega_n)^2\}^2 + \{2\zeta\omega/\omega_n\}^2]^{1/2}}$$

and

$$\tan \phi = \frac{2\zeta\omega/\omega_n}{1 - (\omega/\omega_n)^2}.$$

The curves below show the variation of X and ϕ with ω.

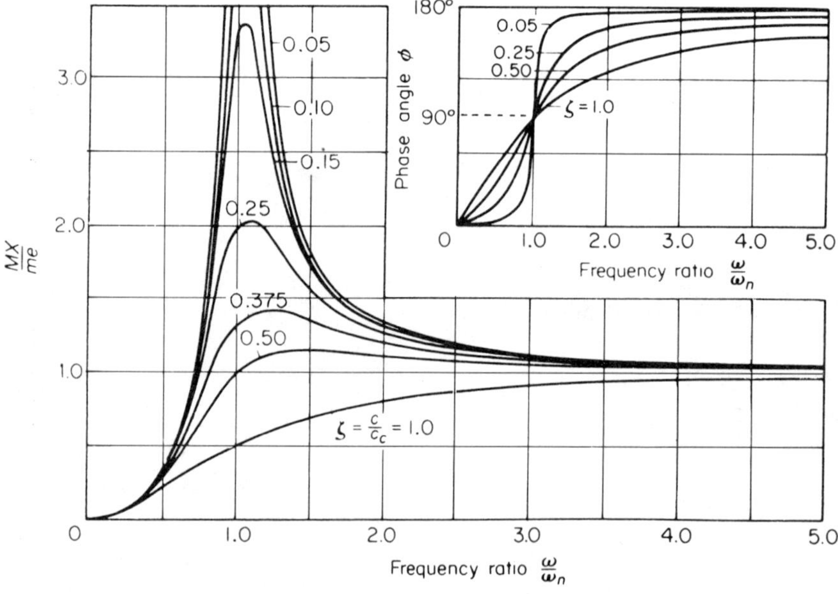

Displacement excitation

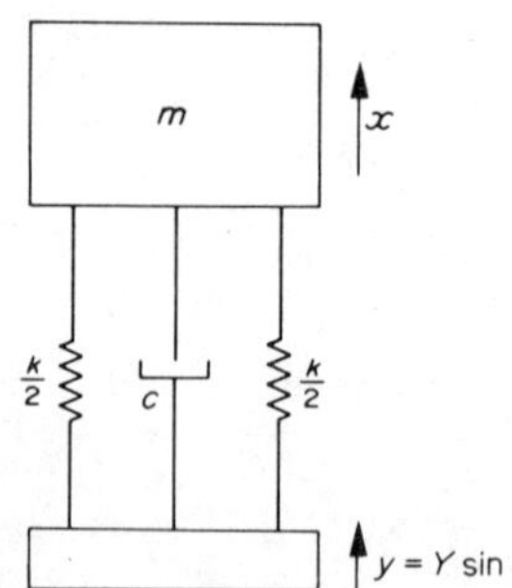

The ratio of the peak amplitudes of the mass and the base is

$$\frac{X}{Y} = \left[\frac{1 + (2\zeta\omega/\omega_n)^2}{\{1 - (\omega/\omega_n)^2\}^2 + \{2\zeta\omega/\omega_n\}^2} \right]^{1/2}$$

and their phase difference is given by

$$\tan \phi = \frac{2\zeta(\omega/\omega_n)^3}{1 - (\omega/\omega_n)^2 + (2\zeta\omega/\omega_n)^2}$$

These results are shown in the curves below. The amplitude
ratio is also the ratio of the transmitted force to the exciting
force in the vibration isolator with constants k and c
separating mass m from ground.

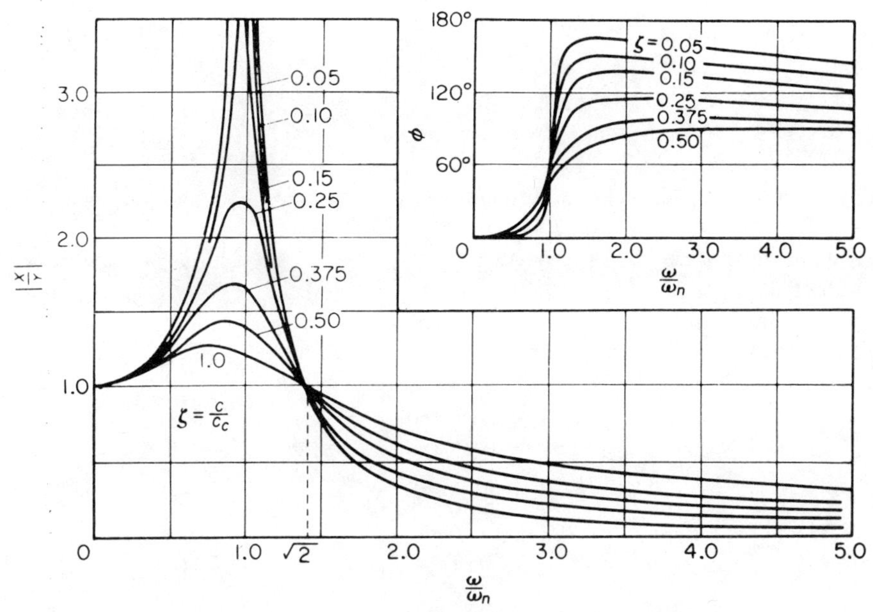

Vibration of beams of uniform section with uniformly distributed load

The natural frequencies of a beam of length l are given by

$$\omega_n = k\sqrt{\frac{EI}{ml^4}}$$

where m is the mass per unit length.

The table below gives the value of k for the 1st, 2nd and 3rd harmonics with different support conditions.

Support	1	2	3
	3·52	22·0	61·7
	9·87	39·5	88·8
	15·4	50·0	104
(also Free–Free)	22·4	61·7	121

Electricity

Electromagnetism

In the following, $\mathbf{E}$, $\mathbf{D}$ are electric field strength and flux density; $\mathbf{H}$, $\mathbf{B}$ are magnetic field strength and flux density; $\mathbf{J}$, i are current density and current, ρ, ρ_s volume and surface charge density; $\mathbf{A}$, ϕ are vector and scalar potentials; V, S, l are volume, area and length; $\mathbf{a}_r$, $\mathbf{n}$ are radial and normal unit vectors; t, n denote tangential and normal; and brackets denote retarded values.

The Lorentz force per unit charge is $\mathbf{E} + \mathbf{v} \times \mathbf{B}$

Maxwell's equations

Integral form

$$\oint_C \mathbf{E} \, . \, d\mathbf{l} = - \int_S \dot{\mathbf{B}} \, . \, d\mathbf{S}$$

$$\oint_C \mathbf{H} \, . \, d\mathbf{l} = \int_S (\mathbf{J} + \dot{\mathbf{D}}) \, . \, d\mathbf{S}$$

$$\oint_S \mathbf{B} \, . \, d\mathbf{S} = 0$$

$$\oint_S \mathbf{D} \, . \, d\mathbf{S} = \int_V \rho \, dV$$

Differential form

$$\text{curl } \mathbf{E} = -\dot{\mathbf{B}}$$

$$\text{curl } \mathbf{H} = \mathbf{J} + \dot{\mathbf{D}}$$

$$\text{div } \mathbf{B} = 0$$

$$\text{div } \mathbf{D} = \rho$$

Equation of continuity

$$\oint_S \mathbf{J} \, . \, d\mathbf{S} = - \int_V \dot{\rho} \, dV \qquad \text{div } \mathbf{J} = -\dot{\rho}$$

with $\mathbf{D} = \epsilon \mathbf{E}$ and $\mathbf{B} = \mu \mathbf{H}$

Constitutive equations

$$\mathbf{D} = \epsilon \mathbf{E}; \ \mathbf{B} = \mu \mathbf{H}; \ \mathbf{J} = \sigma \mathbf{E}$$

Potential functions

$$\mathbf{B} = \nabla \times \mathbf{A}$$

$$\mathbf{E} = -\nabla \phi - \dot{\mathbf{A}}$$

$$\nabla \, . \, \mathbf{A} = -\mu \epsilon \dot{\phi}$$

$$\nabla^2 \phi = \mu \epsilon \ddot{\phi} - \rho / \epsilon$$

$$\nabla^2 \mathbf{A} = \mu \epsilon \ddot{\mathbf{A}} - \mu \mathbf{J}$$

$$\phi = \frac{1}{4\pi\epsilon} \int_V \frac{[\rho]}{r} \, dV$$

$$\mathbf{A} = \frac{\mu}{4\pi} \int_V \frac{[\mathbf{J}]}{r} \, dV$$

$$\mathbf{A} = \frac{\mu}{4\pi} \int_C \frac{[i]}{r} \, dl$$

Boundary conditions

$$E_{t1} = E_{t2} \qquad\qquad H_{t1} = H_{t2}$$
$$D_{n1} - D_{n2} = \rho_s \qquad B_{n1} = B_{n2}$$

At the surface of a perfect conductor: $\mathbf{D} = \rho_s \mathbf{n}$ and $\mathbf{n} \times \mathbf{H} = \mathbf{J}_s$

Biot-Savart law: $d\mathbf{H} = \dfrac{i \, d\mathbf{l} \times \mathbf{a}_r}{4\pi r^2}$

Poisson's equation: $\nabla^2 \phi = -\rho / \epsilon$

Laplace's equation: $\nabla^2 \phi = 0$

Linear passive circuits

In the following Z is impedance and Y admittance.

Star-delta and delta-star transformation

Star to delta:

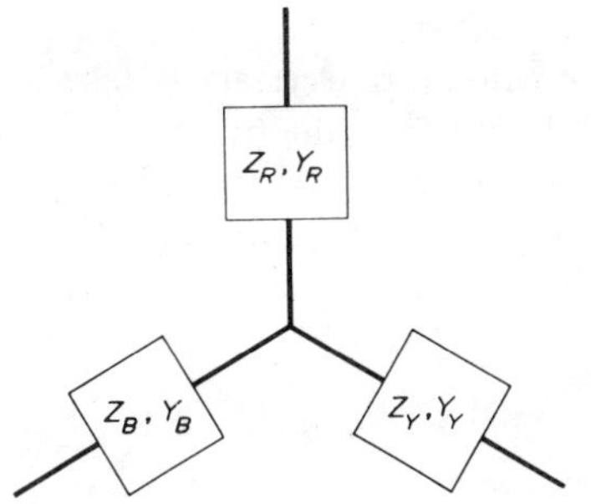

$$Y_R' = \frac{Y_Y Y_B}{Y_R + Y_Y + Y_B}.$$

Delta to star

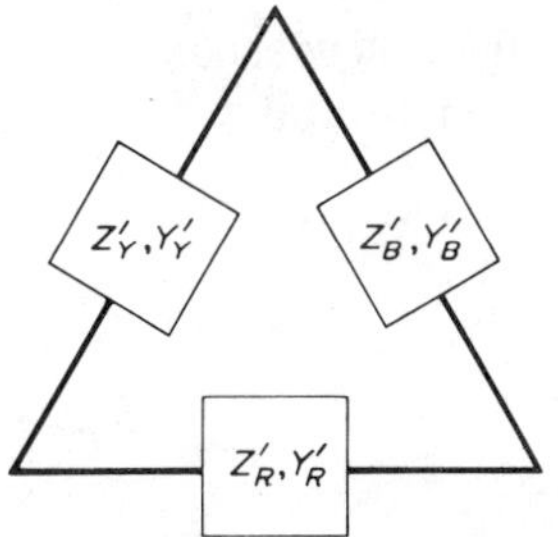

$$Z_R = \frac{Z_Y' Z_B'}{Z_R' + Z_Y' + Z_B'}.$$

Self-inductance of two coils

If the self-inductances of the individual coils are L_1 and L_2 and they are placed so that their mutual inductance is M, then the self-inductance of the combination, when connected in series, is $L_1 + L_2 \pm 2M$ dependent upon the relative directions of current flow. When connected in parallel the self-inductance of the combination is

$$(L_1 L_2 - M^2)/(L_1 + L_2 \mp M).$$

Reciprocity theorem

If a current is produced at point a in a network by a source acting at a point b, then the same current would be produced at point b by the source acting at point a.

Resonance, Q-factor and bandwidth

Series resonant circuit

The resonant frequency (minimum Z, or real Z) is

$$\omega_0 = \frac{1}{\sqrt{(LC)}}$$

and the damped natural frequency is

$$\omega_n = \sqrt{\left(\frac{1}{LC} - \frac{R^2}{4L^2} \right)}$$

$$Q = \frac{\omega_0 L}{R} \qquad \text{so} \qquad \omega_n = \omega_0 \sqrt{\left\{ 1 - \left(\frac{1}{2Q} \right)^2 \right\}}$$

Q is the magnification or

$$\frac{\text{voltage developed across } L \text{ or } C}{\text{voltage across whole circuit}}$$

or

$$\frac{2\pi \, (\text{energy stored in } L \text{ or } C)}{\text{energy dissipated per cycle}}$$

Impedance just off resonance ($\omega_0 \pm \Delta\omega$) is given by

$$Z = R(1 + \mathrm{j}2fQ)$$

where

$$f = \frac{\Delta\omega}{\omega_0} \ll 1$$

If $2fQ = \pm 1$, $|Z| = \sqrt{2}R$ and for constant voltage the current is then $1/\sqrt{2}$ of its maximum value and the power is halved. These points at which $2fQ = \pm 1$ are known as half-power points and the total bandwidth between them is $2\Delta\omega$. Hence

$$\text{half-power bandwidth} = \frac{2\Delta\omega}{\omega_0} = 2f = \frac{1}{Q}.$$

Parallel resonant circuit

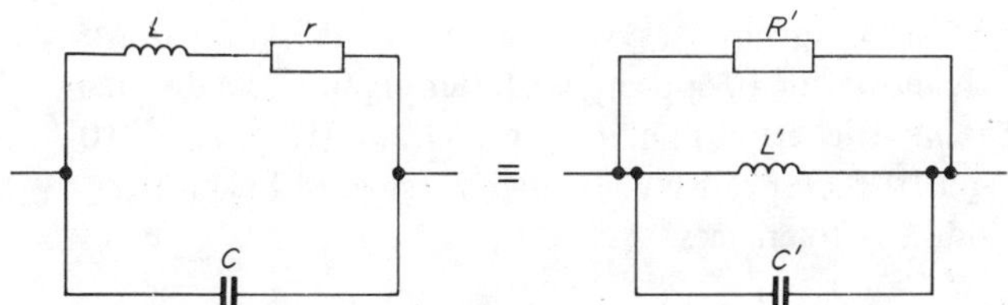

These circuits are equivalent with $L' = L$, $C' = C$ and $R' = Q^2 r$ if $Q \gg 1$. The right-hand version gives results similar to those for the series resonant circuit.

Q becomes $Q' = \dfrac{R'}{\omega_0 L'}$, and represents the current magnification, and the admittance is

$$Y' = \frac{1}{R'} \, (1 + \mathrm{j}2fQ')$$

Resistor and capacitor values

Resistors

The basic colour code is as follows.

0	Black	5	Green
1	Brown	6	Blue
2	Red	7	Violet
3	Orange	8	Grey
4	Yellow	9	White

Older types used these colours to mark the body, the tip
and either a spot or a band. This system was replaced by
one in which three bands of colour appear.

The interpretation is as follows:

Body or first band (closest to end): first digit
Tip or second band: second digit
Spot or third band: power of 10

A gold third band indicates 10^{-1}, silver 10^{-2}.

A fourth band indicates either (*a*) accuracy or tolerance:
10% silver, 5% gold, 2% red and 1% brown or (*b*) the last
band of a basic four-band system in which the first three
indicate digits and the fourth the power of 10.

A salmon pink fifth band indicates a high-stability resistor.

B.S. 1852 specifies an alternative system in which the
digits appear as such and the power of 10 appears as a letter
in place of a decimal point. The letters R, K and M indicate
10^0, 10^3 and 10^6, respectively. A final letter indicates
tolerance as follows: B = ±0·1%, C = ±0·25%, D = ±0·5%,
F = ±1%, G = ±2%, J = ±5%, K = ±10%, M = ±20%, N = ±30%
(e.g. 4R7J = 4·7Ω±5% and 68KK = 68kΩ± 10%).

Preferred values

The following series is that of 10% tolerance resistors:

10, 12, 15, 18, 22, 27, 33, 39, 47, 56, 68, 82

and are normally available from 2·2Ω to 10 MΩ.

Capacitors

B.S. 1852 specifies a system similar to the last for resistors.
The power of 10 appears as a letter in place of a decimal
point. The letters p, n, μ and m indicate 10^{-12}, 10^{-9}, 10^{-6}
and 10^{-3}, respectively. The same system of final letters
indicates tolerances.

Power in a.c. circuits

If voltage **V** and current **I** are expressed as complex numbers

Power = Re **VI*** = Re **V*I** = |**V**|| **I** | cos ϕ

(ϕ is argument of **I** relative to **V**).

The reactive power = Im **V*I**   if reactive power is
         (= |**V**||**I**| sin ϕ)   defined as +ve for
                         leading current.
             = Im **VI***   if reactive power is
         (= − |**V**||**I**| sin ϕ)   defined as +ve for
                         lagging current.

Power measurement in three-phase circuits

It is normal to quote line voltage V and line current I rather
than phase values but phase angles relate to the conditions in
a phase.

$$\text{Total power} = \sqrt{3}\, VI \cos \phi$$

for balanced star or delta connections.

In a system fed through n wires, it is necessary to take
$n - 1$ power readings for the sum to give the true total
power. Therefore, for a delta-connected load and a 3-wire
star load, two wattmeter readings are sufficient. If the loads
are balanced then

$$\tan \phi = \sqrt{3}\left(\frac{W_1 - W_2}{W_1 + W_2}\right)$$

Symmetrical components and similar transformations

The following matrix relation gives the various sequence
components in terms of the unbalanced quantities.

$$\begin{bmatrix} V_1 \\ V_2 \\ V_0 \end{bmatrix} = \tfrac{1}{3} \begin{bmatrix} 1 & a & a^2 \\ 1 & a^2 & a \\ 1 & 1 & 1 \end{bmatrix} \begin{bmatrix} V_R \\ V_Y \\ V_B \end{bmatrix}$$

where a is the operator $-\tfrac{1}{2} + j\sqrt{3}/2$.

The matrix can be inverted to give

$$\begin{bmatrix} V_R \\ V_Y \\ V_B \end{bmatrix} = \begin{bmatrix} 1 & 1 & 1 \\ a^2 & a & 1 \\ a & a^2 & 1 \end{bmatrix} \begin{bmatrix} V_1 \\ V_2 \\ V_0 \end{bmatrix}$$

Other forms give 012 sequence rather than 120, or $1/\sqrt{3}$
multipliers in both equations.

Two-port or four-terminal networks

Alternative conventions for the direction of I_2 give the
arrangements below.

$$\begin{bmatrix} V_1 \\ I_1 \end{bmatrix} = \begin{bmatrix} A & B \\ C & D \end{bmatrix} \begin{bmatrix} V_2 \\ I_2 \end{bmatrix} \qquad \begin{bmatrix} V_1 \\ I_1 \end{bmatrix} = \begin{bmatrix} A' & B' \\ C' & D' \end{bmatrix} \begin{bmatrix} V_2 \\ I_2 \end{bmatrix}$$

The convention in which positive I_2 flows out is that used
for transmission lines.

The *iterative impedance* Z_T is a function of the parameters of the network such that when Z_T is connected to
the output terminals the impedance at the input terminals
is also Z_T.

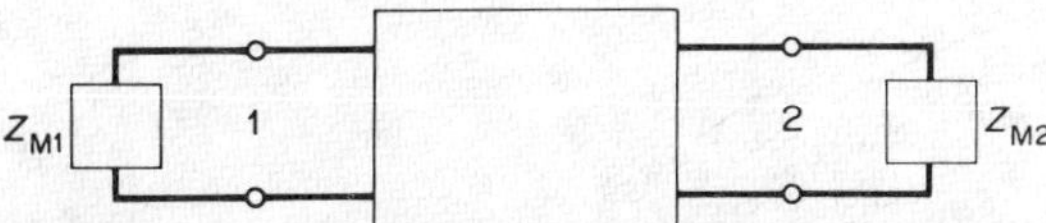

The *image impedances* Z_{M1} and Z_{M2} are functions of the parameters such that when connected as shown the impedance to the right at terminals 1 is Z_{M1} and the impedance to the left at terminals 2 is Z_{M2}.

$$Z_{M1} = \sqrt{\frac{AB}{CD}} \qquad Z_{M2} = \sqrt{\frac{BD}{AC}}$$

Transmission lines

In the following, R, G, L, C are the resistance, conductance, inductance and capacitance per unit length.

The propagation constant, γ, is

$$\sqrt{\{(R + j\omega L)(G + j\omega C)\}} = \alpha + j\beta$$

The characteristic impedance, Z_0, is

$$\sqrt{\left(\frac{R + j\omega L}{G + j\omega C}\right)}$$

The voltage and current at a point distant x from the sending end are given by

$$\begin{bmatrix} V_x \\ I_x \end{bmatrix} = \begin{bmatrix} \cosh \gamma x & -Z_0 \sinh \gamma x \\ -\frac{1}{Z_0} \sinh \gamma x & \cosh \gamma x \end{bmatrix} \begin{bmatrix} V_S \\ I_S \end{bmatrix}$$

The matrix can be inverted with $x = l$ to yield

$$\begin{bmatrix} V_S \\ I_S \end{bmatrix} = \begin{bmatrix} \cosh \gamma l & Z_0 \sinh \gamma l \\ \frac{1}{Z_0} \sinh \gamma l & \cosh \gamma l \end{bmatrix} \begin{bmatrix} V_R \\ I_R \end{bmatrix}$$

If the line is terminated with an impedance Z_L the input impedance is

$$Z_S = Z_0 \frac{Z_L + Z_0 \tanh \gamma l}{Z_0 + Z_L \tanh \gamma l}$$

The reflection coefficient for voltage is

$$\frac{Z_L - Z_0}{Z_L + Z_0} = \rho = \frac{\text{voltage of reflected wave}}{\text{voltage of incident wave}}$$

The voltage standing-wave ratio is

$$\text{V.S.W.R.} = \frac{1 + |\rho|}{1 - |\rho|}$$

Attenuation, wavelength, and phase velocity

The voltage at distance x can be written in the form

$$\begin{aligned} V &= A\,e^{+\gamma x} + B\,e^{-\gamma x} \\ &= A\,e^{\alpha x}\,e^{j\beta x} + B\,e^{-\alpha x}\,e^{-j\beta x} \end{aligned}$$

where A, B are determined by the end conditions, and the two terms represent travelling waves in the positive and negative x-directions. α represents an attenuation and β a change of phase with distance. $\beta\lambda = 2\pi$ where λ is wavelength.

The phase velocity is $v_p = \omega/\beta$ and the group velocity is $v_g = d\omega/d\beta$.

For a *lossless line*

$$R = G = 0; \qquad \alpha = 0;$$
$$\gamma = j\,\omega\sqrt{LC}; \qquad Z_0 = \sqrt{\frac{L}{C}}; \qquad v_p = \frac{1}{\sqrt{LC}};$$

If $\dfrac{L}{R} = \dfrac{C}{G}$ the line is *distortionless* since the attenuation and velocity of propagation are then independent of frequency, and

$$\gamma = \alpha + j\beta = \sqrt{RG} + j\omega\sqrt{LC}$$
$$Z_0 = \sqrt{\frac{R}{G}} = \sqrt{\frac{L}{C}}$$

The Smith Chart allows calculations for reflections at discontinuities

$$\rho = \frac{\mathbf{Z}/\mathbf{Z_0} - 1}{\mathbf{Z}/\mathbf{Z_0} + 1}$$

$\dfrac{R}{Z_0}$ and $\dfrac{X}{Z_0}$ are scaled as shown. $|\rho|$ is measured from centre (boundary radius = 1). $\angle\rho$ is measured anticlockwise.

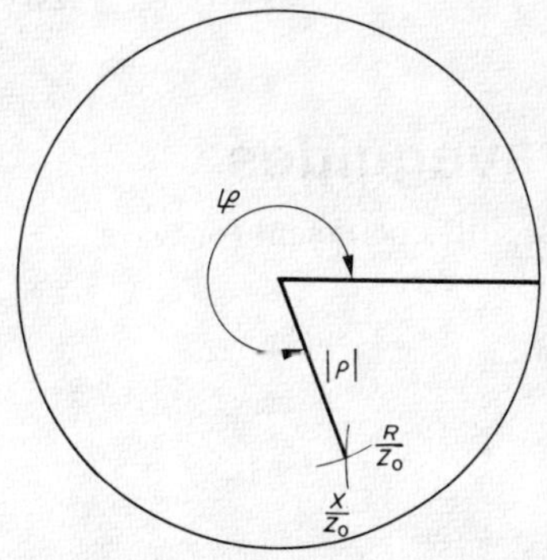

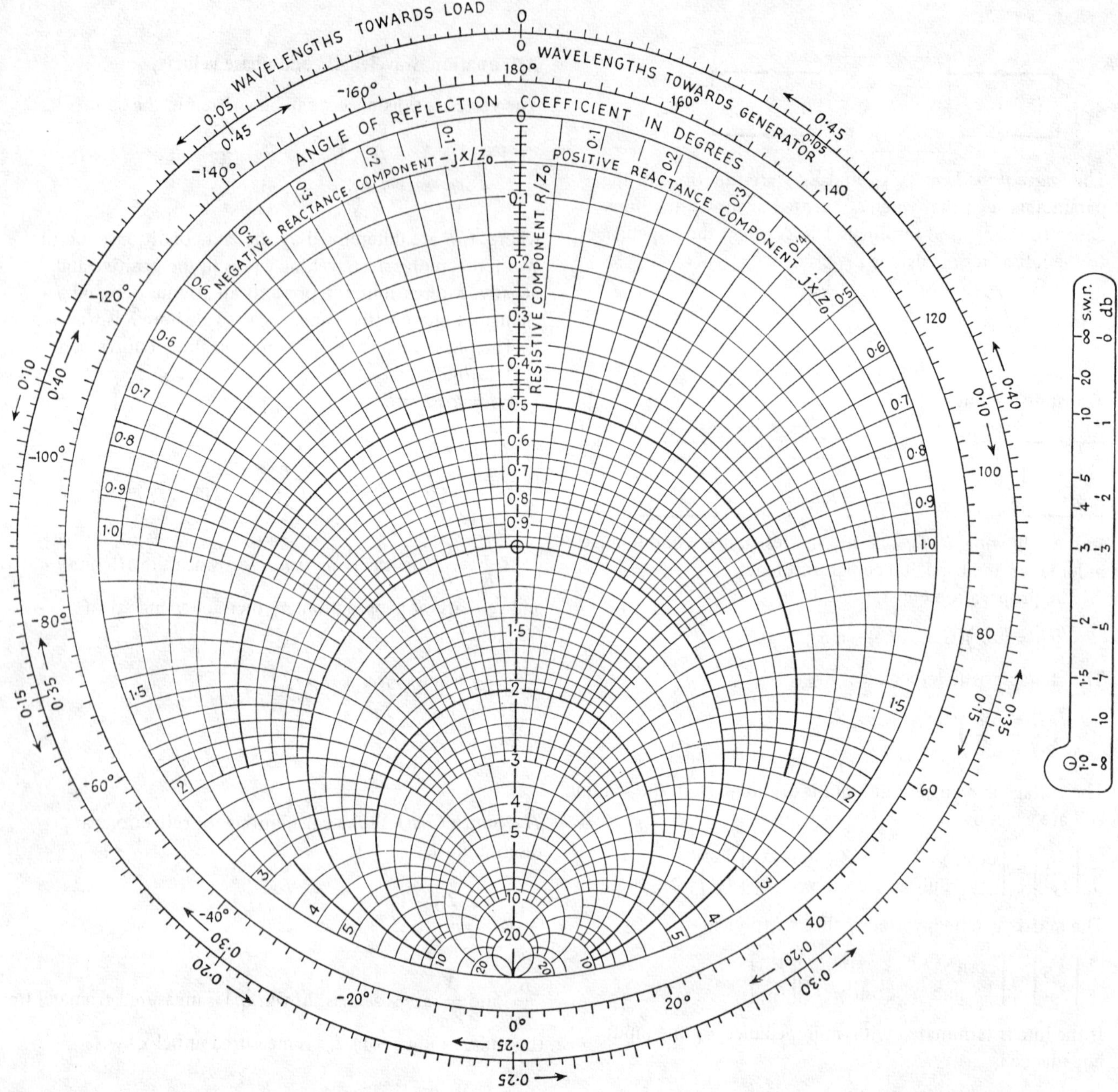

Rectangular waveguides

For any waveguide the phase-shift coefficient (or wave number) is

$$\beta = (\beta_0^2 - \beta_c^2)^{1/2}$$

where

$$\beta_0^2 = \omega^2 \epsilon \mu$$

and for a rectangular waveguide in a TE or TM mode

$$\beta_c^2 = \left(\frac{m\pi}{a}\right)^2 + \left(\frac{n\pi}{b}\right)^2$$

In this m and n denote the number of half-cycles along the x and y coordinates, for which the waveguide internal dimensions are a and b. At cut-off $\beta_0 = \beta_c$, and for evanescence $\beta_0 < \beta_c$. The waveguide wavelength λ_g is given by $2\pi/\beta$.

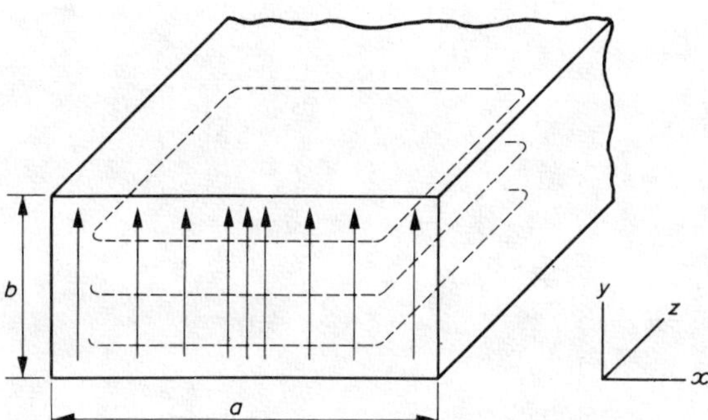

Electric (——) and magnetic (- - -) fields at a particular
instant for a rectangular waveguide in the TE_{10} mode
($m = 1, n = 0$).

Data for standard rectangular waveguides, TE_{10} mode

f_c, λ Frequency, free-space wavelength at cut-off (GHz, mm)

$\Delta f, W$ Recommended frequency range and power rating (GHz, MW)*

A Theoretical attenuation per 100 m for pure copper at $1 \cdot 5 f_c$ (db)†

The dimensions given are, respectively, x- and y-direction internal sizes, wall thickness and tolerance in *inches*.

f_c	λ	Δf	W	A	Dimensions	Designation British	American ‡	Frequency-band designation
0·908	330·2	1·12–1·70	13·47	0·497	6·500 x 3·250 x 0·080 ± 0·008	WG 6	RG-69/U	
1·157	259·1	1·45–2·20	8·29	0·728	5·100 x 2·550 x 0·080 ± 0·008	WG 7	–	
1·372	218·4	1·70–2·60	5·90	0·938	4·300 x 2·150 x 0·080 ± 0·006	WG 8	RG-104/U	L
1·686	177·8	2·20–3·30	3·91	1·275	3·500 x 1·750 x 0·080 ± 0·005	WG 9	–	
1·736	172·7	2·20–3·30	3·80	1·313	3·400 x 1·700 x 0·080 ± 0·005	WG 9A	RG-112/U	
2·078	144·3	2·60–3·95	2·43	1·822	2·840 x 1·340 x 0·080 ± 0·004	WG 10	RG-48/U	S
2·488	120·5	3·30–4·90	1·69	2·381	2·372 x 1·122 x 0·064 ± 0·003	WG 11	–	
2·577	116·3	3·30–4·90	1·60	2·460	2·29 x 1·145 x 0·064 ± 0·003	WG 11A	–	
3·152	95·10	3·95–5·85	1·04	3·430	1·872 x 0·872 x 0·064 ± 0·003	WG 12	RG-49/U	C
3·711	80·78	4·90–7·05	0·806	4·120	1·590 x 0·795 x 0·064 ± 0·002	WG 13	–	
4·310	69·70	5·85–8·20	0·544	5·570	1·372 x 0·622 x 0·064 ± 0·002	WG 14	RG-50/U	
5·260	57·00	7·05–10·0	0·355	7·671	1·122 x 0·497 x 0·064 ± 0·002	WG 15	RG-51/U	
6·557	45·72	8·20–12·4	0·229	10·63	0·900 x 0·400 x 0·050 ± 0·001	WG 16	RG-52/U	X
7·869	38·10	10·0–15·0	0·178	12·90	0·750 x 0·375 x 0·050 ± 0·001	WG 17	–	
9·488	31·60	12·4–18·0	0·123	17·09	0·622 x 0·311 x 0·040 ± 0·001	WG 18	RG-91/U	J
11·571	25·91	15·0–22·0	0·083	23·00	0·510 x 0·255 x 0·040 ± 0·001	WG 19	–	
14·051	21·34	18·0–26·5	0·048	35·76	0·420 x 0·170 x 0·040 ± 0·0008	WG 20	RG-53/U	
17·357	17·27	22·0–33·0	0·037	42·00	0·340 x 0·170 x 0·040 ± 0·0008	WG 21	–	
21·077	14·22	26·5–40·0	0·025	56·70	0·280 x 0·140 x 0·040 ± 0·0008	WG 22	RG-96/U	Q
26·344	11·38	33·0–50·0	0·016	78·80	0·224 x 0·112 x 0·040 ± 0·0008	WG 23	RG-97/U	
31·391	9·550	40·0–60·0	0·010	103·0	0·188 x 0·094 x 0·040 ± 0·0008	WG 24	–	
39·877	7·518	50·0–75·0	0·007	147·0	0·148 x 0·074 x 0·040 ± 0·0008	WG 25	RG-98/U	
48·369	6·198	60·0–90·0	0·005	197·0	0·122 x 0·061 x 0·040 ± 0·0008	WG 26	RG-99/U	O
59·014	5·080	75·0–110	0·003	266·0	0·100 x 0·050 x 0·040 ± 0·0008	WG 27	–	
73·767	4·064	90·0–140	0·002	370·0	0·080 x 0·040 x 0·040 ± 0·0008	WG 28	RG-138/U	
90·791	3·302	110–162	–	–	0·065 x 0·0325 ± 0·00025	WG 29	RG-136/U	
115·75	2·590	140–210	–	–	0·051 x 0·0255 ± 0·00025	WG 30	RG-135/U	
137·27	2·184	170–240	–	–	0·0430 x 0·0215 ± 0·0002	WG 31	RG-137/U	
173·49	1·728	220–310	–	–	0·0340 x 0·017 ± 0·0002	WG 32	RG-139/U	

* The rating W is based on a breakdown field of 3 kV/mm for air, a V.S.W.R. of 2 and a power safety factor of 2.

† For the attenuation at a frequency f other than $1 \cdot 5 f_c$, these figures should be multiplied by $0 \cdot 421 \{(f/f_c)^2 + 1\}/\sqrt{[(f/f_c)\{(f/f_c)^2 - 1\}]}$.

‡ These are American Services Designations; the American Radio and TV Manufacturers Association uses the designation WRX where X is the x-dimension in hundredths of an inch, or the nearest integer.

Resonant cavities

In the table, λ is the resonant wavelength and δ the skin depth, given by $\sqrt{(2/\omega\mu\sigma)}$ for material of conductivity σ and permeability μ at angular frequency ω.

Resonator type	λ	Q
TE_{101}	$2\sqrt{2}a$	$\dfrac{0{\cdot}353\lambda}{\delta} \cdot \dfrac{1}{1 + 0{\cdot}177\lambda/h}$
Circular cylinder TM_{010}	$2{\cdot}61a$	$\dfrac{0{\cdot}383\lambda}{\delta} \cdot \dfrac{1}{1 + 0{\cdot}192\lambda/h}$
Sphere	$2{\cdot}28a$	$0{\cdot}318\lambda/\delta$
Co-axial TEM	$4h$	For optimum Q $b/a = 3{\cdot}6$ and $Z_0 = 77\ \Omega$ $\dfrac{\lambda}{4\delta + 7{\cdot}2\,h\delta/b}$

Radiation and aerials

The Poynting vector is

$$S = \mathbf{E} \times \mathbf{H}$$

and for orthogonal fields in an isotropic non-conducting medium of permeability μ and permittivity ϵ has the value

$$S = EH = E^2\sqrt{(\epsilon/\mu)}$$
$$= H^2\sqrt{(\mu/\epsilon)}$$

An *isotropic radiator* emitting a mean power P produces a mean S of $P/4\pi r^2$ at distance r, and the r.m.s. electric field in free space is then

$$E = \sqrt{(30P)}/r$$

The *gain* of an aerial is the ratio of the power it emits per steradian in a given direction to the power per steradian emitted by a reference aerial of the same total power. Usually, the direction chosen is that of maximum power density and the reference aerial is an isotropic radiator.

The directivity may be measured either by the maximum gain or by the *beam width*, the angle contained between points at which the power density is half of the maximum.

The *radiation resistance* R_r of an aerial is such that the aerial radiates power $I^2 R_r$ when fed with r.m.s. current I. The *aperture* of a receiving aerial is the ratio of the power received to the Poynting vector of the incident field. The effective aperture of an aerial is greatest when it is matched; for a lossless aerial of gain G it is then given by $\lambda^2 G/4\pi$,

where λ is the wavelength. The power received by a matched aerial is

$$P = V^2/4 R_r$$

where V is the integral of the induced electric field along its length.

Non-isotropic radiators

	Current distri-bution	Radiation resistance	Maxi-mum gain	Beam width	Aperture
Hertzian dipole	constant	$80\pi^2 (l/\lambda)^2$	1·5	90°	$3\lambda^2/8\pi$
Half-wave dipole	half-cosine	73·1 Ω	1·64	78°	$30\lambda^2/73\pi$

(Here l is the total length of the aerial and λ the wavelength.)

Poles and zeros

A transfer function of a linear finite lumped-parameter system can be expressed as the ratio of two polynomials in the complex variable s arising from generalized impedances.

The transfer function may be written

$$\frac{K(s - z_1)(s - z_2)\ldots}{(s - p_1)(s - p_2)\ldots}$$

where $z_1 z_2 \ldots p_1 p_2 \ldots$, etc., are the zeros and poles of the function and can be plotted as points on an Argand diagram.

Examples

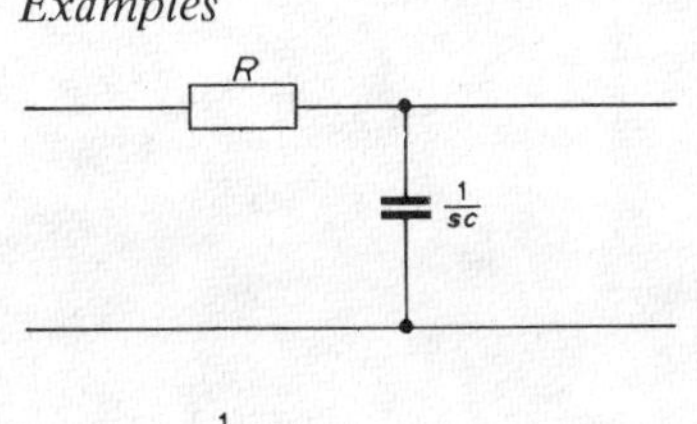

$$\frac{V_{out}}{V_{in}} = \frac{1}{RC\left(s + \dfrac{1}{RC}\right)}$$

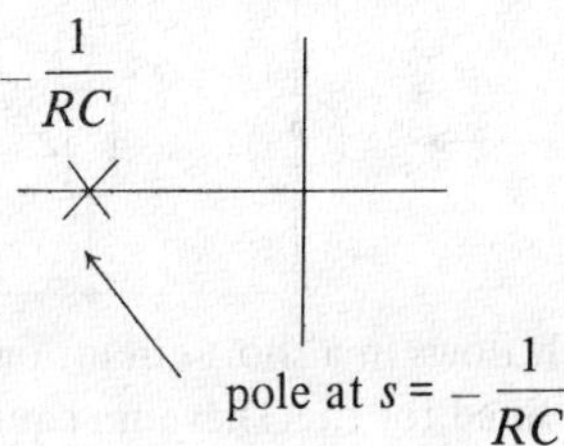

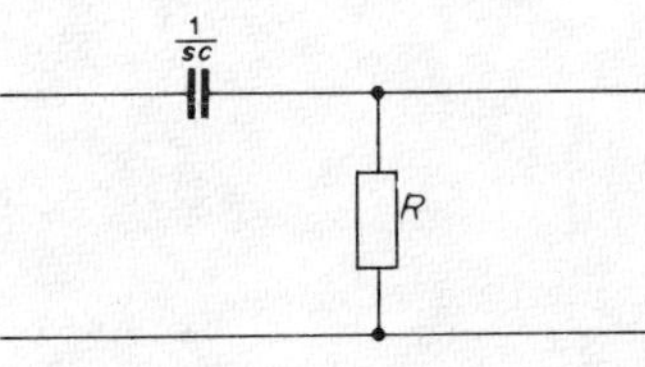

$$\frac{V_{out}}{V_{in}} = \frac{s}{s + \dfrac{1}{RC}}$$

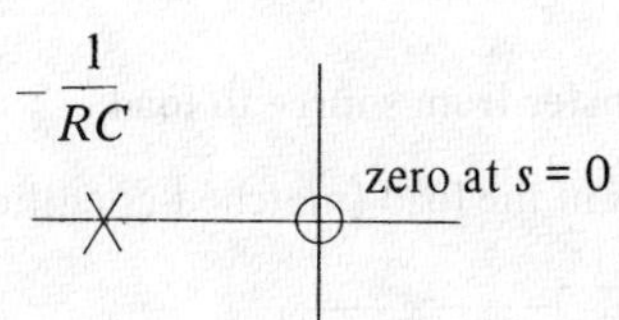

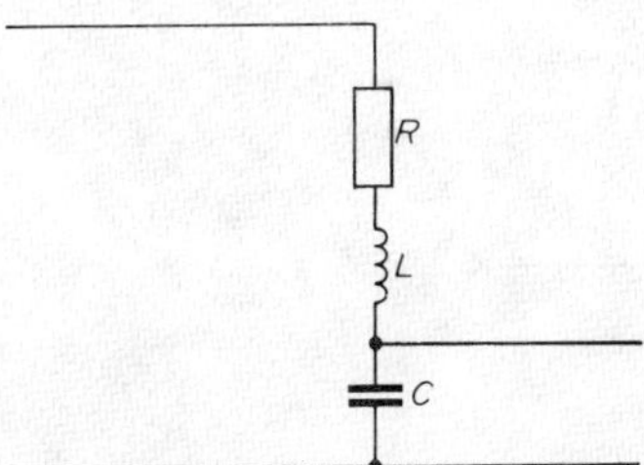

$$\frac{V_{out}}{V_{in}} = \frac{1}{LC\left(s^2 + \dfrac{R}{L}s + \dfrac{1}{LC}\right)}$$

The response to a sinusoidal input and the transient response can be deduced from the positions of the poles and zeros.

Linear active circuits

Superposition principle

The response of a linear system to a number of simultaneously
applied excitations is equal to the sum of the responses
taken one at a time. When any one source is being considered
all the others are de-activated; de-activation means that
independent voltage sources are replaced by short circuits
and independent current sources by open circuits.

Thévenin's theorem and equivalent circuit

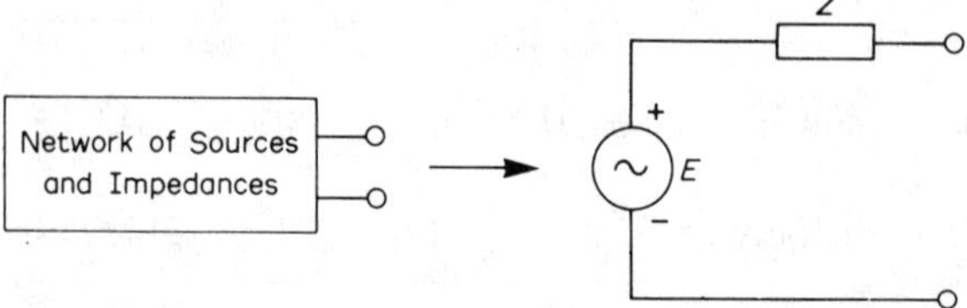

A two-terminal network containing sources and impedances
can always be replaced, as far as any load is concerned, by a
voltage source and impedance as shown. The value of E is the
voltage which is measured at the terminals when open-
circuited. The value of Z is the impedance presented at the
open-circuited terminals when all the sources are de-activated.

Norton's theorem and equivalent circuit

This is the equivalent to Thévenin's theorem in terms of a
current source.

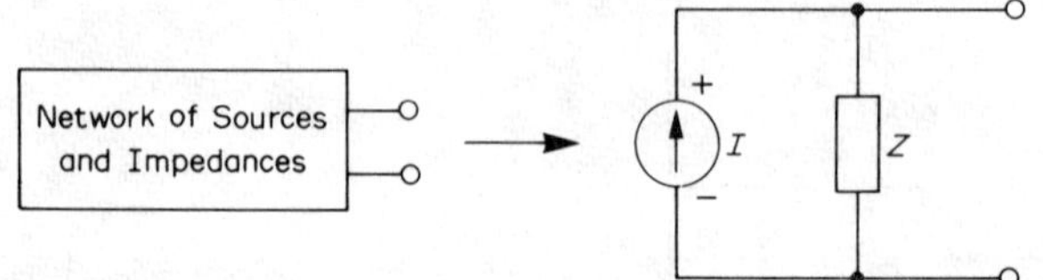

I is the current which flows in a short-circuit on the
terminals and Z is as defined for the Thévenin equivalent
circuit.

Maximum power transfer from source to load

For maximum power in the load (matched condition)

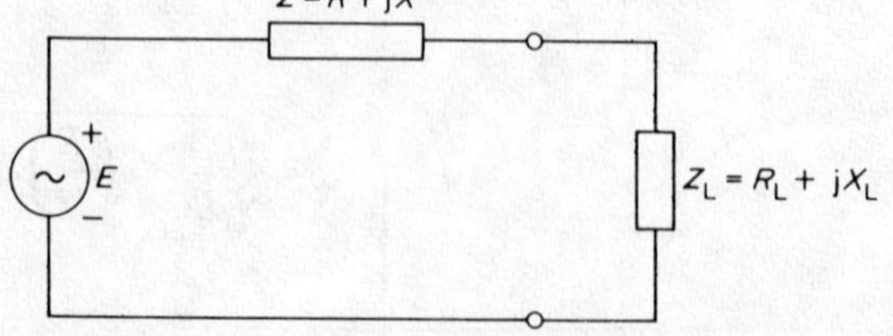

$$Z_L = Z^*$$

or

$$R_L = R, \qquad X_L = -X$$

(The overall efficiency is not necessarily then 50% if Z is a
Thévenin impedance.)

Small-signal equivalent circuits for valves

Valve, triode or pentode

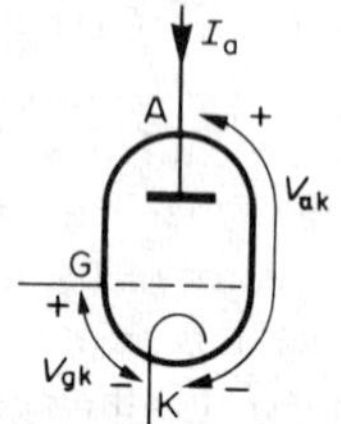

$$g_m = \left(\frac{\partial I_a}{\partial V_{gk}}\right)_{V_{ak}}$$

$$r_a = \left(\frac{\partial V_{ak}}{\partial I_a}\right)_{V_{gk}}$$

$$\mu = g_m r_a = -\left(\frac{\partial V_{ak}}{\partial V_{gk}}\right)_{I_a}$$

Voltage generator

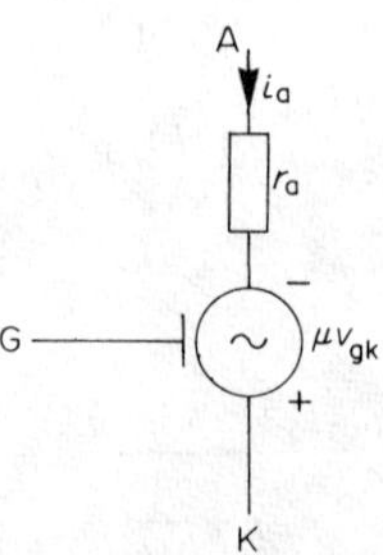

Current generator

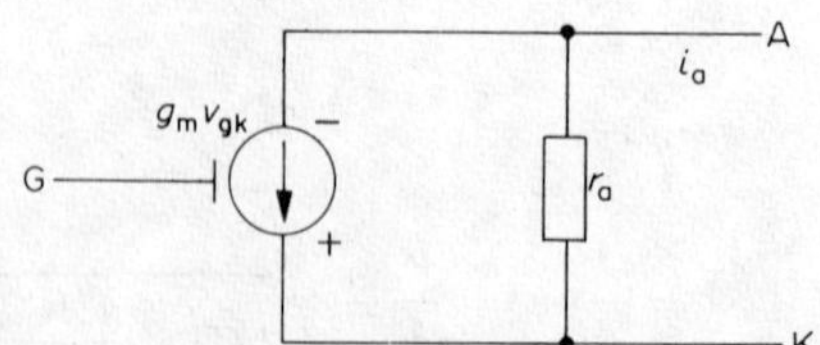

Transistor equivalent circuits

Hybrid parameters

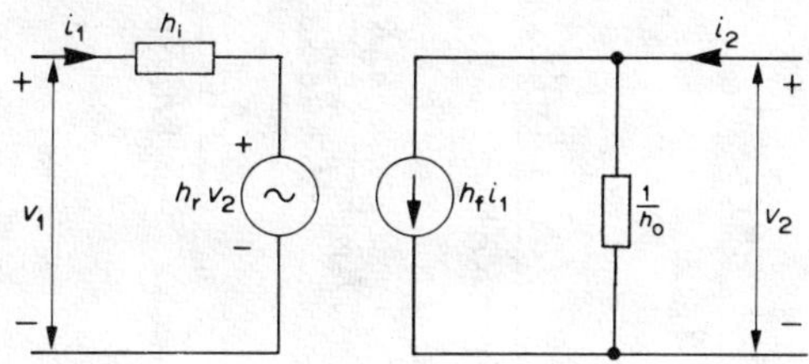

$$\begin{bmatrix} v_1 \\ i_2 \end{bmatrix} = \begin{bmatrix} h_{11} & h_{12} \\ h_i & h_r \\ \hline h_f & h_o \\ h_{21} & h_{22} \end{bmatrix} \begin{bmatrix} i_1 \\ v_2 \end{bmatrix}$$

$h_{11} = h_i$ = input impedance with output short circuited to a.c.

$h_{12} = h_r$ = reverse voltage transfer with input open circuited to a.c.

$h_{21} = h_f$ = forward current ratio with output short circuited to a.c.

$h_{22} = h_0$ = output admittance with input open circuited to a.c.

Conventions

Lower-case subscripts refer to small signal values e.g.

$$h_f = \frac{i_2}{i_1} = \frac{\partial I_2}{\partial I_1}$$

Capital subscripts refer to large signal (or d.c.) values e.g.

$$h_F = \frac{I_2}{I_1}$$

A second subscript letter b, e or c can be added to the first to indicate which terminal of the transistor is common to input and output.

Relationships between *h*-parameters for different connections

Common base	Common emitter	Common collector
h_{ib}	$h_{ie} = \dfrac{h_{ib}}{1 - \alpha}$	$h_{ic} = \dfrac{h_{ib}}{1 - \alpha}$
h_{rb}	$h_{re} = \dfrac{h_{ib}h_{ob}}{1 - \alpha} - h_{rb}$	$h_{rc} = 1$

Common base	Common emitter	Common collector
$h_{fb} = -\alpha$	$h_{fe} = \dfrac{\alpha}{1 - \alpha} = \beta$	$h_{fc} = \dfrac{-1}{1 - \alpha}$
h_{ob}	$h_{oe} = \dfrac{h_{ob}}{1 - \alpha}$	$h_{oc} = \dfrac{h_{ob}}{1 - \alpha}$

r-parameters and T equivalent circuit

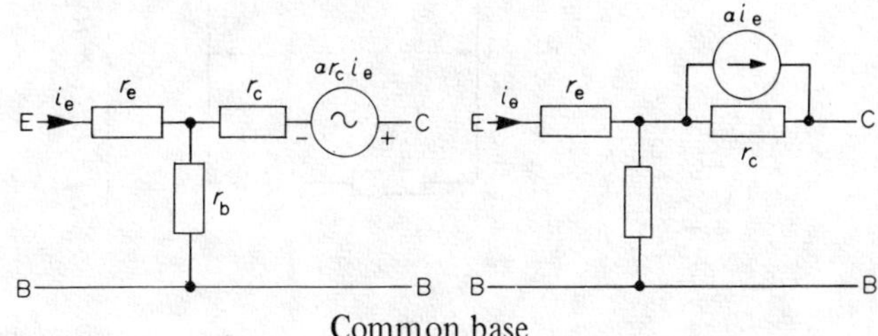

Common base

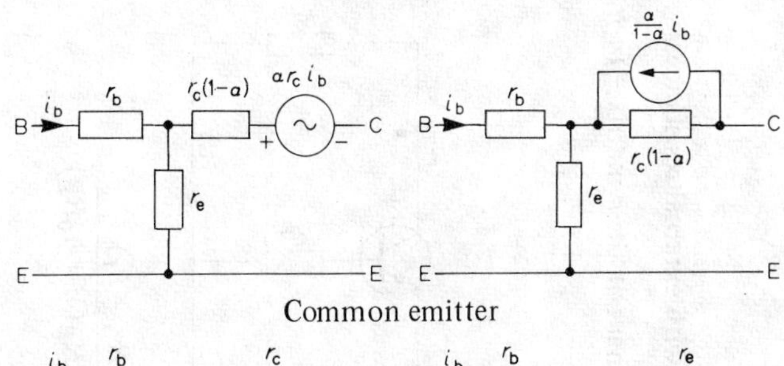

Common emitter

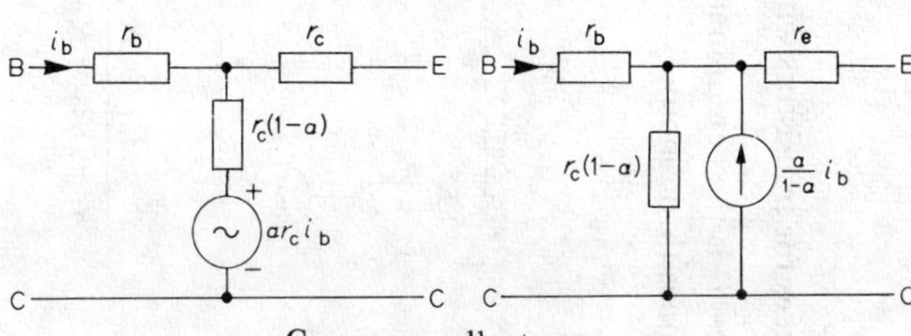

Common collector

Values of *r*-parameters in terms of *h*-parameters

$$\alpha = \frac{h_{fe}}{1 + h_{fe}} = -h_{fb} \qquad \beta = h_{fe} = \frac{-h_{fb}}{1 + h_{fb}}$$

$$r_e = \frac{h_{re}}{h_{oe}} = h_{ib} - \frac{h_{rb}}{h_{ob}}(1 + h_{fb})$$

$$r_b = h_{ie} - \frac{h_{re}}{h_{oe}}(1 + h_{fe}) = \frac{h_{rb}}{h_{ob}}$$

$$r_c = \frac{1 + h_{fe}}{h_{oe}} = \frac{1 - h_{rb}}{h_{ob}}$$

Voltage and current gains, input and output resistances for transistors with external resistances.

These circuits are intended for use at low frequencies. D is the determinant of the h matrix $= (h_i h_0 - h_f h_r)$, $M = D + h_f - h_r + 1$ and h_i etc. are *common-base* values throughout.

Common-base arrangements

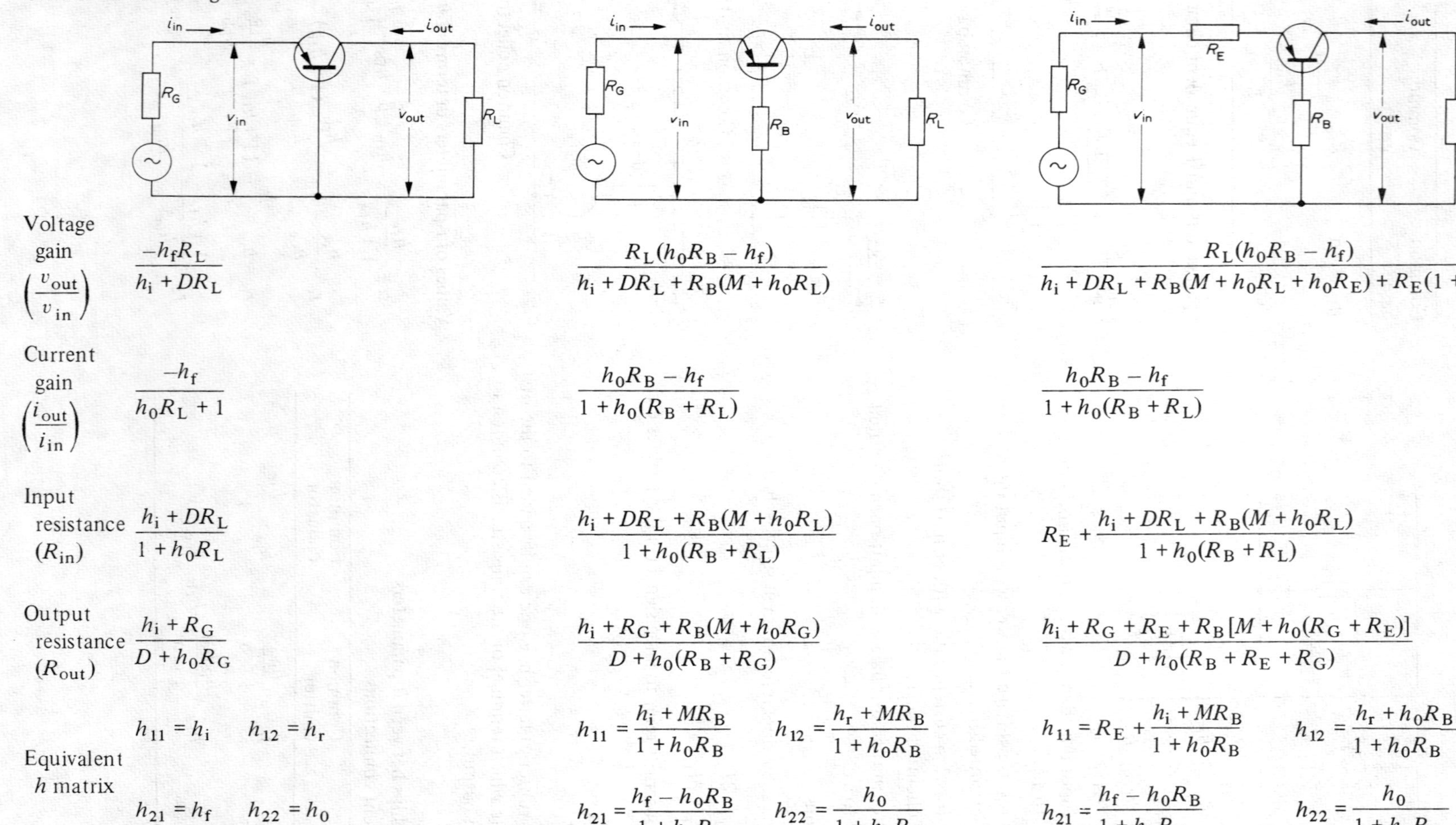

Voltage gain $\left(\dfrac{v_{\text{out}}}{v_{\text{in}}}\right)$	$\dfrac{-h_f R_L}{h_i + DR_L}$	$\dfrac{R_L(h_0 R_B - h_f)}{h_i + DR_L + R_B(M + h_0 R_L)}$	$\dfrac{R_L(h_0 R_B - h_f)}{h_i + DR_L + R_B(M + h_0 R_L + h_0 R_E) + R_E(1 + h_0 R_L)}$
Current gain $\left(\dfrac{i_{\text{out}}}{i_{\text{in}}}\right)$	$\dfrac{-h_f}{h_0 R_L + 1}$	$\dfrac{h_0 R_B - h_f}{1 + h_0(R_B + R_L)}$	$\dfrac{h_0 R_B - h_f}{1 + h_0(R_B + R_L)}$
Input resistance (R_{in})	$\dfrac{h_i + DR_L}{1 + h_0 R_L}$	$\dfrac{h_i + DR_L + R_B(M + h_0 R_L)}{1 + h_0(R_B + R_L)}$	$R_E + \dfrac{h_i + DR_L + R_B(M + h_0 R_L)}{1 + h_0(R_B + R_L)}$
Output resistance (R_{out})	$\dfrac{h_i + R_G}{D + h_0 R_G}$	$\dfrac{h_i + R_G + R_B(M + h_0 R_G)}{D + h_0(R_B + R_G)}$	$\dfrac{h_i + R_G + R_E + R_B[M + h_0(R_G + R_E)]}{D + h_0(R_B + R_E + R_G)}$

Equivalent h matrix:

$$h_{11} = h_i \qquad h_{12} = h_r$$
$$h_{21} = h_f \qquad h_{22} = h_0$$

$$h_{11} = \frac{h_i + MR_B}{1 + h_0 R_B} \qquad h_{12} = \frac{h_r + MR_B}{1 + h_0 R_B}$$
$$h_{21} = \frac{h_f - h_0 R_B}{1 + h_0 R_B} \qquad h_{22} = \frac{h_0}{1 + h_0 R_B}$$

$$h_{11} = R_E + \frac{h_i + MR_B}{1 + h_0 R_B} \qquad h_{12} = \frac{h_r + h_0 R_B}{1 + h_0 R_B}$$
$$h_{21} = \frac{h_f - h_0 R_B}{1 + h_0 R_B} \qquad h_{22} = \frac{h_0}{1 + h_0 R_B}$$

Common-emitter arrangements

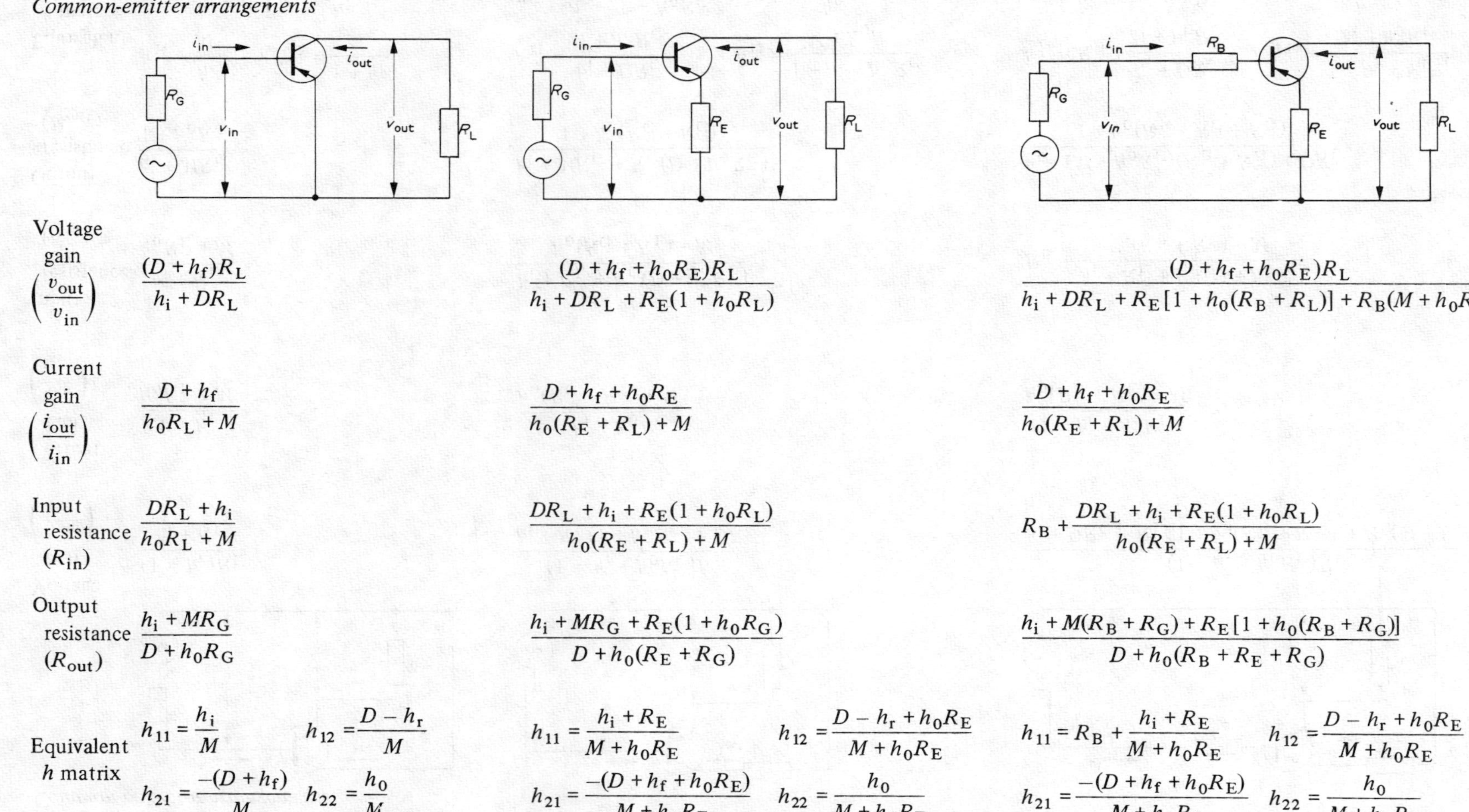

Voltage gain $\left(\dfrac{v_{out}}{v_{in}}\right)$	$\dfrac{(D+h_f)R_L}{h_i+DR_L}$	$\dfrac{(D+h_f+h_0R_E)R_L}{h_i+DR_L+R_E(1+h_0R_L)}$	$\dfrac{(D+h_f+h_0R_E)R_L}{h_i+DR_L+R_E[1+h_0(R_B+R_L)]+R_B(M+h_0R_L)}$
Current gain $\left(\dfrac{i_{out}}{i_{in}}\right)$	$\dfrac{D+h_f}{h_0R_L+M}$	$\dfrac{D+h_f+h_0R_E}{h_0(R_E+R_L)+M}$	$\dfrac{D+h_f+h_0R_E}{h_0(R_E+R_L)+M}$
Input resistance (R_{in})	$\dfrac{DR_L+h_i}{h_0R_L+M}$	$\dfrac{DR_L+h_i+R_E(1+h_0R_L)}{h_0(R_E+R_L)+M}$	$R_B+\dfrac{DR_L+h_i+R_E(1+h_0R_L)}{h_0(R_E+R_L)+M}$
Output resistance (R_{out})	$\dfrac{h_i+MR_G}{D+h_0R_G}$	$\dfrac{h_i+MR_G+R_E(1+h_0R_G)}{D+h_0(R_E+R_G)}$	$\dfrac{h_i+M(R_B+R_G)+R_E[1+h_0(R_B+R_G)]}{D+h_0(R_B+R_E+R_G)}$
Equivalent h matrix	$h_{11}=\dfrac{h_i}{M}\quad h_{12}=\dfrac{D-h_r}{M}$ $h_{21}=\dfrac{-(D+h_f)}{M}\quad h_{22}=\dfrac{h_0}{M}$	$h_{11}=\dfrac{h_i+R_E}{M+h_0R_E}\quad h_{12}=\dfrac{D-h_r+h_0R_E}{M+h_0R_E}$ $h_{21}=\dfrac{-(D+h_f+h_0R_E)}{M+h_0R_E}\quad h_{22}=\dfrac{h_0}{M+h_0R_E}$	$h_{11}=R_B+\dfrac{h_i+R_E}{M+h_0R_E}\quad h_{12}=\dfrac{D-h_r+h_0R_E}{M+h_0R_E}$ $h_{21}=\dfrac{-(D+h_f+h_0R_E)}{M+h_0R_E}\quad h_{22}=\dfrac{h_0}{M+h_0R_E}$

Common-collector arrangements

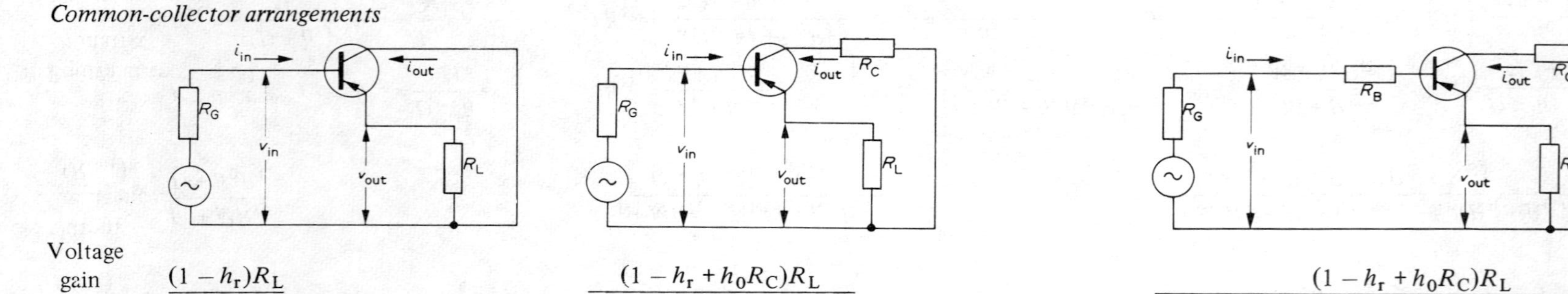

Voltage gain $\left(\dfrac{v_{out}}{v_{in}}\right)$	$\dfrac{(1-h_r)R_L}{h_i+R_L}$	$\dfrac{(1-h_r+h_0R_C)R_L}{h_i+R_L(1+h_0R_C)+DR_C}$	$\dfrac{(1-h_r+h_0R_C)R_L}{h_i+DR_C+R_L[1+h_0(R_B+R_C)]+R_B(M+h_0R_C)}$
Current gain $\left(\dfrac{i_{out}}{i_{in}}\right)$	$\dfrac{1-h_r}{h_0R_L+M}$	$\dfrac{1-h_r+h_0R_C}{h_0(R_C+R_L)+M}$	$\dfrac{1-h_r+h_0R_C}{h_0(R_C+R_L)+M}$
Input resistance (R_{in})	$\dfrac{h_i+R_L}{h_0R_L+M}$	$\dfrac{h_i+R_L+R_C(D+h_0R_L)}{h_0(R_C+R_L)+M}$	$R_B+\dfrac{h_i+R_L+R_C(D+h_0R_L)}{h_0(R_C+R_L)+M}$
Output resistance (R_{out})	$\dfrac{h_i+MR_G}{1+h_0R_G}$	$\dfrac{h_i+MR_G+R_C(D+h_0R_G)}{1+h_0(R_C+R_G)}$	$\dfrac{h_i+(M+h_0R_C)(R_B+R_G)+DR_C}{1+h_0(R_B+R_C+R_G)}$

Equivalent h matrix:

$$h_{11}=\frac{h_i}{M}\qquad h_{12}=\frac{1+h_f}{M}$$
$$h_{21}=\frac{h_r-1}{M}\qquad h_{22}=\frac{h_0}{M}$$

$$h_{11}=\frac{h_i+DR_C}{M+h_0R_C}\qquad h_{12}=\frac{1+h_f+h_0R_C}{M+h_0R_C}$$
$$h_{21}=\frac{h_r-1-h_0R_C}{M+h_0R_C}\qquad h_{22}=\frac{h_0}{M+h_0R_C}$$

$$h_{11}=R_B+\frac{h_i+DR_C}{M+h_0R_C}\qquad h_{12}=\frac{1+h_f+h_0R_C}{M+h_0R_C}$$
$$h_{21}=\frac{h_r-1-h_0R_C}{M+h_0R_C}\qquad h_{22}=\frac{h_0}{M+h_0R_C}$$

Field-effect transistors

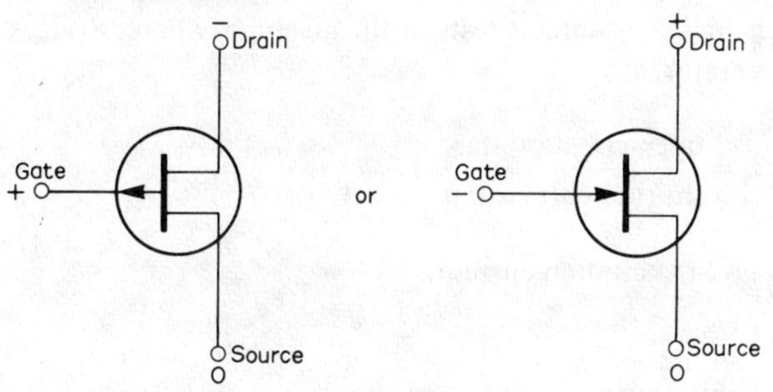

or

Insulated-gate field-effect transistors; metal-oxide semiconductor transistors (I.G.F.E.T.; M.O.S.T. or M.O.S.F.E.T.)

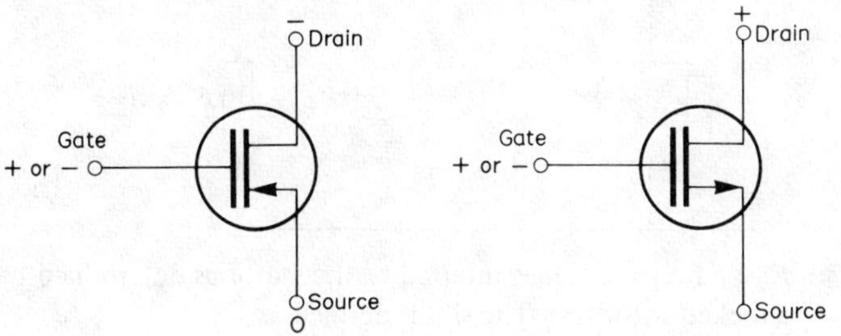

Low-frequency equivalent circuit

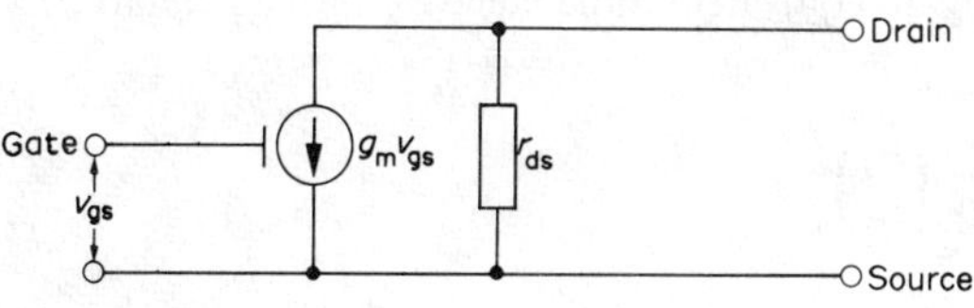

Higher frequencies

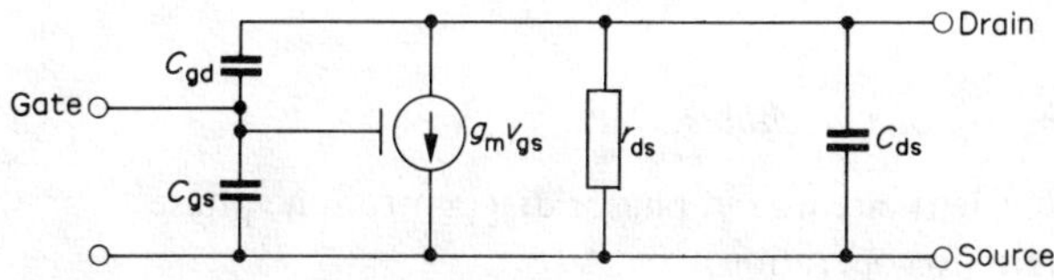

Electrical machines

In the following Φ is flux per pole, P number of poles, n number of turns in series per phase, N speed (rev/s) and f frequency (Hz).

Regulation =

$$\frac{\text{change in voltage (or speed) between no load and full load}}{\text{rated voltage (or speed)}}$$

D.C. machines

$$\text{E.M.F.} = \frac{\Phi ZNP}{A}$$

where Z is the total number of conductors and A the number of parallel paths in which they are arranged.

$$\text{Torque} = \frac{\Phi ZPI}{2\pi A}$$

where I is the total armature current.

For ordinary windings $A = P$ if lap-wound, and $A = 2$ if wave-wound (whatever the value of P).

A.C. machines

$$\text{E.M.F.} = 4\cdot44\,\Phi nf \text{ (r.m.s.)}$$

$$\text{Synchronous speed} = 120\,f/P \text{ rev/min}$$
$$= 60\,f/p \text{ rev/min}$$
$$= 2\pi f/p \text{ rad/s}$$

where p is the number of pole-pairs.

Transformers

Equivalent circuits:

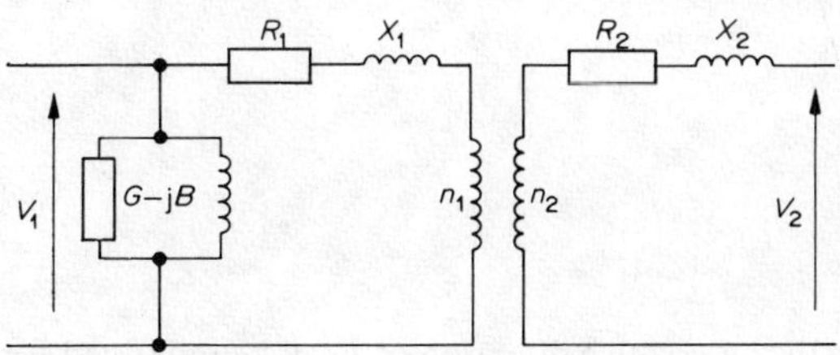

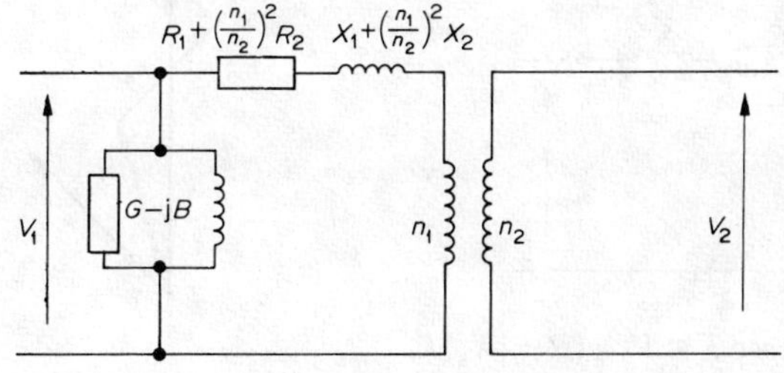

An impedance Z_2 in the secondary may be considered to be replaced by an equivalent impedance Z_2' in the primary, where

$$Z_2' = Z_2 \left(\frac{n_1}{n_2}\right)^2$$

For an ideal transformer ($G, B, R_1, X_1, R_2, X_2 = 0$)

$$\frac{V_1}{V_2} = \frac{n_1}{n_2}; \qquad I_2 n_2 = -I_1 n_1$$

Synchronous machines

Equivalent circuit and phasor diagram for one phase (alternator operation):

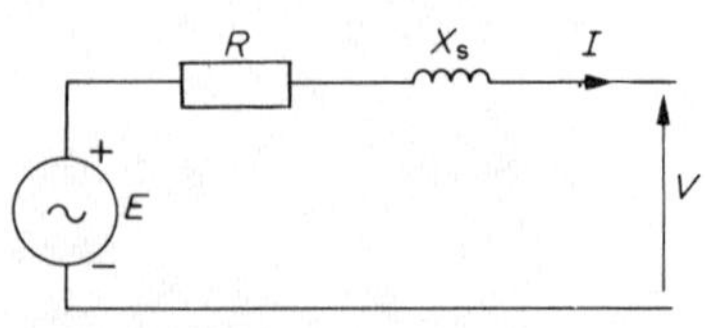
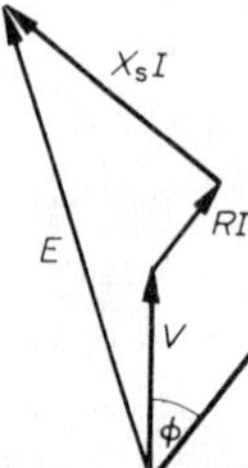

$$V = E - (R + jX_s)I$$

where R is armature resistance and X_s is synchronous reactance (including leakage and armature reaction). Usually $R \ll X_s$.

For operation as a motor the same equivalent circuit and sign convention for I can be used; the phasor diagram is then:

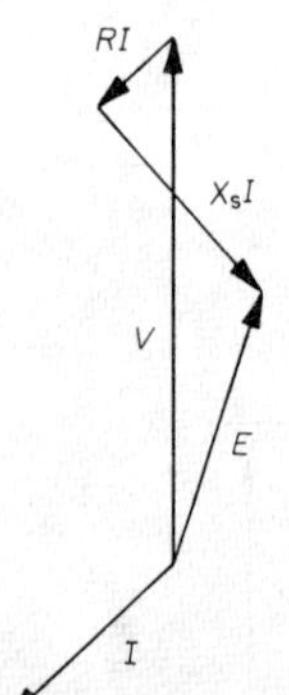

Alternatively the positive direction of I can be redefined for I', to give:

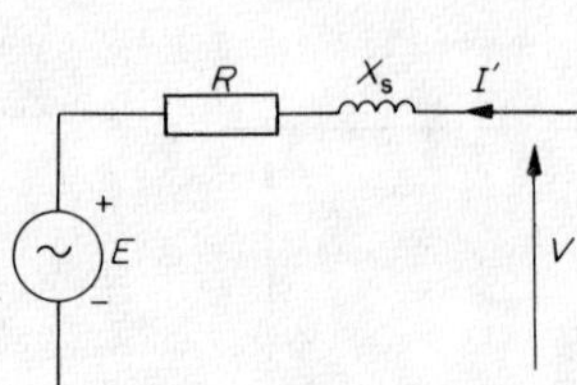
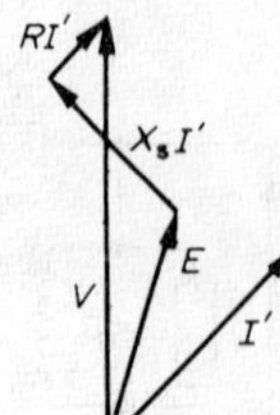

in which case $E = V - (R + jX_s)I'$.

The synchronous reactance X_s can be determined from open- and short-circuit tests of the machine, when driven as a generator, as

$$X_s = \frac{\text{open-circuit voltage}}{\text{short-circuit current}}$$

at a given excitation current.

Asynchronous or induction motor

Equivalent circuit:

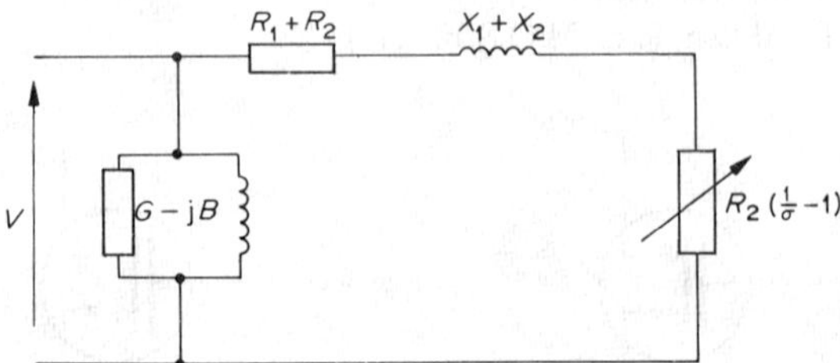

R_2 and X_2 are values referred to the stator as determined by a locked-rotor test. The slip is defined as

$$\sigma = \frac{\omega_s - \omega}{\omega_s}$$

where

ω_s = synchronous speed

ω = actual speed

The slip is also given by

$$\sigma = \frac{\text{rotor copper loss}}{\text{gross power input to rotor}}$$

The torque per phase is

$$T = \frac{V^2 R_2 \sigma/\omega_s}{(R_1\sigma + R_2)^2 + \sigma^2 X^2}$$

where $X = X_1 + X_2$. The maximum torque is

$$T_{max} = \frac{V^2}{2X\omega_s}$$

at the slip

$$\sigma = R_2/X$$

provided that $R_1 \ll X$.

Solid-state electronic properties

In the following, m, e and N_e are the free mass, charge and number density of electrons; h is Planck's constant and k Boltzmann's constant.

Free electrons

In a medium containing free electrons with a collision time τ, an electromagnetic wave of angular frequency ω propagates with a wave-number

$$\beta = \omega \sqrt{\left\{ \mu\epsilon \left(1 + \frac{j\sigma}{\omega\epsilon(1 - j\omega\tau)} \right) \right\}}$$

where μ, ϵ, σ are the permeability, permittivity and conductivity of the medium. For $\omega\tau \gg 1$,

$$\beta = \omega\sqrt{\{\mu\epsilon (1 - \omega_p^2/\omega^2)\}}$$

where ω_p is the plasma frequency, given by $\sqrt{(N_e e^2/m\epsilon)}$ or $\sqrt{(\sigma/\epsilon\tau)}$.

For free electrons confined by an infinite potential barrier, the density of energy states per unit volume in the energy interval E to $E + dE$ is

$$Z(E)\, dE = 4\pi \left(\frac{2m}{h^2}\right)^{3/2} E^{1/2}\, dE$$

The Fermi-Dirac function for the probability of occupation of a state of energy E, at absolute temperature T, is

$$F(E) = \{e^{(E - E_F)/kT} + 1\}^{-1}$$

where E_F is the Fermi energy. The Fermi energy for free electrons is

$$E_F = \frac{h^2}{2m} \left(\frac{3N_e}{8\pi}\right)^{2/3}$$

Semiconductors

The number of electrons in the conduction band of a semiconductor with energy gap E_g is

$$N_e = 2 \left(\frac{2\pi m_e^* kT}{h^2}\right)^{3/2} e^{-(E_g - E_F)/kT}$$

where m_e^* is the effective mass of an electron at the bottom of the band; this assumes that for the electrons $E - E_F \gg kT$.

The number of holes (effective mass m_h^*) in the valence band is

$$N_h = 2 \left(\frac{2\pi m_h^* kT}{h^2}\right)^{3/2} e^{-E_F/kT}$$

The Fermi energy for an intrinsic semiconductor is

$$E_F = \tfrac{1}{2}E_g + \tfrac{3}{4}kT \ln (m_h^*/m_e^*)$$

The *rectifier equation* for the forward current I across a junction with forward voltage bias V is

$$I = I_0(e^{eV/kT} - 1)$$

where I_0 is the reverse saturation current.

Dielectrics

The *Clausius-Mossotti* equation for the polarizability of a non-polar molecule of number density N is

$$\alpha = \frac{\epsilon_r - 1}{\epsilon_r + 2} \frac{3\epsilon_0}{N}$$

where ϵ_r is the dielectric constant of the material.

The *Debye equations* for the frequency dependence of complex permittivity $\epsilon' - j\epsilon''$ are

$$\epsilon' = \epsilon_\infty + \frac{\epsilon_s - \epsilon_\infty}{1 + \omega^2\tau^2}$$

$$\epsilon'' = \frac{\omega\tau}{1 + \omega^2\tau^2} (\epsilon_s - \epsilon_\infty)$$

where ϵ_s, ϵ_∞ are the static and high-frequency permittivities and τ the dipole relaxation time.

Miscellaneous

Gauges for wire and sheet metal

The sizes given refer to the thickness of sheet or the diameter of wire.

I.S.O. (metric) sizes

The following sizes in millimetres form the R40 series; of these the first and alternate sizes are the R20 series, of which the first and alternate sizes are the R10 series. The order of preference is R10, R20, R40.

R10, R20	0·020	0·040	0·080	0·160	0·315	0·63	1·25	2·5	5·0	10·0	20·0
	0·021	0·042	0·085	0·170	0·335	0·67	1·32	2·65	5·3	10·6	21·2
R20	0·022	0·045	0·090	0·180	0·355	0·71	1·40	2·8	5·6	11·2	22·4
	0·024	0·048	0·095	0·190	0·375	0·75	1·50	3·0	6·0	11·8	23·6
R10, R20	0·025	0·050	0·100	0·200	0·40	0·80	1·60	3·15	6·3	12·5	25·0
	0·026	0·053	0·106	0·212	0·425	0·85	1·70	3·35	6·7	13·2	
R20	0·028	0·056	0·112	0·224	0·45	0·90	1·80	3·55	7·1	14·0	
	0·030	0·060	0·118	0·236	0·475	0·95	1·90	3·75	7·5	15·0	
R10, R20	0·032	0·063	0·125	0·250	0·50	1·00	2·00	4·0	8·0	16·0	
	0·034	0·067	0·132	0·265	0·53	1·06	2·12	4·25	8·5	17·0	
R20	0·036	0·071	0·140	0·28	0·56	1·12	2·24	4·5	9·0	18·0	
	0·038	0·075	0·150	0·30	0·60	1·18	2·36	4·75	9·5	19·0	

Standard wire gauge

The following gauge numbers correspond to the sizes given in thousandths of an inch; equivalent sizes in millimetres are also given to two or three significant figures.

SWG	in x 10^3	mm	SWG	in x 10^3	mm	SWG	in x 10^3	mm	SWG	in x 10^3	mm
7/0	500	12·7	9	144	3·66	24	22	0·559	39	5·2	0·132
6/0	464	11·8	10	128	3·25	25	20	0·508	40	4·8	0·122
5/0	432	11·0	11	116	2·95	26	18	0·457	41	4·4	0·112
4/0	400	10·2	12	104	2·64	27	16·4	0·417	42	4·0	0·102
3/0	372	9·45	13	92	2·34	28	14·8	0·376	43	3·6	0·091
00	348	8·84	14	80	2·03	29	13·6	0·345	44	3·2	0·081
0	324	8·23	15	72	1·83	30	12·4	0·315	45	2·8	0·071
1	300	7·62	16	64	1·63	31	11·6	0·295	46	2·4	0·061
2	276	7·01	17	56	1·42	32	10·8	0·274	47	2·0	0·051
3	252	6·40	18	48	1·22	33	10·0	0·254	48	1·6	0·041
4	232	5·89	19	40	1·02	34	9·2	0·234	49	1·2	0·030
5	212	5·38	20	36	0·914	35	8·4	0·213	50	1·0	0·025
6	192	4·88	21	32	0·813	36	7·6	0·193			
7	176	4·47	22	28	0·711	37	6·8	0·173			
8	160	4·06	23	24	0·610	38	6·0	0·152			

Standard screw threads

In the following, the diameter given is the nominal diameter which is the basic value of the major diameter (over the crests of an external thread).

I.S.O. metric

The table below gives only the first-choice threads in the coarse and fine series. (The constant-pitch series includes threads up to 300 mm diameter with pitches of 0·35, 0·5, 0·75, 1, 1·25, 1·5, 2, 3, 4 and 6 mm.) The dimensions are in millimetres. The theoretical thread depths are approximately 0·613 and 0·541 of the pitch for external and internal threads respectively.

Diameter	Pitch Coarse	Pitch Fine	Diameter	Pitch Coarse	Pitch Fine
1·6	0·35	–	16	2	1·5
2	0·4	–	20	2·5	1·5
2·5	0·45	0·35*	24	3	2
3	0·5	0·35*	30	3·5	2
4	0·7	0·5*	36	4	3
5	0·8	0·5*	42	4·5	3
6	1	0·75*	48	5	3
8	1·25	1	56	5·5	4
10	1·5	1·25	64	6	4
12	1·75	1·25	†		

* These threads form part of the constant-pitch series.
† Threads of diameters greater than 70 mm in the coarse and fine series are taken from the constant-pitch series, with pitches of 6 mm (coarse) or 4 mm (fine) and first-choice diameters 72, 80, 90, 100, 110, 125, 140, 160, 180, 200, 220, 250 and 280 mm.

Unified

The diameters are in inches. There is also an extra-fine series, and a constant-pitch series with threads up to 6 in diameter and pitches corresponding to 32, 28, 20, 16, 12, 8, 6 and 4 threads per inch. The threads are identical in form to the I.S.O. metric threads.

Diameter	Threads per inch Coarse	Fine	Diameter	Threads per inch Coarse	Fine
$\frac{1}{4}$	20	28	$1\frac{3}{8}$	6	12
$\frac{5}{16}$	18	24	$1\frac{1}{2}$	6	12
$\frac{3}{8}$	16	24	$1\frac{3}{4}$	5	
$\frac{7}{16}$	14	20			
$\frac{1}{2}$	13	20	2	$4\frac{1}{2}$	
			$2\frac{1}{4}$	$4\frac{1}{2}$	
$\frac{9}{16}$	12	18	$2\frac{1}{2}$	4	
$\frac{5}{8}$	11	18	$2\frac{3}{4}$	4	
$\frac{3}{4}$	10	16	3	4	
$\frac{7}{8}$	9	14			
1	8	12	$3\frac{1}{4}$	4	
			$3\frac{1}{2}$	4	
$1\frac{1}{8}$	7	12	$3\frac{3}{4}$	4	
$1\frac{1}{4}$	7	12	4	4	

Whitworth

The diameters are in inches; the two series are British Standard Whitworth (BSW) and British Standard Fine (BSF). Sizes which are not recommended are excluded here. The theoretical thread depth is approximately 0·640 of the pitch.

Diameter	Threads per inch BSW	BSF	Diameter	Threads per inch BSW	BSF
$\frac{3}{16}$	24		2	$4\frac{1}{2}$	7
$\frac{1}{4}$	20	26	$2\frac{1}{4}$	4	6
$\frac{5}{16}$	18	22	$2\frac{1}{2}$	4	6
$\frac{3}{8}$	16	20	$2\frac{3}{4}$	$3\frac{1}{2}$	6
$\frac{7}{16}$	14	18	3	$3\frac{1}{2}$	5
$\frac{1}{2}$	12	16	$3\frac{1}{4}$		5
$\frac{9}{16}$		16	$3\frac{1}{2}$	$3\frac{1}{4}$	$4\frac{1}{2}$
$\frac{5}{8}$	11	14	$3\frac{3}{4}$		$4\frac{1}{2}$
$\frac{3}{4}$	10	12	4	3	$4\frac{1}{2}$
$\frac{7}{8}$	9	11	$4\frac{1}{4}$		4
1	8	10	$4\frac{1}{2}$	$2\frac{7}{8}$	4
$1\frac{1}{8}$	7	9	5	$2\frac{3}{4}$	4
$1\frac{1}{4}$	7	9	$5\frac{1}{2}$	$2\frac{5}{8}$	4
$1\frac{1}{2}$	6	8	6	$2\frac{1}{2}$	4
$1\frac{3}{4}$	5	7			

British Association (B.A.)

The dimensions are in millimetres. No. 0 B.A. is not recommended and only even numbers are first-choice. The theoretical thread depth is 0·6 of the pitch.

No.	Diameter	Pitch	No.	Diameter	Pitch
0	6·0	1·00	13	1·2	0·25
1	5·3	0·90	14	1·0	0·23
2	4·7	0·81	15	0·90	0·21
3	4·1	0·73	16	0·79	0·19
4	3·6	0·66	17	0·70	0·17
5	3·2	0·59	18	0·62	0·15
6	2·8	0·53	19	0·54	0·14
7	2·5	0·48	20	0·48	0·12
8	2·2	0·43	21	0·42	0·11
9	1·9	0·39	22	0·37	0·10
10	1·7	0·35	23	0·33	0·09
11	1·5	0·31	24	0·29	0·08
12	1·3	0·28	25	0·25	0·07

References

The following is a selection of sources from which more extensive tables and data for engineering may be found.

Mathematics

1. Comrie, L. J. *Six-figure Mathematical Tables* (2 vols.), 3rd ed. Edinburgh, Chambers, 1963–65.
2. Korn, G. A. and Korn, T. M. *Mathematical Handbook for Scientists and Engineers.* New York, McGraw-Hill, 1961.
3. Abramowitz, M. and Stegun, L. A. *Handbook of Mathematical Functions.* New York, Dover, 1965.
4. British Association, continued by Royal Society, *Mathematical Tables.* Cambridge, C.U.P., various.
5. Fletcher, A. *et al., An Index of Mathematical Tables* (2 vols.), 2nd ed. Oxford, Blackwell, 1962.

Properties of matter

6. Smithells, C. J. *Metals Reference Book*, 5th ed. London, Butterworth, 1976.
7. Ross, R. B. *Metallic Materials Specification Handbook* 2nd ed. London, Chapman and Hall, 1972.
8. Touloukian, Y.S. *Thermophysical Properties of High Temperature Solid Materials* (6 vols.), New York, Macmillan, 1967.
9. Society of the Plastics Industry (U.S.A.), *Plastics Engineering Handbook.* New York, Reinhold, 1960.
10. *The Radiochemical Manual.* Amersham, Bucks, Radiochemical Centre, 1962.
11. *UK Steam Tables in SI Units.* London, Arnold, 1970.
12. Keenan, J. H. and Kaye, J. *Gas Tables.* New York, Wiley, 1948.
13. Predvoditelev, A. C. *et al., Tables of Thermodynamic Functions of Air* (6000–20 000 K, 10^{-3}–10^3 atm) (2 vols.). Moscow 1959, trans. Assoc. Tech. Services Inc., New Jersey, 1962.

Elasticity and structures; mechanics

14. Roark, R. J. *Formulas for Stress and Strain*, 4th ed. New York, McGraw-Hill, 1965.
15. Comrie J. *Civil Engineering Reference Book* (4 vols.), 2nd ed. London, Butterworth, 1961.
16. Flügge, W. *Handbook of Engineering Mechanics.* New York, McGraw-Hill, 1962.
17. *Handbook on Structural Steelwork.* London, The British Constructional Steelwork Association and The Constructional Steel Research and Development Organisation, 1971.

Electricity

18. Say, M. G. *Electrical Engineer's Reference Book,* 13th ed. London, Butterworth, 1973.
19. Ball, A. M. *Radio Valve and Transistor Data,* 9th ed. London, Iliffe, 1970.
20. Turner, L. W. *Electronic Engineer's Reference Book,* 4th ed. London, Butterworth, 1974.
21. Harvey, A. F. *Microwave Engineering.* New York, Academic Press, 1963.

General and miscellaneous

22. Kaye, G. W. C. and Laby, T. H. *Physical and Chemical Constants*, 14th ed. London, Longmans, 1973
23. Weast, R. C. *Handbook of Chemistry and Physics,* 56th ed. Cleveland, Ohio, CRC Press, 1975.
24. Prockter, C. E. *Kempe's Engineers Year Book* (2 vols.). West Wickham, Kent, Morgan-Grampian, yearly.
25. Houghton, B. *Mechanical Engineering: the sources of information.* London, Bingley, 1970.
26. *Handbook No. 18: Metric Standards for Engineering.* London, B.S.I., 1966.
27. *British Standards Yearbook.* London, B.S.I., yearly.